© 2025 World Scientific Publishing Company
https://doi.org/10.1142/9789819802104_fmatter

Preface

Stochastic differential equations and stochastic partial differential equations, or, briefly, S(P)DEs, and their applications belong to the modern branches of mathematics. They have applications in many disciplines, such as natural sciences, computer science, engineering fields, and finance. Fractional stochastic processes can be used to model short- and long-range dependencies with random influences. Fractal time derivatives are applied to describe effects over large frequency ranges. Over the past 50 years, the theory of rough paths and associated ordinary and partial differential equations has been developed. The rough path theory has the advantage of being able to give a pathwise meaning to S(P)DEs.

In this volume, a comprehensive overview is presented, covering both the theory and numerics of S(P)DEs driven by fractional Brownian motion using fractional white noise calculus, rough equations, or time-fractional derivatives. Applications to optimal control, finance, and physics are given.

This book is intended for graduate students, researchers, and practitioners in the field of stochastics, numerics, physics, and finance, as well as scientists with a strong background in mathematics and undergraduate students in applied mathematics.

The topics presented in this volume are:

- state-of-the-art numerical schemes for solving rough differential equations,
- mild solutions to semilinear rough partial differential equations,

- optimal control for a stochastic Schrödinger equation perturbed by multiplicative fractional noise,
- calibration of non-semimartingale models through an adjoint approach,
- strong convergence rates of an exponential integrator and finite element method for time-fractional S(P)DEs.

In Chapter 1, Redmann and Werner first present the basics of rough path theory and describe finite-dimensional rough differential equations. Numerical schemes for these equations are introduced. These include Taylor schemes, Runge–Kutta schemes, the log-ODE method, and the Runge–Kutta–log-ODE method. The authors discuss advantages as well as drawbacks of the methods and compare their performance in numerical experiments for high-dimensional rough differential equations where the path is generated by fractional Brownian motion with the Hurst index $\frac{1}{3} < H \leq \frac{1}{2}$.

Semilinear rough partial differential equations with time-inhomogeneous coefficients are considered by Tappe in Chapter 2. The state space is a Banach space, and the driving signal is an α-Hölder rough path, where $\alpha \in (\frac{1}{3}, \frac{1}{2}]$, and with values in a further Banach space. The solution is defined in the mild sense. Therefore, rough convolution integrals are introduced in the sense of the Gubinelli integral. An existence and uniqueness theorem is proven. Applications to S(P)DEs driven by infinite-dimensional Wiener processes and infinite-dimensional fractional Brownian motions are presented as well.

In Chapter 3, a minimization problem for a nonlinear stochastic Schrödinger equation driven by a fractional Brownian motion with the Hurst index $H \in (0,1)$ is discussed by Grecksch and Lisei. The state equation is defined in the variational sense. A separation approach is used, so that the solution is given by the product of the solution of a controlled pathwise problem and the solution of an SDE. A general cost function is introduced for the optimal control problem. This allows the quadratic case to be considered as an example. Finite-dimensional Galerkin approximations and a linearization method are presented and used to derive ε-optimal controls.

In Chapter 4, Bender and Thiel design and analyze a Monte Carlo algorithm for calibrating a financial model, in which certain quantities (e.g., volatility) are represented in terms of a stochastic differential equation driven by a continuous p-variation process for $p \in (1,2)$ (e.g., a fractional Brownian motion with the Hurst parameter larger than $\frac{1}{2}$). The p-variation process can be correlated with the Brownian motion, which drives the stock prices, in order to capture the so-called leverage effect. The key tool is an adjoint gradient representation via a new type of anticipating backward stochastic differential equation, which is formulated in terms of the Russo–Vallois forward integral. The rates of convergence for an Euler approximation of this adjoint equation are provided. Finally, the results are illustrated using a case study, calibrating a fractional Heston model to market data.

Noupelah and Tambue consider in Chapter 5 a semilinear S(P)DE driven by a Hilbert space valued Wiener process and a compensated Poisson random measure, where a time-fractional derivative in the sense of Caputo occurs. The linear operator is not necessarily self-adjoint. The solution process is defined as a mild solution, and its unique existence is proven. The problem is discretized in time by a variant of the exponential integrator scheme and in space by the finite element method. A mean square error estimate is given. The convergence order depends on the regularity of the initial condition and the power of the time-fractional derivative. Finally, the results are illustrated using a simulation study.

The editors would like to thank all the authors for their interesting contributions, topics which are current and stimulating sources of investigation in the field of fractional stochastic processes and their applications.

Wilfried Grecksch
and
Hannelore Lisei

About the Editors

Wilfried Grecksch studied mathematics at the Dresden University of Technology, Germany. After two years in industry, he worked as an assistant, lecturer, and associate professor at the Technical University of Merseburg until 1991. He earned his doctorate from Dresden University of Technology in 1976 and completed his habilitation at Merseburg in 1980. In 1992, Wilfried Grecksch was appointed professor of stochastics at Martin Luther University Halle-Wittenberg, where he served as a rector from 2000 to 2006. He has been a professor emeritus since 2016. His research focuses on stochastic analysis, with a particular emphasis on stochastic partial differential equations and their optimal control.

Hannelore Lisei received her PhD degree in mathematics from Martin Luther University Halle-Wittenberg, Germany, and her habilitation in mathematics from Babeş-Bolyai University, Cluj-Napoca, Romania. She is currently an associate professor at this university, where she teaches courses, conducts seminars, and pursues research in probability theory and statistics. Her research focus is mainly on stochastic analysis, and her mathematical contributions are primarily related to the existence, approximation, and optimization results for stochastic equations with applications in physics, such as Navier–Stokes equations and nonlinear Schrödinger equations. She has also published results on various problems in deterministic variational calculus and co-authored papers with applications to the modeling of chemical processes.

© 2025 World Scientific Publishing Company
https://doi.org/10.1142/9789819802104_fmatter

About the Contributors

Christian Bender obtained his PhD from the University of Konstanz, Germany, in 2003. Before joining Saarland University as a full professor for applied mathematics in 2009, he was a postdoctoral researcher at the Weierstrass Institute in Berlin and an assistant professor at TU Braunschweig. His research interests span a wide range of topics in stochastic analysis, computational stochastics, and mathematical finance. These include stochastic calculus for fractional Brownian motion and related processes, as well as Monte Carlo algorithms for nonlinear option pricing.

Aurelien Junior Noupelah is a lecturer and researcher in the field of applied mathematics. He holds a PhD from the University of Dschang, Cameroon, and has over five years of experience mentoring students and conducting research in stochastic and numerical analysis. His research is dedicated to developing numerical methods for solving stochastic partial differential equations (SPDEs). His focus is on using fractional calculus to approach problems involving systems with memory, specifically those modeled by SPDEs with fractional components. His research improves predictive modeling in oil and gas recovery, geothermal energy extraction, and resource management. Aurelien's research advances numerical accuracy by bridging theory and application, resulting in better decision-making across various industries.

Martin Redmann is a professor of applied stochastics at the Martin Luther University Halle-Wittenberg, Germany. His research focuses on stochastic/rough differential equations, their numerics, and model reduction.

Antoine Tambue is a full professor in mathematics at the Department of Computer Science, Electrical Engineering, and Mathematical Sciences, Western Norway University of Applied Sciences (HVL), Norway. In 2010, he obtained a PhD in mathematics at Heriot-Watt University, UK, via an interdisciplinary collaborative project (Bridging the Gaps Between Engineering and Mathematics) between the Department of Mathematics and the Institute of Petroleum Engineering (since 2019, Institute of GeoEnergy Engineering). His main research interests are in stochastic calculus and its numerics, numerical analysis, scientific computing, stochastic optimal control, computational finance, and computational statistics.

Stefan Tappe studied mathematics at the University of Paderborn, and he prepared his PhD thesis as a research fellow at Humboldt University of Berlin under the supervision of Prof. Uwe Küchler. After completing his PhD thesis in 2005, he served as a postdoctoral researcher at the Ludwig Maximilian University of Munich in Germany, at the Vienna Institute of Finance in Austria, and at ETH Zürich in Switzerland. In 2011, Dr. Tappe accepted an offer for a junior professorship at the University of Hannover in Germany. Afterward, he held several deputy professorships and research stints in Germany. Currently, Dr. Tappe is affiliated with the Albert Ludwig University of Freiburg, where he holds a research position, funded by the DFG (German Research Foundation). His research interests include stochastic analysis, stochastic partial differential equations, and financial mathematics.

Matthias Thiel studied mathematics at Saarland University, Germany. In 2023, he obtained his PhD under the supervision of Christian Bender. The chapter on calibration of non-semimartingale models is based on the results of his doctoral dissertation.

Justus Werner studied mathematics at the Martin Luther University Halle-Wittenberg, Germany. Since 2023, he has been a PhD student at this university. His research interests are in rough paths and related numerical methods.

Contents

Preface	v
About the Editors	ix
About the Contributors	xi

Chapter 1. State-of-the-Art Numerical Schemes for Solving Rough Differential Equations   1
Martin Redmann and Justus Werner

1.1.	Introduction	1
1.2.	Foundation of Rough Path Theory	3
1.3.	Overview of Numerical Schemes	9
	1.3.1. Taylor schemes	9
	1.3.2. Runge–Kutta schemes	12
	1.3.3. Log-ODE method	14
	1.3.4. RK–log-ODE method	16
1.4.	Numerical Examples	17
1.5.	Conclusion	22
Acknowledgments		22
References		23

Chapter 2. Mild Solutions to Semilinear Rough Partial Differential Equations   25
Stefan Tappe

2.1.	Introduction	25
2.2.	Frequently Used Notation	27

2.3.	Rough Path Theory	28
2.4.	Strongly Continuous Semigroups	35
2.5.	Compositions of Regular Paths with Functions	41
2.6.	Compositions of Controlled Rough Paths with Functions	42
	2.6.1. Compositions with time-dependent smooth functions	43
	2.6.2. Compositions with bilinear operators	46
	2.6.3. Compositions with time-dependent linear operators	52
2.7.	Regular Convolution Integrals	60
2.8.	Rough Convolution Integrals	62
2.9.	Rough Partial Differential Equations	67
	2.9.1. Solution concepts	67
	2.9.2. The space for the fixed point problem	72
	2.9.3. Auxiliary results	76
	2.9.4. Existence and uniqueness of local mild solutions	79
	2.9.5. Further auxiliary results	80
	2.9.6. Existence and uniqueness of global mild solutions	85
2.10.	Stochastic Partial Differential Equations Driven by Infinite-Dimensional Wiener Processes	88
	2.10.1. Infinite-dimensional Wiener process as a rough path	88
	2.10.2. Coincidence of the two integrals	93
	2.10.3. Stochastic partial differential equations	97
2.11.	Stochastic Partial Differential Equations Driven by Infinite-Dimensional Fractional Brownian Motion	99
Funding Statement		101
References		101

Chapter 3. Fractional Noise-Perturbed Nonlinear Schrödinger Equations: Stochastic Minimization Problems   **105**

Wilfried Grecksch and Hannelore Lisei

3.1.	Introduction	105
3.2.	Preliminaries	109

Chapter 1

State-of-the-Art Numerical Schemes for Solving Rough Differential Equations

Martin Redmann[*] and Justus Werner[†]

*Faculty of Natural Sciences II, Institute of Mathematics,
Martin Luther University Halle-Wittenberg,
D-06099 Halle (Saale), Germany
[*]martin.redmann@mathematik.uni-halle.de
[†]justus.werner@mathematik.uni-halle.de*

In this chapter, we give an introduction to rough paths. Subsequently, we consider various approaches to solving rough differential equations numerically, discuss the advantages and drawbacks of each individual scheme, and compare their performance in numerical experiments.

1.1. Introduction

The use of stochastic differential equations (SDEs) as a sophisticated mathematical tool for modeling real-life applications has grown in popularity because incorporating uncertainties makes the models more realistic. However, it is often not possible to determine solutions for SDEs in an analytically closed form. Therefore, numerical schemes are important tools to approximate their solutions. There are various classes of numerical methods for solving SDEs.

The most popular ones are the Taylor schemes, such as the Euler–Maruyama and Milstein schemes, as well as the Runge–Kutta methods (see Refs. [1–5]). Additionally, some approaches from Lie theory (see Refs. [6–8]) have also been proposed. In practice, the implementation of all these techniques is limited to low strong approximation orders since high-order schemes struggle with efficiently calculating the involved iterated integrals. This makes the implementation of high-order schemes efficient only in certain special cases.

A deterministic (or path-wise) approach to SDEs is given by the rough path theory, developed by Lyons. In this case, the driver is a more general Hölder continuous function instead of a continuous semimartingale. Rough differential equations (RDEs) are driven by objects which can be interpreted as vectors of iterated integrals and therefore make higher-order numerical methods relevant again. While the most widely studied numerical schemes available for RDEs are of the Taylor type [9,10], there is little literature on Runge–Kutta-type schemes [11]. Many of these approaches rely on the availability of iterated integrals which are usually not known analytically. Therefore, methods using only the path information of the driver have been considered [11–13] for a simpler implementation. However, they are restricted to low orders of convergence. Also, the Lie theory approach, namely the log-ODE method [14–17], has been studied extensively, building upon the Taylor schemes involving iterated integrals. Unfortunately, the Taylor methods rely on nested derivatives, which are computationally expensive, making them inefficient in, e.g., large-scale settings. We aim to modify the Lie theory *ansatz* and exploit the advantages of the Runge–Kutta schemes in this context, which incur a lower computational cost since they are derivative-free. The contribution of this chapter is to introduce a novel Runge–Kutta–log-ODE method, which is based on a new Runge–Kutta scheme that extends the Runge–Kutta *ansatz* from SDE theory [1]. Both new techniques require considerably reduced computational effort compared to their well-known Taylor counterparts and explicitly depend on second-order iterated integrals. In particular, this explicit dependence is an advantage when compared to the class of methods used in Ref. [11]. In this work, the focus is rather on applying the new approaches and testing their theoretical properties through numerical experiments. Therefore, it can be seen as a good starting point for further theoretical studies of new methods with high potential.

This chapter is structured as follows. In Section 1.2, we introduce the basics of rough path theory and describe RDEs. Section 1.3 contains an overview of various numerical schemes, including our newly proposed approach. We conclude with Section 1.4, in which we conduct numerical experiments for high-dimensional RDEs and discuss the performance of each method.

1.2. Foundation of Rough Path Theory

First, we introduce the essentials of rough path theory, including tensor algebra and signatures.

Definition 1.1 ([16], Definition A.1). We say that $T(\mathbb{R}^k) := \bigoplus_{i=0}^{\infty}(\mathbb{R}^k)^{\otimes i}$ is the tensor algebra of $\mathbb{R}^k$ and $T((\mathbb{R}^k)) = \{\mathbf{a} = (a_0, a_1, \ldots) : a_n \in (\mathbb{R}^k)^{\otimes n}\ \forall n \geq 0\}$ is the set of the formal series of tensors of $\mathbb{R}^k$. Similarly, we define the truncated tensor algebra $T^N(\mathbb{R}^k) := \mathbb{R} \oplus \mathbb{R}^k \oplus (\mathbb{R}^k)^{\otimes 2} \oplus \cdots \oplus (\mathbb{R}^k)^{\otimes N}$ for $N \in \mathbb{N}$. Moreover, $T(\mathbb{R}^k)$, $T((\mathbb{R}^k))$, and $T^N(\mathbb{R}^k)$ can be endowed with the operations of addition and multiplication.

Given $\mathbf{a} = (a_0, a_1, \ldots)$ and $\mathbf{b} = (b_0, b_1, \ldots)$, we have

$$\begin{aligned}\mathbf{a} + \mathbf{b} &= (a_0 + b_0, a_1 + b_1, \ldots), \\ \mathbf{a} \otimes \mathbf{b} &= (c_0, c_1, \ldots),\end{aligned} \quad (1.1)$$

where for $n \geq 0$, the nth term $c_n \in (\mathbb{R}^k)^{\otimes n}$ can be written using the usual tensor product as

$$c_n := \sum_{i=0}^{n} a_i \otimes b_{n-i}.$$

In rough path theory, we consider two parameter functions with values in $T^N(\mathbb{R}^k)$. Instead of choosing two arbitrary parameters in $[0, T]$, it is often useful to order the parameters via the simplex $\Delta_T := \{(s, t) \in [0, T]^2 : s < t\}$. We follow this approach notationwise.

Definition 1.2 ([16], Definition A.2). The signature $S_{\cdot,\cdot}(X) : \Delta_T \to T((\mathbb{R}^k))$ of a path $X : [0, T] \to \mathbb{R}^k$ of bounded variation over the interval $[s,t]$ is defined as the following collection of iterated (Riemann–Stieltjes) integrals:

$$S_{s,t}(X) := \left(1, X^{(1)}_{s,t}, X^{(2)}_{s,t}, \cdots\right) \in T((\mathbb{R}^k)),$$

where for $n \geq 1$,

$$X^{(n)}_{s,t} := \int \cdots \int_{s<u_1<\cdots<u_n<t} dX_{u_1} \otimes \cdots \otimes dX_{u_n} \in (\mathbb{R}^k)^{\otimes n}.$$

Similarly, we can define the depth-N (or truncated) signature of the path X on $[s,t]$ as

$$S^N_{s,t}(X) := \left(1, X^{(1)}_{s,t}, \cdots, X^{(N)}_{s,t}\right) \in T^N(\mathbb{R}^k).$$

Definition 1.3. We define $\pi_n \colon T^N(\mathbb{R}^k) \to (\mathbb{R}^k)^{\otimes n}$ as the projection map onto $(\mathbb{R}^k)^{\otimes n}$ for $n = 0, 1, \ldots, N$.

Next, we introduce a suitable norm for signatures and, later, for rough paths.

Definition 1.4. Let $\mathbf{X} \colon \Delta_T \to T^{\lfloor p \rfloor}(\mathbb{R}^k)$, where $\lfloor \cdot \rfloor$ denotes the floor function. For $p \geq 1$, we introduce the p-variation norm

$$\|\mathbf{X}\|_{p\text{-var}} := \max_{1 \leq n \leq \lfloor p \rfloor} \sup_{\mathcal{D}} \left(\sum_{t_i \in \mathcal{D}} \|\pi_n(\mathbf{X}_{t_i,t_{i+1}})\|^{\frac{p}{n}} \right)^{\frac{n}{p}}$$

and the induced p-variation metric between two continuous paths $\mathbf{Z}^1$ and $\mathbf{Z}^2$ with values in $T^{\lfloor p \rfloor}(\mathbb{R}^k)$ as

$$d_p(\mathbf{Z}^1, \mathbf{Z}^2) := \max_{1 \leq n \leq \lfloor p \rfloor} \sup_{\mathcal{D}} \left(\sum_{t_i \in \mathcal{D}} \|\pi_n(\mathbf{Z}^1_{t_i,t_{i+1}}) - \pi_n(\mathbf{Z}^2_{t_i,t_{i+1}})\|^{\frac{p}{n}} \right)^{\frac{n}{p}},$$

where the supremum is taken over all partitions $\mathcal{D}$ of $[0,T]$ and the norms $\|\cdot\|$ must satisfy

$$\|a \otimes b\| \leq C \|a\| \|b\|,$$

for $a \in (\mathbb{R}^k)^{\otimes n}$, $b \in (\mathbb{R}^k)^{\otimes m}$ and a constant $C \geq 0$. For example, we can take $\|\cdot\|$ to be the projective or injective tensor norms (see

Propositions 2.1 and 3.1 in Ref. [18]). Additionally, we define a metric for the two cases $p = 0$ and $p = \infty$ in the following way:

$$d_{\infty;[0,T]}(\mathbf{Z}^1, \mathbf{Z}^2) := \sup_{t \in [0,T]} d(\mathbf{Z}^1_{0,t}, \mathbf{Z}^2_{0,t}),$$

$$d_{0;[0,T]}(\mathbf{Z}^1, \mathbf{Z}^2) := \sup_{0 \leq s < t \leq T} d(\mathbf{Z}^1_{s,t}, \mathbf{Z}^2_{s,t}),$$

where d is the so-called Carnot–Caratheodory metric (see Theorem 7.32 in Ref. [10]).

Next, we consider an important object in rough path theory, namely the geometric p-rough path.

Definition 1.5 ([16], Theorem B.1). For $p \geq 1$, we say that $\mathbf{X}\colon \Delta_T \to T^{\lfloor p \rfloor}(\mathbb{R}^k)$ is a geometric p-rough path if $\mathbf{X}$ is a continuous path in the tensor algebra $T^{\lfloor p \rfloor}(\mathbb{R}^n)$ and there exists a sequence (x^n) of continuous finite variation paths $x^n : [0,T] \to \mathbb{R}^k$ whose truncated signatures converge to $\mathbf{X}$ in the p-variation metric

$$d_p\big(S^{\lfloor p \rfloor}(x^n), \mathbf{X}\big) \to 0 \tag{1.2}$$

as $n \to \infty$.

The following identity, known as Chen's relation, tells us precisely how to "patch together" rough paths over adjacent intervals.

Lemma 1.1. *If $\mathbf{X}$ is a geometric p-rough path, then*

$$\mathbf{X}_{s,t} = \mathbf{X}_{s,u} \otimes \mathbf{X}_{u,t} \tag{1.3}$$

holds for $0 \leq s < u < t \leq T$.

Proof. In Ref. [10, Theorem 7.11], this statement is proven for $S^{\lfloor p \rfloor}(x^n)$ instead of $\mathbf{X}$ (see (1.2)). A careful passage to the limit concludes the proof. □

If not stated otherwise, hereinafter we always assume that $p \geq 1$.

Example 1.1. Assume that $X = W.(\omega)\colon [0,T] \to \mathbb{R}^k$ is a path of a standard Brownian motion W. Since the paths of W are almost surely of finite p-variation for $p > 2$, we need to make sense of $\mathbb{W}$ in $\mathbf{X} = (W, \mathbb{W})$, taking values in $T^2(\mathbb{R}^k)$. The two most common approaches are to define $\mathbb{W}$ via either Itô or Stratonovich integrals. Note that the construction of this second-order information is not path-wise, although rough analysis is a path-wise *ansatz*. While the approach involving Itô integrals leads to a non-geometric rough path, the approach via Stratonovich integrals indeed results in a random geometric p-rough path, $\mathbf{X} = (W, \mathbb{W}^{\text{Strat}})$, where for a path $W_\cdot(\omega)\colon [0,T] \to \mathbb{R}^k$,

$$\mathbb{W}^{\text{Strat}}_{s,t}(\omega) = \left(\int_s^t W_{s,u} \otimes \circ dW_u\right)(\omega)$$

is based on a Stratonovich integral. A potential smooth approximation in (1.2) is the well-known Wong–Zakai approximation [19].

Since we now have the definition of a rough path as a potential driver, the aim is to consider RDEs and construct a solution concept.

We begin by introducing the notion of smoothness for vector fields.

Definition 1.6 ([10], Definition 10.2). A map $f\colon E \to F$ between two normed spaces E and F is called γ-Lipschitz (in symbols $f \in \text{Lip}(\gamma)$) if f is bounded with $\lfloor \gamma \rfloor$-bounded Fréchet derivatives, where the last Fréchet derivative $D^{\lfloor \gamma \rfloor} f$ is Hölder continuous with exponent $\gamma - \lfloor \gamma \rfloor$. Then, the following norm is finite:

$$\|\cdot\|_{\text{Lip}(\gamma)} := \max_{0 \leq k \leq \lfloor \gamma \rfloor} \|D^k f\|_\infty \vee \|D^{\lfloor \gamma \rfloor} f\|_{(\gamma - \lfloor \gamma \rfloor)\text{-Höl}}.$$

There are at least two concepts to define solutions of RDEs, which are equivalent in our framework. We start with the classical approach of Lyons [20], which is similar to the SDE case. The idea is to define a rough integral and subsequently introduce a differential equation involving such an integral.

Definition 1.7 ([10], Definition 10.44). Let $\mathbf{X}\colon \Delta_T \to T^{\lfloor p \rfloor}(\mathbb{R}^k)$ be a geometric p-rough path[a] and $f = (f_1, \ldots, f_k)$ be a collection of maps $f_j\colon \mathbb{R}^{n_1} \to \mathbb{R}^k$. We say that $\mathbf{Z}\colon \Delta_T \to T^{\lfloor p \rfloor}(\mathbb{R}^{n_1})$ is a rough integral of f along $\mathbf{X}$ if there exists a sequence $(x^n)_n$ with $x^n\colon [0,T] \to \mathbb{R}^k$

---

[a]This definition also holds in the case of weak geometric p-rough paths, which is a more general concept.

and x^n is of bounded variation such that
$$\forall n : x_0^n = X_0,$$
$$\lim_{n\to\infty} d_{0;[0,T]}(S^{\lfloor p \rfloor}(x^n), \mathbf{X}) = 0,$$
$$\sup_n \|S^{\lfloor p \rfloor}(x^n)\|_{p\text{-var}} < \infty,$$
where $X = \pi_1(\mathbf{X})$ and
$$\lim_{n\to\infty} d_\infty \left(S^{\lfloor p \rfloor} \left(\int f(x_u^n) \, dx_u^n \right), \mathbf{Z} \right) = 0.$$

The following results provide the conditions under which the rough integral is well defined.

Theorem 1.1 ([10], Theorem 10.47). *Assume that:*

- *for $f = (f_1, \ldots, f_k)$, $f_i \in \text{Lip}(\gamma - 1)$ with $i = 1, \ldots, k$ and $\gamma > p \geq 1$;*
- *$\mathbf{X} \colon \Delta_T \to T^{\lfloor p \rfloor}(\mathbb{R}^k)$ is a geometric p-rough path.*

Then, for all $s < t \in [0, T]$, there exists a unique rough-path integral of f along $\mathbf{X}$. The indefinite integral $\int f(X) d\mathbf{X}$ is a geometric rough path: There exists a constant C depending only on p and γ such that for all $s < t$ in $[0, T]$,

$$\left\| \int f(X) d\mathbf{X} \right\|_{p\text{-var};[s,t]} \leq C \|f\|_{\text{Lip}(\gamma-1)} \left(\|\mathbf{X}\|_{p\text{-var};[s,t]} \vee \|\mathbf{X}\|_{p\text{-var};[s,t]}^p \right)$$

with $X = \pi_1(\mathbf{X})$.

In the following, we write $\mathbf{X}_t$ instead of $\mathbf{X}_{0,t}$ for simplicity of notation. Now, a solution to the RDE
$$dY_t = f(Y_t) d\mathbf{X}_t,$$
$$Y_0 = y_0,$$
can then be defined via the integral equation
$$Y_t = y_0 + \int_0^t f(Y_s) d\mathbf{X}_s,$$
where it turns out that $\int_0^t f(Y_s) d\mathbf{X}_s$ is well defined as a rough integral under suitable regularity conditions on f. Note that we actually

need a more general notion of integrals as opposed to that given in Definition 1.7 in order to make sense of RDEs. However, we omit this extension to improve the readability of this chapter. The other approach, which we will use in the remainder of the chapter, is to define a solution as an appropriate limit of solutions to Stieltjes differential equations.

Definition 1.8 ([10], Definition 10.17). Let $\mathbf{X}\colon \Delta_T \to T^{\lfloor p \rfloor}(\mathbb{R}^k)$ be a geometric p-rough path. We say that $Y \in C([0,T], \mathbb{R}^{n_1})$ is a solution to the RDE

$$dY_t = f(Y_t) d\mathbf{X}_t,$$
$$Y_0 = y_0,$$

if there exists a sequence $(x^n)_n$ of bounded variations functions such that $\lim_{n\to\infty} d_{0;[0,T]}(S_{\lfloor p \rfloor}(x^n), \mathbf{X}) = 0$ and $\sup_n \|S_{\lfloor p \rfloor}(x^n)\|_{p\text{-var}} < \infty$ hold as well as solutions Y^n to the Stieltjes differential equations

$$dY_t^n = f(Y_t^n) dx_t^n,$$
$$Y_0^n = y_0,$$

exist such that

$$Y^n \to Y \quad \text{uniformly on } [0,T] \text{ as } n \to \infty.$$

Remark 1.1. Both concepts of solutions are equivalent in our framework (see Remark 10.19 in Ref. [10]). While the second concept might be more intuitive, the approach of Lyons can immediately be generalized for more general rough paths $\mathbf{X}$.

Finally, we give an existence and uniqueness result for the solution of RDEs.

Theorem 1.2 ([10], Theorem 10.26). *Assume that:*

- $f = (f_1, \ldots, f_k)$ *is a collection of* $\mathrm{Lip}(\gamma)$-*vector fields* $f_j \colon \mathbb{R}^n \to \mathbb{R}^k$ *for* $\gamma > p \geq 1$;
- $\mathbf{X}\colon \Delta_T \to T^{\lfloor p \rfloor}(\mathbb{R}^k)$ *is a geometric p-rough path;*
- $y_0 \in \mathbb{R}^n$ *is considered the initial condition at time zero.*

Then, there exists a unique solution $Y\colon [0,T] \to \mathbb{R}^n$ to the RDE

$$dY_t = f(Y_t)d\mathbf{X}_t,$$
$$Y_0 = y_0. \tag{1.4}$$

A main advantage of rough path theory is that $Y(y_0, f, \mathbf{X})$ under the above assumptions is locally Lipschitz continuous in the initial value, the vector field, and the driver. This fact is used to prove the uniqueness of a solution of (1.4).

Remark 1.2. It is possible to recover the solution of (1.2) as the rough path $\mathbf{Y}$ itself. In general, we only consider the first level $\pi_1(\mathbf{Y}) = Y$ of the solution.

1.3. Overview of Numerical Schemes

In this section, we present schemes that can be used to compute numerical solutions to RDEs. Let $\mathbf{X}\colon \Delta_T \to T^{\lfloor p \rfloor}(\mathbb{R}^k)$ be a geometric p-rough path and $f = (f_1, \ldots, f_k)\colon \mathbb{R}^n \to \mathbb{R}^{n \times k}$. We aim to approximate a function $Y\colon [0,T] \to \mathbb{R}^n$ that satisfies

$$dY_t = f(Y_t)d\mathbf{X}_t = \sum_{i=1}^{k} f_i(Y_t)d\mathbf{X}_t^i, \tag{1.5}$$
$$Y_0 = y_0.$$

Additionally, we assume that $f_i \in \operatorname{Lip}(\gamma)$ for $\gamma > p$ and $i = 1, \ldots, k$ to ensure the existence of a unique solution according to Theorem 1.2.

1.3.1. Taylor schemes

A special case of this class of methods was first introduced in Ref. [9] and is extensively studied in Ref. [10] in full generality.[b] Notation-wise, we follow the approach of Ref. [16].

We begin by examining the origin of the Taylor schemes. Let $X\colon [0,T] \to \mathbb{R}$ be continuously differentiable, and assume that

---

[b]What we call Taylor schemes are studied there under the name of (step-N) Euler schemes.

$f\colon \mathbb{R}^n \to \mathbb{R}^n$ is smooth. We denote D_f as the Jacobian of f. Now, we exploit the first-order Taylor expansion of f, i.e., $f(Y_s) \approx f(Y_a) + D_f(Y_a)(Y_s - Y_a)$, in order to find a Taylor approximation of the solution to (1.5) around the point a with $0 < t - a \ll 1$. In this context, we neglect terms of higher order and obtain

$$\begin{aligned}
Y_t &= Y_a + \int_a^t f(Y_s)dX_s \\
&\approx Y_a + \int_a^t f(Y_a) + D_f(Y_a)(Y_s - Y_a)dX_s \\
&= Y_a + \int_a^t \left(f(Y_a) + D_f(Y_a)\int_a^s f(Y_u)dX_u\right)dX_s \\
&\approx Y_a + \int_a^t \left(f(Y_a) + D_f(Y_a)f(Y_a)\int_a^s dX_u\right)dX_s \\
&= Y_a + f(Y_a)\int_a^t dX_s + D_f(Y_a)f(Y_a)\int_a^t \int_a^s dX_u\, dX_s \\
&= Y_a + f(Y_a)\pi_1(S_{a,t}(X)) + D_f(Y_a)f(Y_a)\pi_2(S_{a,t}(X)).
\end{aligned} \quad (1.6)$$

This gives a second-order Taylor-like expansion of the solution, which in the case of SDEs is known as the Milstein scheme. The levels of the signature play a crucial role as polynomials on paths. Higher-order Taylor expansions result in expressions using higher-order signature terms.

Indeed, this connection can be made rigorous, which leads to a precise definition of the Taylor schemes. We begin by describing the role of the vector fields, which are usually known as elementary differentials.

Definition 1.9 ([16], Definition A.6). For $f\colon \mathbb{R}^n \to \mathbb{R}^{n\times k}$, we define $f^{\circ m}\colon \mathbb{R}^n \to L((\mathbb{R}^k)^{\otimes m}, \mathbb{R}^n)$ recursively for $m \in \mathbb{N}$ by

$$f^{\circ 0}(y) := y,$$
$$f^{\circ 1}(y) := f(y),$$
$$f^{\circ m+1}(y) := D(f^{\circ m})(y)f(y),$$

for $y \in \mathbb{R}^n$, where $D(f^{\circ m})$ denotes the Fréchet derivative of $f^{\circ m}$.

Now, we can describe the Taylor schemes for more general drivers and orders covering the motivation in (1.6).

Definition 1.10 ([16], Definition A.7). Let $\mathbf{X}\colon \Delta_T \to T^{\lfloor p \rfloor}(\mathbb{R}^k)$ be a geometric p-rough path, $f \in C^{N-1}(\mathbb{R}^n)$, and $N = \lfloor p \rfloor$. The Taylor operator and the associated RDE approximation are given by

$$\mathrm{Taylor}(Y_s, f, \mathbf{X}_{s,t}) := \sum_{k=0}^{N} f^{\circ k}(Y_s)\pi_k(\mathbf{X}_{s,t}) \approx Y_t. \qquad (1.7)$$

This local approximation is repeated m times on a partition $D = \{0 = t_0 < t_1 < \cdots < t_m = T\}$. This enables us to define a numerical solution for the Nth-level Taylor method via

$$y_i^{N\text{-Tay}} = \mathrm{Taylor}(y_{i-1}^{N\text{-Tay}}, f, \mathbf{X}_{t_i, t_{i+1}}), \quad (i = 1, \ldots, m),$$

where $y_i^{N\text{-Tay}} \approx Y_{t_i}$ and $y_0^{N\text{-Tay}} = y_0$.

Example 1.2. The reader might be familiar with these types of Taylor schemes from SDE theory. The strong Taylor approximations in Ref. [2] are very similar. In particular, the case of $N = 1$ is the Euler–Maruyama scheme for rough paths, while the case of $N = 2$ is the Milstein scheme for rough paths.

We continue with results on convergence and corresponding rates. In the following, we make use of the Euclidean norm, denoted by $|\cdot|$ hereafter.

Theorem 1.3 ([10], Theorem 10.30). *Assume that:*

- $\mathbf{X}\colon \Delta_T \to T^{\lfloor p \rfloor}(\mathbb{R}^k)$ *is a geometric p-rough path;*
- $f = (f_1, \ldots, f_k)$ *is a collection of $\mathrm{Lip}(\gamma)$-vector fields on $\mathbb{R}^n$ for $\gamma > p \geq 1$;*
- $D = \{0 = t_0 < t_1 < \cdots < t_m = T\}$ *is a fixed partition of $[0, T]$, where $h = \max_i t_{i+1} - t_i$.*

Set $N := \lfloor \gamma \rfloor \geq \lfloor p \rfloor$. Then, there exists a constant $C = C(p, \gamma) > 0$ so that

$$|Y_T - y_m^{N\text{-Tay}}| \leq C h^{\alpha},$$

where $\alpha = \frac{(N+1)}{p} - 1$ and h is sufficiently small.

Remark 1.3. These convergence rates represent the worst-case rates, and in some special cases, better rates can arise. For example, choose the random rough path $\mathbf{X} = (W, \mathbb{W}^{\text{Strat}})$, which is the Stratonovich lift of a standard Brownian motion W. In this case, Theorem 1.3 gives a convergence rate of $\alpha = \frac{2+1}{2+\varepsilon_1} - 1 = 0.5 - \varepsilon_2$ for some arbitrary small $\varepsilon_1, \varepsilon_2 > 0$ and almost all paths of W; however, from Ref. [21], we know that the Milstein scheme converges a.s. path-wise with an order of $1 - \varepsilon$.

1.3.2. *Runge–Kutta schemes*

In the theory of ordinary differential equations, the Runge–Kutta schemes are preferred over the usual Taylor schemes. The reason for this is that the elementary differentials (see Definition 1.9) are expensive to compute from a numerical perspective. Another advantage of the Runge–Kutta methods is that they are derivative-free.

In Ref. [11], a Runge–Kutta approach is presented that discretizes very general RDEs with non-geometric drivers. It is shown that one can achieve an arbitrary good rate analogous to Theorem 1.3 using a proper choice of the coefficients within the numerical scheme (and given enough regularity of the vector fields). Although this Runge–Kutta *ansatz* works perfectly in theory, it has the disadvantage that the associated coefficients depend implicitly on iterated integrals, and these dependencies are hard to identify in general. To address the difficulty of implementing of this method, Ref. [11] presents another approach in which the driver is geometric. In detail, the geometric p-rough path is discretized using the lift of a piecewise linear approximation, and the Runge–Kutta schemes are used to solve the resulting ODEs. This leads to schemes involving path information only. In contrast to the approach in Ref. [11], we aim to introduce the Runge–Kutta schemes for RDEs with explicit second-order information. To the best of the authors' knowledge, there exist no Runge–Kutta scheme for RDEs that shows explicit dependence on higher levels. This means we can establish a higher-order scheme that is implementable. To do so, we exploit a Runge–Kutta scheme with explicit second-order information designed for SDEs [1] and transfer it to the much more general case of RDEs. We investigate the scheme numerically, including numerical evidence for the convergence of the scheme and an experiment on the order of convergence. The theoretical analysis is beyond the scope of this chapter.

We begin by defining a Runge–Kutta operator which shall replace the Taylor operator in (1.7). The definition of the Runge–Kutta operator is based on Section 6.1 in Ref. [1]. We specify the method by choosing fixed coefficients for the Runge–Kutta operator, which fulfill certain order conditions in the SDE case.

Definition 1.11. Let $\mathbf{X}\colon \Delta_T \to T^{\lfloor p \rfloor}(\mathbb{R}^k)$ be a geometric p-rough path with $p < 3$. Then, the Runge–Kutta operator and the associated RDE approximation are defined by

$$Y_t \approx \text{Runge–Kutta}(Y_s, f, \mathbf{X}_{s,t})$$

$$:= Y_s + \sum_{l=1}^{k} f_l(Y_s)\pi_1(\mathbf{X}_{s,t})_l + \frac{\sqrt{t-s}}{2} \sum_{l=1}^{k} \left(f_l(Z_l^1) - f_l(Z_l^2) \right),$$

where $\pi_1(\mathbf{X}_{s,t})_l$ is the lth element of $\pi_1(\mathbf{X}_{s,t})$ and

$$Z_l^1 = Y_s + \frac{1}{\sqrt{t-s}} \sum_{j=1}^{k} f_j(Y_s)\pi_2(\mathbf{X}_{s,t})_{j,l},$$

$$Z_l^2 = Y_s - \frac{1}{\sqrt{t-s}} \sum_{j=1}^{k} f_j(Y_s)\pi_2(\mathbf{X}_{s,t})_{j,l},$$

with $\pi_2(\mathbf{X}) \in (\mathbb{R}^k)^{\otimes 2} \simeq \mathbb{R}^{k \times k}$ and $\pi_2(\mathbf{X})_{j,l}$ being the element in the jth row and lth column.

This local approximation is repeated m times on a partition $D = \{0 = t_0 < t_1 < \cdots < t_m = T\}$. We define the numerical solution of the Runge–Kutta method via

$$y_i^{RK} = \text{Runge–Kutta}(y_{i-1}^{RK}, f, \mathbf{X}_{t_i, t_{i+1}}), \quad (i = 1, \ldots, m),$$

where $y_0^{RK} = y_0$. The intuition is that $y_i^{RK} \approx Y_{t_i}$.

Remark 1.4. While the Taylor scheme is a class of numerical methods depending on $N \in \mathbb{N}$ in Definition 1.11, we give the Runge–Kutta method for the concrete case of $N = 2$. In Ref. [1], the Runge–Kutta methods for SDEs are given for $N \in \mathbb{N}$. In future publications, the authors plan to investigate this case for RDEs as well. The $N = 1$ Runge–Kutta scheme coincides with the first-level Taylor method.

Remark 1.5. The motivation for the Taylor method (1.6) is based on the Taylor series. Similarly, the Runge–Kutta operator can be motivated by an expansion of the solution using the so-called Butcher series, or B-series.

1.3.3. Log-ODE method

The method was introduced in Ref. [14], analyzed in Ref. [15], and received recent attention in Ref. [17]. Again, we mainly follow the notation of Ref. [16]. The log-ODE method is an approach derived from Lie theory and therefore has the advantage of respecting the geometry of the problem. If the solution of (1.5) lies in a certain manifold, then it is possible to construct an approximation using the log-ODE method which also lies in the same manifold.

We begin by introducing the basic Lie theory of rough paths.

Definition 1.12 ([16], Definition A.3). For $\mathbf{a} = (a_0, a_1, \ldots) \in T((\mathbb{R}^k))$ with $a_0 > 0$, define $\log(\mathbf{a})$ to be an element of $T((\mathbb{R}^k))$ given by the following series:

$$\log(\mathbf{a}) := \log(a_0) + \sum_{l=1}^{\infty} \frac{(-1)^l}{l} \left(1 - \frac{\mathbf{a}}{a_0}\right)^{\otimes l}, \tag{1.8}$$

where $\mathbf{1} = (1, 0, 0, \ldots)$ is the unit element of $T((\mathbb{R}^k))$ and $\log(a_0)$ is viewed as $\log(a_0)\mathbf{1}$.

Next, we define the truncated logarithm map.

Definition 1.13 ([16], Definition A.4). For $\mathbf{a} = (a_0, a_1, \ldots) \in T((\mathbb{R}^k))$ with $a_0 > 0$, define $\log^N(\mathbf{a})$ to be an element of $T^N(\mathbb{R}^k)$ defined from the logarithm map (1.8) as

$$\log^N(\mathbf{a}) := P_N(\log(\tilde{a})),$$

where $\tilde{a} = (a_0, a_1, \ldots, a_N, 0, 0, \ldots) \in T((\mathbb{R}^k))$ and P_N denotes the orthogonal projection map from $T((\mathbb{R}^k))$ onto $T^N(\mathbb{R}^k)$.

Finally, we introduce the log-signature.

Definition 1.14 ([16], Definition A.5). The log-signature of a path $X: [0, T] \to \mathbb{R}^n$ of bounded variation over the interval $[s, t]$ is defined as $\mathrm{LogSig}_{s,t}(X) := \log(S_{s,t}(X))$, where $S_{s,t}(X)$ denotes the signature of X. Likewise, the depth-N log-signature of X is defined for each $N \in \mathbb{N}$ as $\mathrm{LogSig}^N_{s,t}(X) := \log^N(S^N_{s,t}(X))$.

Remark 1.6. While for a path X of bounded variation the signature $S_{s,t}(X)$ (and likewise $S^N_{s,t}(X)$) is an element of the Lie group, the log-signature $\mathrm{LogSig}_{s,t}(X)$ (and likewise $\mathrm{LogSig}^N_{s,t}(X)$) is an element of the Lie algebra. For further information on the Lie theory of rough paths, the authors suggest Chapter 7 in Ref. [10].

With these definitions, we are ready to introduce the log-ODE method.

Definition 1.15 ([16], Definition A.8). Let $\mathbf{X}\colon \Delta_T \to T^{\lfloor p \rfloor}(\mathbb{R}^k)$ be a geometric p-rough path. We consider the ODE

$$\frac{dz}{du} = \text{Taylor}(z(u), f, \text{LogSig}_{s,t}^N(\mathbf{X})),$$
$$z(0) = Y_s, \tag{1.9}$$

where $u \in [0, 1]$. Now, (1.9) is constructed for the purpose of approximating the RDE solution Y in (1.5) at time point t in the sense that

$$Y_t \approx z(1).$$

Remark 1.7. The construction of the vector fields in (1.9) is similar to the Taylor methods in Definition 1.10. The difference is the use of the log-signature instead of the signature.

Remark 1.8. The procedure in Definition 1.15 is usually repeated m times on a partition $D = \{0 = t_0 < t_1 < \cdots < t_m = T\}$. We define the numerical solution of the Nth-level log-ODE method by

$$y_{i+1}^{N\text{-Log-ODE}} := z(1),$$

where z is the solution of

$$\frac{dz}{du} = \text{Taylor}(z(u), f, \text{LogSig}_{t_i, t_{i+1}}^N(\mathbf{X})),$$
$$z(0) = y_i^{N\text{-Log-ODE}},$$

and $y_0^{N\text{-Log-ODE}} = y_0$. Then, $y_i^{N\text{-Log-ODE}} \approx Y_{t_i}$.

We continue with a result on the convergence order of the log-ODE method.

Theorem 1.4 ([16], Theorem B.1). *Assume that:*

- $\mathbf{X}\colon \Delta_T \to T^{\lfloor p \rfloor}(\mathbb{R}^k)$ *is a geometric p-rough path;*
- $f = (f_1, \ldots, f_k)$ *is a collection of* $\mathrm{Lip}(\gamma)$-*vector fields on* $\mathbb{R}^n$ *for* $\gamma > p \geq 1$;
- $D = \{0 = t_0 < t_1 < \cdots < t_m = T\}$ *is a fixed partition of* $[0, T]$, *where* $h = \max\limits_i t_{i+1} - t_i$.

Set $N := \lfloor \gamma \rfloor \geq \lfloor p \rfloor$. *Then, there exists a constant* $C = C(p, \gamma, \|f\|_{\mathrm{Lip}(\gamma)}) > 0$ *so that*

$$|Y_T - y_m^{N\text{-Log-ODE}}| \leq Ch^\alpha,$$

where $\alpha = \frac{N+1}{p} - 1$ *and* h *is sufficiently small.*

We see that the order of convergence for a fixed N for both the log-ODE method and the Taylor method is the same.

Remark 1.9. The assumptions in the above theorem ensure that the right-hand side of the ODE (1.9) is globally bounded and Lipschitz continuous. The above error estimate also holds when the vector field f is linear ([16], Remark B.8).

1.3.4. RK–log-ODE method

On the basis of the introduced Runge–Kutta operator, the authors propose a log-ODE method, which is referred to as the RK-log-ODE method in the following. Since we introduced the Runge–Kutta method only for $N = 2$, we will examine the RK–log-ODE method for $N = 2$ as well.

Definition 1.16. Let $\mathbf{X}\colon \Delta_T \to T^{\lfloor p \rfloor}(\mathbb{R}^k)$ be a geometric p-rough path and $p < 3$, $N = \lfloor p \rfloor$. We consider the ODE

$$\frac{dz}{du} = \text{Runge–Kutta}(z(u), f, \text{LogSig}_{s,t}^N(\mathbf{X})), \qquad (1.10)$$
$$z(0) = Y_s,$$

where $u \in [0, 1]$.

Now, (1.10) is constructed for the purpose of approximating the RDE solution Y in (1.5) at time point t in the sense that

$$Y_t \approx z(1).$$

Let $D = \{0 = t_0 < t_1 < \cdots < t_m = T\}$ be a partition of $[0,T]$. We define the numerical solution of the RK–log-ODE method by

$$y_{i+1}^{\text{RK–Log-ODE}} := \tilde{z}(1),$$

where $\tilde{z}$ is the solution of

$$\frac{d\tilde{z}}{du} = \text{Runge–Kutta}(\tilde{z}(u), f, \text{LogSig}_{t_i, t_{i+1}}^{N}(\mathbf{X})),$$

$$\tilde{z}(0) = y_i^{\text{RK–Log-ODE}},$$

and $y_0^{\text{RK–Log-ODE}} = y_0$. Then, $y_i^{\text{RK–Log-ODE}} \approx Y_{t_i}$.

1.4. Numerical Examples

Before we present our results, we provide an outline of the experiments. Let $\mathbf{X} \colon \Delta_T \to T^{\lfloor p \rfloor}(\mathbb{R}^k)$ be a geometric p-rough path and $f = (f_1, \ldots, f_k) \colon \mathbb{R}^n \to \mathbb{R}^{n \times k}$. We aim to approximate the solution $Y \colon [0,T] \to \mathbb{R}^n$ of the RDE

$$\begin{aligned} dY_t &= f(Y_t)d\mathbf{X}_t, \\ Y_0 &= y_0, \end{aligned} \qquad (1.11)$$

where $t \in [0,T]$, with $T = 1$ and

$$(f(Y_t))_{ij} = \begin{cases} a_i \cdot \cos(j \cdot Y_t^i), & j \text{ even,} \\ a_i \cdot \sin(j \cdot Y_t^i), & j \text{ odd.} \end{cases}$$

The coefficients $a_i \in \mathbb{R}$ are sampled independently from a uniform distribution $U([-1,1])$. We solve this RDE for random initial values y_0 (also uniformly distributed on $[-1,1]^n$) and for varying, growing dimensions n and k.

At first, we comment on the geometric p-rough path $\mathbf{X}$. As the underlying function, we choose a path of a fractional Brownian motion (fBm) B^H with Hurst index $H = 0.4$, i.e.,

$X = \pi_1(\mathbf{X}) = B^H(\omega)$. From analysis, it is known that a function of the Hölder regularity α is of finite p-variation with $p > \frac{1}{\alpha}$. Since the paths of the fBm are almost surely Hölder continuous with $\alpha < H$, X is of finite p-variation for all $p > 2.5$ (see Ref. [22]).

We approximate $\mathbf{X}_{s,t}$ using a sequence of $(x^n)_n$ of finite variation. This is possible by Definition 1.5. In Ref. [23], it is proven that the Wong–Zakai approximation, i.e., a piecewise linear approximation, is a possible choice in Definition 1.5 for a fBm with $\frac{1}{4} < H < \frac{1}{2}$. Since X is of finite p-variation for all $p > 2.5$, we need to compute the first $N = \lfloor p \rfloor = 2$ levels of the signature or log-signature, respectively. While the first level $\pi_1(\mathbf{X}_{s,t})$ is nothing more than the increment $X_{s,t}$, the second level cannot be computed directly and is approximated via $\pi_2(S_{s,t}(x^n))$, which we computed directly using the following lemma.

Lemma 1.2. *Let x^n be piecewise linear for a partition $[s = \tau_0 < \cdots < \tau_n = t]$. Then,*

$$\pi_2(S_{s,t}(x^n)) = \frac{1}{2} \sum_{l=1}^{n} \pi_1(S_{\tau_{l-1},\tau_l}(x^n)) \otimes \pi_1(S_{\tau_{l-1},\tau_l}(x^n))$$

$$+ \sum_{l=1}^{n} \sum_{i=1}^{l} \pi_1(S_{\tau_{i-1},\tau_i}(x^n)) \otimes \pi_1(S_{\tau_{l-1},\tau_l}(x^n)).$$

Proof. The statements holds true in the case of $n = 1$ since

$$\int_{\tau_{l-1}}^{\tau_l} \int_{\tau_{l-1}}^{u} dx_i^n(r) dx_j^n(u) = \frac{1}{2}(x_i^n(\tau_l) - x_i^n(\tau_{l-1}))(x_j^n(\tau_l) - x_j^n(\tau_{l-1}))$$

for $i, j = (1, \ldots, k)$.

The general case follows via induction over n assuming that the statement holds true for $n - 1$. With Chen's identity (1.3) and the arithmetic of the tensor algebra (1.1), it follows that

$$\pi_2(S_{s,t}(x^n)) = \pi_2(S_{\tau_0,\tau_{n-1}}(x^n)) + \pi_2(S_{\tau_{n-1},\tau_n}(x^n))$$
$$+ \pi_1(S_{\tau_0,\tau_{n-1}}(x^n)) \otimes \pi_1(S_{\tau_{n-1},\tau_n}(x^n))$$

$$= \frac{1}{2}\sum_{l=1}^{n-1} \pi_1(S_{\tau_{l-1},\tau_l}(x^n)) \otimes \pi_1(S_{\tau_{l-1},\tau_l}(x^n))$$

$$+ \sum_{l=1}^{n-1}\sum_{i=1}^{l} \pi_1(S_{\tau_{i-1},\tau_i}(x^n)) \otimes \pi_1(S_{\tau_{l-1},\tau_l}(x^n))$$

$$+ \frac{1}{2}\pi_1(S_{\tau_{n-1},\tau_n}(x^n)) \otimes \pi_1(S_{\tau_{n-1},\tau_n}(x^n))$$

$$+ \pi_1(S_{\tau_0,\tau_{n-1}}(x^n)) \otimes \pi_1(S_{\tau_{n-1},\tau_n}(x^n))$$

$$= \frac{1}{2}\sum_{l=1}^{n} \pi_1(S_{\tau_{l-1},\tau_l}(x^n)) \otimes \pi_1(S_{\tau_{l-1},\tau_l}(x^n))$$

$$+ \sum_{l=1}^{n}\sum_{i=1}^{l} \pi_1(S_{\tau_{i-1},\tau_i}(x^n)) \otimes \pi_1(S_{\tau_{l-1},\tau_l}(x^n)).$$

This concludes the proof. □

Then, we compute the second level of the log-signature $\text{LogSig}_{s,t}^N(\mathbf{X})$ via

$$\pi_2(\text{LogSig}_{s,t}^N(\mathbf{X}_{s,t})) = \pi_2(\mathbf{X}_{s,t}) - \frac{1}{2}(\pi_1(\mathbf{X}_{s,t}) \otimes \pi_1(\mathbf{X}_{s,t})).$$

Note that we choose a representation of $\text{LogSig}_{s,t}^N(\mathbf{X})$ on the basis of the truncated tensor algebra $T^N(\mathbb{R}^k)$ rather than on the basis of the Lie algebra.

All implementations are made in Python. There are some more implementation details to clarify. The derivatives in Definitions 1.10 and 1.15 are computed using forward automatic differentiation, where we use the function `torch.func.jvp`. The ODEs in Definitions 1.15 and 1.16 are solved using a single step of the Runge–Kutta 23 scheme from the scipy library using the `solve_ivp` function. The underlying fBm is sampled on a partition of $[0, T]$ with twice as many grid points as the partition of the numerical solutions. For the computation of a reference solution of (1.11), we use Heun's third-order method from Ref. [11] with $m = 2^{15}$ grid points.

Now, we apply the proposed methods. Let "TM" denote the second-level Taylor method (Definition 1.10), and let "Log-ODE"

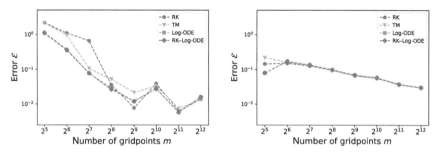

Fig. 1.1. Error vs. grid points for RDE (1.11) with $n = 2$ and $k = 2$ for two different paths.

denote the second-level log-ODE method (Definition 1.15). Finally, let "RK" and "RK–Log-ODE" denote the Runge–Kutta (Definition 1.11) and Runge–Kutta–log-ODE methods (Definition 1.16), respectively. We begin with an investigation of the error

$$\mathcal{E} = |Y_T - y_m|$$

for a partition $0 = t_0 < \cdots < t_m = T$. We analyze the error versus the number of grid points in log-log diagrams. The step size of each method equals $\frac{1}{m}$. Figure 1.1 suggests that the RK and RK–Log-ODE indeed converge. From the convergence analysis of TM (Theorem 1.3) and Log-ODE (Theorem 1.4), we expect to obtain at least an order of convergence of $\alpha = \frac{N+1}{p} - 1 = \frac{3}{2.5} - 1 = 0.2$. Theoretically, the expected order of convergence (EOC) should remain unperturbed either by the use of ODE solvers for the ODEs (1.9) and (1.10), as they have a much higher convergence rate of 3, or by the numerical approximation of $\mathbf{X}$ because this rate, according to Ref. [23], is $2H - 0.5 = 0.3$, which is also greater than $\alpha = 0.2$.

Since the development of the error suffers from the roughness of the fBm, it is not always possible to observe a clear EOC. Therefore, in Figure 1.2, we average the error over 200 independent trajectories of the fBm in order to smooth out perturbations in the EOC. Figure 1.2 suggests a similar convergence behavior for TM, RK, Log-ODE, and RK–Log-ODE. The actual EOC seems to be higher than the theoretically predicted worst-case rate of 0.2. The authors expect the EOC to be $\alpha = 0.3$. This is natural because the total error of the approximation would then be determined by the error of the discretization of the rough path, i.e., $\mathbf{X}_{t_i, t_{i+1}} \approx S^2_{t_i, t_{i+1}}(x^n)$, where

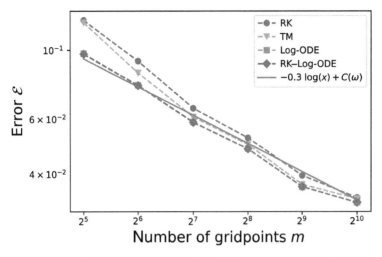

Fig. 1.2. Error vs. grid points for RDE (1.11) with $n = k = 2$.

Table 1.1. Running times (in s) of numerical schemes for RDE (1.11) with varying dimensions n and k but a fixed $m = 2^8$.

	$n = 2$ $k = 2$	$n = 10$ $k = 10$	$n = 100$ $k = 20$	$n = 1000$ $k = 20$
TM	0.25	1.2	3	19
RK	0.01	0.1	0.5	1.6
Log-ODE	1	19	160	789
RK–Log-ODE	0.08	1.9	12	45

this approximation converges with a rate of 0.3. We refer to Remark 1.3 once more, where a similar effect was pointed out.

We complete the section with an investigation of the computational effort required for each method to solve the RDE (1.11). In Table 1.1, we compare the running times for different dimensions and a fixed grid size, while in Table 1.2, we fix the dimensions and vary the grid size. There are two important observations to make from the experiments on the running time. First, the methods based on the Runge–Kutta operator heavily outperform their Taylor counterparts, namely RK is faster than TM and RK–Log-ODE performs better than Log-ODE. Second, one can conclude by simply looking

Table 1.2. Running times (in s) of numerical schemes for RDE (1.11) with varying grid points m but fixed dimensions $n = 2$ and $k = 2$.

	$m = 2^{10}$	$m = 2^{12}$	$m = 2^{14}$	$m = 2^{16}$
TM	1.57	6.3	25	101
RK	0.09	0.35	1.4	5.7
Log-ODE	6.6	29	103	415
RK–Log-ODE	0.45	3.6	7.1	28

at the running times that the TM and RK methods outperform their Log-ODE counterparts. This was not the intention of the experiment. Usually Log-ODE (and RK–Log-ODE) is used on a coarse grid, which saves time, while TM (and RK) works on a very finely partitioned grid. Since we compared all the methods for the same partition size, this outcome is expected.

1.5. Conclusion

In this chapter, we proposed two efficient Runge–Kutta approaches for potentially high-dimensional RDEs driven by geometric p-rough paths with $p < 3$. Compared to well-known methods such as the Taylor schemes (Definition 1.10) and the log-ODE method (Definition 1.15), numerical evidence (see Tables 1.1 and 1.2) suggests a significantly reduced computational cost for the proposed Runge–Kutta scheme (Definition 1.11) and the RK–Log-ODE method (Definition 1.16), especially in the case of high dimensions. However, numerical evidence (see Figures 1.1 and 1.2) also indicates the same order of convergence for all the methods. A numerical analysis of the proposed schemes is an open question, as is the generalization of the proposed schemes to general geometric p-rough paths for $p \geq 3$, which the authors intend to tackle in the future.

Acknowledgments

The work on this study began while MR and JW were supported by the DFG individual grant "Low-order approximations for large-scale problems arising in the context of high-dimensional PDEs and spatially discretized SPDEs" — project number 499366908. Both authors would like to thank Prof. Wilfried Grecksch for his helpful suggestions on the manuscript and Dr. Christian Bayer for discussions on related topics.

References

[1] A. Rößler, Runge–Kutta methods for the strong approximation of solutions of stochastic differential equations. *SIAM Journal on Numerical Analysis.* **48**(3), 922–952 (2010). https://doi.org/10.1137/09076636X.

[2] P. Kloeden and E. Platen, *Numerical Solution of Stochastic Differential Equations.* Stochastic Modelling and Applied Probability. Vol. 23, Springer, Berlin (2011). https://books.google.de/books?id=BCvtssom1CMC.

[3] K. Burrage and P. M. Burrage, Order conditions of stochastic Runge–Kutta methods by B-series. *SIAM Journal on Numerical Analysis.* **38**(5), 1626–1646 (2000). https://doi.org/10.1137/S0036142999363206.

[4] K. Debrabant and A. Kværnø, B-series analysis of stochastic Runge–Kutta methods that use an iterative scheme to compute their internal stage values. *SIAM Journal on Numerical Analysis.* **47**, 181–203 (2008). https://doi.org/10.1137/070704307.

[5] G. Milstein and M. Tretyakov, *Stochastic Numerics for Mathematical Physics.* Springer, Berlin (2004). https://doi.org/10.1007/978-3-662-10063-9.

[6] F. Castell, Asymptotic expansion of stochastic flows. *Probability Theory and Related Fields.* **96**, 225–239 (1993). https://doi.org/10.1007/BF01192134.

[7] F. Castell and J. Gaines, The ordinary differential equation approach to asymptotically efficient schemes for solution of stochastic differential equations (1996). http://www.numdam.org/item/AIHPB_1996__32_2_231_0/.

[8] F. Castell and J. Gaines, An efficient approximation method for stochastic differential equations by means of the exponential Lie series. *Mathematics and Computers in Simulation.* **38**(1), 13–19 (1995). https://doi.org/10.1016/0378-4754(93)E0062-A.

[9] A. M. Davie, Differential equations driven by rough paths: An approach via discrete approximation (2007). https://arxiv.org/abs/0710.0772.

[10] P. K. Friz and N. B. Victoir, *Multidimensional Stochastic Processes as Rough Paths: Theory and Applications.* Cambridge Studies in Advanced Mathematics. Cambridge University Press, Cambridge (2010).

[11] M. Redmann and S. Riedel, Runge-Kutta methods for rough differential equations. *Journal of Stochastic Analysis.* **3**(4) (2022). https://repository.lsu.edu/josa/vol3/iss4/6.

[12] A. Deya, A. Neuenkirch, and S. Tindel, A Milstein-type scheme without Lévy area terms for SDEs driven by fractional Brownian motion. *Annales de l'Institut Henri Poincaré, Probabilités et Statistiques.* **48**(2), 518–550 (2012). http://dx.doi.org/10.1214/10-AIHP392.

[13] J. Hong, C. Huang, and X. Wang, Symplectic Runge–Kutta methods for Hamiltonian systems driven by Gaussian rough paths. *Applied Numerical Mathematics.* **129**, 120–136 (2018). https://doi.org/10.1016/j.apnum.2018.03.006.

[14] T. Lyons, Rough paths, signatures and the modelling of functions on streams (2014). https://arxiv.org/abs/1405.4537.

[15] Y. Boutaib, L. G. Gyurkó, T. Lyons, and D. Yang, Dimension-free Euler estimates of rough differential equations (2013). https://arxiv.org/abs/1307.4708.

[16] J. Morrill, P. Kidger, C. Salvi, J. Foster, and T. J. Lyons, Neural CDEs for long time series via the log-ODE method (2020). *CoRR,* https://arxiv.org/abs/2009.08295.

[17] C. Bayer, S. Breneis, and T. Lyons, An adaptive algorithm for rough differential equations (2023). https://arxiv.org/abs/2307.12590.

[18] R. A. Ryan, *Introduction to Tensor Products of Banach Spaces.* Springer Monographs in Mathematics. Springer, London (2002).

[19] E. Wong and M. Zakai, On the convergence of ordinary integrals to stochastic integrals. *The Annals of Mathematical Statistics.* **36**(5), 1560–1564 (1965). http://www.jstor.org/stable/2238444.

[20] T. J. Lyons, Differential equations driven by rough signals. *Revista Matemática Iberoamericana.* **14**(2), 215–310 (1998). http://eudml.org/doc/39555.

[21] P. Kloeden and A. Neuenkirch, The pathwise convergence of approximation schemes for stochastic differential equations. *LMS Journal of Computation and Mathematics.* **10** (2007). https://doi.org/10.1112/S1461157000001388.

[22] G. Shevchenko, Fractional Brownian motion in a nutshell (2014). https://arxiv.org/abs/1406.1956.

[23] P. Friz and S. Riedel, Convergence rates for the full Gaussian rough paths (2012). https://arxiv.org/abs/1108.1099.

Chapter 2

Mild Solutions to Semilinear Rough Partial Differential Equations

Stefan Tappe

*Department of Mathematical Stochastics,
Albert Ludwig University of Freiburg,
Ernst-Zermelo-Straße 1, D-79104 Freiburg, Germany
stefan.tappe@math.uni-freiburg.de*

We provide an existence and uniqueness result for mild solutions to rough partial differential equations in the framework of the semigroup approach. Applications to stochastic partial differential equations driven by infinite-dimensional Wiener processes and infinite-dimensional fractional Brownian motion are presented as well.

2.1. Introduction

In this chapter, we deal with rough partial differential equations (RPDEs) with time-inhomogeneous coefficients of the form

$$\begin{cases} dY_t = (AY_t + f_0(t, Y_t))dt + f(t, Y_t)d\mathbf{X}_t, \\ Y_0 = \xi. \end{cases} \quad (2.1)$$

The state space of the RPDE (2.1) is a Banach space W, and the driving signal $\mathbf{X} = (X, \mathbb{X}) \in \mathscr{C}^\alpha([0, T], V)$ is a rough path with values in a Banach space V for some index $\alpha \in (\frac{1}{3}, \frac{1}{2}]$. Furthermore, the operator A is the generator of a C_0-semigroup $(S_t)_{t \geq 0}$ on W, and

$f_0 : [0,T] \times W \to W$ and $f : [0,T] \times W \to L(V,W)$ are appropriate mappings. We are interested in the existence of mild solutions, which means that the variations-of-constants formula

$$Y_t = S_t \xi + \int_0^t S_{t-s} f_0(s, Y_s)\, ds + \int_0^t S_{t-s} f(s, Y_s)\, d\mathbf{X}_s, \quad t \in [0,T], \tag{2.2}$$

is satisfied.

By now, there is a well-established theory of rough paths, including rough integration and rough differential equations (RDEs); see the textbook in Ref. [1] and also the lecture notes in Ref. [2]. The goals of the current chapter are to extend the existing RDE results to RPDEs with time-inhomogeneous coefficients of the form (2.1) and to apply these findings to stochastic partial differential equations (SPDEs) driven by infinite-dimensional Wiener processes and to SPDEs driven by infinite-dimensional fractional Brownian motion.

To some extent, RPDEs and various related aspects have already been studied in the literature; we refer, for example, to Refs. [3–9] for results about RPDEs, Refs. [10–12] for RPDE results with a particular focus on fractional Brownian motion, Refs. [13–16] for RPDE results in relation to random dynamical systems, Refs. [17–19] for results about SPDEs driven by Wiener processes, Refs. [20–25] for results concerning SPDEs driven by fractional Brownian motion, and Refs. [26–28] for various related aspects of infinite-dimensional fractional Brownian motion.

One of the principle challenges when dealing with mild solutions to RPDEs is an appropriate definition of the rough convolution

$$\int_0^t S_{t-s} Y_s\, d\mathbf{X}_s, \quad t \in [0,T], \tag{2.3}$$

for controlled rough paths (Y, Y'), which shows up in the variation-of-constants formula (2.2). For this reason, most of the aforementioned papers about RPDE results assume that the semigroup $(S_t)_{t \geq 0}$ is analytic. Considering the scales of Hilbert spaces, this assumption allows to establish a mild sewing lemma and then to provide a definition of the rough convolution (2.3) in the spirit of the Gubinelli integral.

In this chapter, we pursue a different approach. The essential idea is to provide a direct definition of rough convolutions (2.3) by using the well-known Gubinelli integral, without the need for establishing a mild sewing lemma and a new integral definition. This is possible for controlled rough paths $(Y, Y') \in \mathscr{D}_X^{2\alpha}([0, T], L(V, D(A^2)))$, where $D(A^2)$ denotes the second-order domain of the generator; see Section 2.8 for details. Of course, this imposes additional conditions on the coefficients f_0 and f of the RPDE (2.1); however, on the other hand, our results (see, in particular, Theorem 2.3) hold true for arbitrary C_0-semigroups $(S_t)_{t \geq 0}$. Moreover, in contrast to most of the aforementioned papers, we treat RPDEs with time-inhomogeneous coefficients.

The remainder of this chapter is organized as follows. In Section 2.2, we briefly explain some frequently used notation. In Section 2.3, we provide the required background about rough path theory. In Section 2.4, we provide the required results about strongly continuous semigroups. In Section 2.5, we treat compositions of regular paths with functions, and in Section 2.6, we deal with compositions of controlled rough paths with functions. In Section 2.7, we consider regular convolution integrals, and in Section 2.8, we investigate rough convolution integrals. After these preparations, in Section 2.9, we deal with RPDEs; in particular, we establish the announced existence and uniqueness result. Afterward, in Section 2.10, we apply our findings to SPDEs driven by infinite-dimensional Wiener processes, and in Section 2.11, we apply our findings to SPDEs driven by infinite-dimensional fractional Brownian motion.

2.2. Frequently Used Notation

In this section, we briefly explain some frequently used notation. The symbols V and W typically denote Banach spaces equipped with their respective norms, always written as $|\cdot|$. The space $L(V, W)$ of all continuous linear operators $T : V \to W$ is also a Banach space. We also use the notation $L(V) = L(V, V)$. On the product space $V \times W$, we always consider the norm

$$|v, w| = |v| + |w|,$$

which makes $V \times W$ a Banach space. For another Banach space E, we denote by $L^{(2)}(V \times W, E)$ the space of all continuous bilinear operators $T : V \times W \to E$.

For a bounded function $f : V \to W$, we denote by $\|f\|_\infty$ the supremum norm. For $n \in \mathbb{N}$, we denote by $C^n = C^n(V, W)$ the space of all mappings $f : V \to W$ which are n-times continuously Fréchet differentiable. Furthermore, we denote by $C_b^n = C_b^n(V, W)$ the subspace of all $f \in C^n$ such that

$$\|f\|_{C_b^n} := \sum_{k=0}^{n} \|D^k f\|_\infty < \infty.$$

The space $\mathrm{Lip}(V, W)$ denotes the space of all Lipschitz continuous functions, that is, the space of all functions $f : V \to W$ such that

$$|f|_{\mathrm{Lip}} := \sup_{\substack{x,y \in V \\ x \neq y}} \frac{|f(x) - f(y)|}{|x - y|} < \infty.$$

Let $T \in \mathbb{R}_+$ and $\alpha \in (0, 1]$ be arbitrary. We denote by $\mathcal{C}^\alpha([0, T], V)$ the space of all α-Hölder continuous functions, that is, the space of all functions $X : [0, T] \to V$ such that

$$\|X\|_\alpha := \sup_{\substack{s,t \in [0,T] \\ s \neq t}} \frac{|X_{s,t}|}{|t - s|^\alpha} < \infty.$$

Let $X \in \mathcal{C}^\alpha([0, T], V)$ be arbitrary. For a subinterval $I \subset [0, T]$, we use the notation

$$\|X\|_{\alpha;I} := \sup_{\substack{s,t \in I \\ s \neq t}} \frac{|X_{s,t}|}{|t - s|^\alpha}.$$

Moreover, for any $h > 0$, the quantity $\|X\|_{\alpha;h}$ denotes the supremum over all $\|X\|_{\alpha;I}$, where $I \subset [0, T]$ is any subinterval with $|I| \leq h$.

2.3. Rough Path Theory

In this section, we provide the required background about rough path theory. Further details can be found, for example, in Ref. [1] or Ref. [2].

Let V be a Banach space, and let $T \in \mathbb{R}_+$ be a time horizon. In the sequel, a function $X : [0, T] \to V$ will be called a *path*. For a path $X : [0, T] \to V$, we agree on the notation

$$X_{s,t} := X_t - X_s, \quad s, t \in [0, T].$$

Definition 2.1. Let $\alpha \in (\frac{1}{3}, \frac{1}{2}]$ be arbitrary. We define the space $\mathscr{C}^\alpha([0, T], V)$ of all α-Hölder rough paths (over V) as the space of all pairs $\mathbf{X} = (X, \mathbb{X})$ with paths $X : [0, T] \to V$ and $\mathbb{X} : [0, T]^2 \to V \otimes V$ such that

$$\|X\|_\alpha := \sup_{\substack{s,t \in [0,T] \\ s \neq t}} \frac{|X_{s,t}|}{|t-s|^\alpha} < \infty,$$

$$\|\mathbb{X}\|_{2\alpha} := \sup_{\substack{s,t \in [0,T] \\ s \neq t}} \frac{|\mathbb{X}_{s,t}|}{|t-s|^{2\alpha}} < \infty,$$

and *Chen's relation*

$$\mathbb{X}_{s,t} - \mathbb{X}_{s,u} - \mathbb{X}_{u,t} = X_{s,u} \otimes X_{u,t} \quad \text{for all } s, u, t \in [0, T] \tag{2.4}$$

is satisfied.

Remark 2.1. The path $\mathbb{X}$ of a rough path $\mathbf{X} = (X, \mathbb{X})$ is also called the associated *second-order process* or *Lévy area*.

Definition 2.2. For a rough path $\mathbf{X} = (X, \mathbb{X}) \in \mathscr{C}^\alpha([0, T], V)$, for some index $\alpha \in (\frac{1}{3}, \frac{1}{2}]$, we introduce

$$\|\mathbf{X}\|_\alpha := \|X\|_\alpha + \|\mathbb{X}\|_{2\alpha}.$$

Lemma 2.1. *Let $\beta \in (\frac{1}{3}, \frac{1}{2}]$ and $\alpha \in (0, \beta)$ be arbitrary. Then, for each rough path $\mathbf{X} = (X, \mathbb{X}) \in \mathscr{C}^\beta([0, T], V)$, we also have $\mathbf{X} \in \mathscr{C}^\alpha([0, T], V)$ and the estimate*

$$\|\mathbf{X}\|_\alpha \leq \|X\|_\beta T^{\beta-\alpha} + \|\mathbb{X}\|_{2\beta} T^{2(\beta-\alpha)}.$$

Proof. Since $\alpha < \beta$, we have $\mathbf{X} \in \mathscr{C}^\alpha([0, T], V)$. Hence, the calculation

$$\|\mathbf{X}\|_\alpha = \|X\|_\alpha + \|\mathbb{X}\|_{2\alpha}$$
$$\leq \|X\|_\beta T^{\beta-\alpha} + \|\mathbb{X}\|_{2\beta} T^{2(\beta-\alpha)}$$

establishes the proof. $\square$

For what follows, we fix an index $\alpha \in (\frac{1}{3}, \frac{1}{2}]$. The following result is well known.

Lemma 2.2. *The space $\mathscr{C}^\alpha([0,T], V)$, equipped with the metric*

$$d_\alpha(\mathbf{X}, \mathbf{Y}) = |X_0 - Y_0| + |||\mathbf{X} - \mathbf{Y}|||_\alpha,$$

is a complete metric space.

Definition 2.3. We define the space $\mathscr{C}_g^\alpha([0,T], V)$ of *weakly geometric rough paths* as the set of all $(X, \mathbb{X}) \in \mathscr{C}^\alpha([0,T], V)$ such that

$$\mathrm{Sym}(\mathbb{X}_{s,t}) = \frac{1}{2} X_{s,t} \otimes X_{s,t} \quad \text{for all } s, t \in [0,T].$$

We list two well-known auxiliary results about weakly geometric rough paths.

Lemma 2.3. *Let $X : [0,T] \to V$ be a continuous path which is piecewise of class C^1. More precisely, there are $n \in \mathbb{N}$ and $0 = t_0 < t_1 < \cdots < t_n = T$ such that $X|_{(t_{i-1}, t_i)} \in C^1((t_{i-1}, t_i), V)$ for all $i = 1, \ldots, n$. Then, we have $\mathbf{X} = (X, \mathbb{X}) \in \mathscr{C}_g^\alpha([0,T], V)$, where the second-order process $\mathbb{X} : [0,T]^2 \to V \otimes V$ is given by*

$$\mathbb{X}_{s,t} = \int_s^t X_{s,r} \otimes dX_r, \quad s, t \in [0,T].$$

Lemma 2.4. *$\mathscr{C}_g^\alpha([0,T], V)$ is a closed subset of $\mathscr{C}^\alpha([0,T], V)$ with respect to the metric d_α.*

For what follows, we fix a rough path $\mathbf{X} = (X, \mathbb{X}) \in \mathscr{C}^\alpha([0,T], V)$. Furthermore, let W be another Banach space.

Definition 2.4. We introduce the following notions:

(a) We say that a path $Y \in \mathcal{C}^\alpha([0,T], W)$ is *controlled* by X if there exists $Y' \in \mathcal{C}^\alpha([0,T], L(V, W))$ such that $\|R^Y\|_{2\alpha} < \infty$, where the remainder term R^Y is given by

$$R_{s,t}^Y := Y_{s,t} - Y_s' X_{s,t} \quad \text{for all } s, t \in [0,T].$$

(b) This defines the space of *controlled rough paths*, and we write
$$(Y, Y') \in \mathscr{D}_X^{2\alpha}([0,T], W).$$

(c) We call Y' a *Gubinelli derivative* of Y (with respect to X).

We endow the space $\mathscr{D}_X^{2\alpha}([0,T], W)$ with the seminorm
$$\|Y, Y'\|_{X, 2\alpha} := \|Y'\|_\alpha + \|R^Y\|_{2\alpha}.$$

Then, the space $\mathscr{D}_X^{2\alpha}([0,T], W)$ endowed with the norm
$$\|\|Y, Y'\|\|_{X, 2\alpha} := |Y_0| + |Y'_0| + \|Y, Y'\|_{X, 2\alpha}$$

is a Banach space. We also consider the seminorm
$$|Y, Y'|_{X, 2\alpha} := |Y'_0| + \|Y, Y'\|_{X, 2\alpha}.$$

Lemma 2.5. *For each $Y \in \mathcal{C}^{2\alpha}([0,T], W)$, we have $(Y, 0) \in \mathscr{D}_X^{2\alpha}([0,T], W)$ with*
$$\|Y, Y'\|_{X, 2\alpha} = \|Y\|_{2\alpha}.$$

Proof. Since $Y' = 0$, we have $R^Y = Y$, proving the statement. □

Lemma 2.6. *For all $(Y, Y') \in \mathscr{D}_X^{2\alpha}([0,T], W)$, we have*
$$\|Y'\|_\infty \leq C|Y, Y'|_{X, 2\alpha}, \tag{2.5}$$
$$\|Y\|_\alpha \leq C|Y, Y'|_{X, 2\alpha}(\|\|\mathbf{X}\|\|_\alpha + T^\alpha), \tag{2.6}$$
$$\|Y\|_\infty \leq C \|\|Y, Y'\|\|_{X, 2\alpha}(\|\|\mathbf{X}\|\|_\alpha + 2), \tag{2.7}$$

where the constant $C > 0$ depends on α and T. Furthermore, for $T \leq 1$, the estimates (2.5)–(2.7) hold true with $C = 1$.

Proof. For each $t \in [0, T]$, we have
$$|Y'_t| \leq |Y'_0| + |Y'_t - Y'_0| \leq |Y'_0| + \|Y'\|_\alpha t^\alpha,$$

and hence

$$\|Y'\|_\infty \leq |Y'_0| + \|Y'\|_\alpha T^\alpha \leq \max\{1, T^\alpha\} |Y, Y'|_{X, 2\alpha},$$

proving (2.5). Now, let $s, t \in [0, T]$ be arbitrary. Recalling that $Y_{s,t} = Y'_s X_{s,t} + R^Y_{s,t}$, we have

$$|Y_{s,t}| \leq \|Y'\|_\infty |X_{s,t}| + |R^Y_{s,t}|$$
$$\leq \|Y'\|_\infty \|X\|_\alpha |t-s|^\alpha + \|R^Y_{s,t}\|_{2\alpha} |t-s|^{2\alpha}$$
$$\leq \|Y'\|_\infty \|X\|_\alpha |t-s|^\alpha + \|R^Y_{s,t}\|_{2\alpha} T^\alpha |t-s|^\alpha.$$

Therefore, we obtain

$$\|Y\|_\alpha \leq \|Y'\|_\infty \|X\|_\alpha + \|R^Y\|_{2\alpha} T^\alpha$$
$$\leq \max\{1, T^\alpha\} |Y, Y'|_{X, 2\alpha} \|X\|_\alpha + \|R^Y\|_{2\alpha} T^\alpha$$
$$\leq \max\{1, T^\alpha\} |Y, Y'|_{X, 2\alpha} (\|\mathbf{X}\|_\alpha + T^\alpha),$$

showing (2.6). Using this estimate, we deduce that

$$\|Y\|_\infty \leq |Y_0| + \|Y\|_\alpha T^\alpha$$
$$\leq \|Y, Y'\|_{X, 2\alpha} + \max\{1, T^\alpha\} T^\alpha \|Y, Y'\|_{X, 2\alpha} (\|\mathbf{X}\|_\alpha + T^\alpha)$$
$$\leq \|Y, Y'\|_{X, 2\alpha} \left(\max\{1, T^\alpha\} T^\alpha \|\mathbf{X}\|_\alpha + 1 + \max\{1, T^\alpha\} T^{2\alpha} \right),$$

proving (2.7). □

Lemma 2.7. *Let E be another Banach space. Then, for all $(Y, Y') \in \mathscr{D}_X^{2\alpha}([0, T], W)$ and $(Z, Z') \in \mathscr{D}_X^{2\alpha}([0, T], E)$, we have*

$$((Y, Z), (Y, Z)') \in \mathscr{D}_X^{2\alpha}([0, T], W \times E), \tag{2.8}$$

where the Gubinelli derivative is given by $(Y, Z)' = (Y', Z')$, and we have the estimates

$$\|(Y, Z), (Y, Z)'\|_{X, 2\alpha} \leq \|Y, Y'\|_{X, 2\alpha} + \|Z, Z'\|_{X, 2\alpha}, \tag{2.9}$$

$$|(Y, Z), (Y, Z)'|_{X, 2\alpha} \leq |Y, Y'|_{X, 2\alpha} + |Z, Z'|_{X, 2\alpha}, \tag{2.10}$$

$$\|(Y, Z), (Y, Z)'\|_{X, 2\alpha} \leq \|Y, Y'\|_{X, 2\alpha} + \|Z, Z'\|_{X, 2\alpha} \tag{2.11}$$

Proof. Let $s, t \in [0,T]$ be arbitrary. Then, we have

$$|(Y_t, Z_t) - (Y_s, Z_s)| = |(Y_t - Y_s, Z_t - Z_s)| = |Y_t - Y_s| + |Z_t - Z_s|$$
$$\leq (\|Y\|_\alpha + \|Z\|_\alpha)|t - s|^\alpha,$$

showing that $(Y, Z) \in \mathcal{C}^\alpha([0,T], W_1 \times W_2)$. Similarly, we have

$$|(Y'_t, Z'_t) - (Y'_s, Z'_s)| = |(Y'_t - Y'_s, Z'_t - Z'_s)| = |Y'_t - Y'_s| + |Z'_t - Z'_s|$$
$$\leq (\|Y'\|_\alpha + \|Z'\|_\alpha)|t - s|^\alpha,$$

showing that $(Y', Z') \in \mathcal{C}^\alpha([0,T], L(V, W_1 \times W_2))$. Furthermore, we have

$$R^{(Y,Z)}_{s,t} = (Y_{s,t}, Z_{s,t}) - (Y'_s, Z'_s) X_{s,t} = (R^Y_{s,t}, R^Z_{s,t}),$$

and hence

$$|R^{(Y,Z)}_{s,t}| = |R^Y_{s,t}| + |R^Z_{s,t}| \leq (\|R^Y\|_{2\alpha} + \|R^Z\|_{2\alpha})|t - s|^{2\alpha},$$

proving (2.8). Moreover, the previous calculations reveal that

$$\|(Y', Z')\|_\alpha \leq \|Y'\|_\alpha + \|Z'\|_\alpha,$$
$$\|R^{(Y,Z)}\|_{2\alpha} \leq \|R^Y\|_{2\alpha} + \|R^Z\|_{2\alpha}.$$

Therefore, the estimates (2.9)–(2.11) are an immediate consequence. □

Now, let $(Y, Y') \in \mathscr{D}^{2\alpha}_X([0,T], L(V, W))$ be a controlled rough path. Noting that $L(V, L(V, W)) \hookrightarrow L(V \otimes V, W)$, we consider the candidate

$$\int_0^1 Y_s \, d\mathbf{X}_s := \lim_{|\Pi| \to 0} \sum_{[s,t] \in \Pi} (Y_s X_{s,t} + Y'_s \mathbb{X}_{s,t}) \quad (2.12)$$

for the rough integral of Y against $\mathbf{X}$. For the following result, we refer to Ref. [1, Theorem 4.10].

Theorem 2.1 (Gubinelli). *For each controlled rough path* $(Y, Y') \in \mathscr{D}^{2\alpha}_X([0,T], L(V, W))$, *the following statements are true:*

(a) *The W-valued rough integral (2.12) exists.*
(b) *For all $s, t \in [0, T]$, we have*
$$\left| \int_s^t Y_r \, d\mathbf{X}_r - Y_s X_{s,t} - Y'_s \mathbb{X}_{s,t} \right|$$
$$\leq C \big(\|X\|_\alpha \|R^Y\|_{2\alpha} + \|\mathbb{X}\|_{2\alpha} \|Y'\|_\alpha \big) |t-s|^{3\alpha},$$
with a constant $C > 0$ depending on α.
(c) *The map*
$$\mathscr{D}_X^{2\alpha}([0,T], L(V,W)) \to \mathscr{D}_X^{2\alpha}([0,T], W),$$
$$(Y, Y') \mapsto (Z, Z') := \left(\int_0^\cdot Y_s \, d\mathbf{X}_s, Y \right)$$
is a continuous linear operator between Banach spaces, and we have
$$\|Z, Z'\|_{X, 2\alpha} \leq \|Y\|_\alpha + \|Y'\|_\infty \|\mathbb{X}\|_{2\alpha}$$
$$+ C T^\alpha \big(\|X\|_\alpha \|R^Y\|_{2\alpha} + \|\mathbb{X}\|_{2\alpha} \|Y'\|_\alpha \big),$$
with a constant $C > 0$ depending on α.

Lemma 2.8. *Let E be another Banach space such that $E \subset W$ as sets, and there is a constant $C > 0$ such that $|y|_W \leq C |y|_E$ for all $y \in E$. Then, for each controlled rough path $(Y, Y') \in \mathscr{D}_X^{2\alpha}([0,T], L(V,E))$, we also have $(Y, Y') \in \mathscr{D}_X^{2\alpha}([0,T], L(V,W))$ and the identity*
$$(E\text{-}) \int_0^1 Y_s \, d\mathbf{X}_s = (W\text{-}) \int_0^1 Y_s \, d\mathbf{X}_s,$$
where the integrals above denote the rough integrals (2.12) in the Banach spaces $(E, |\cdot|_E)$ and $(W, |\cdot|_W)$.

Proof. The result is easily verified, noting that for any sequence $(y_n)_{n \in \mathbb{N}} \subset E$, the convergence $|y_n - y|_E \to 0$ implies $|y_n - y|_W \to 0$. □

Proposition 2.1. *Let $(Y, Y') \in \mathscr{D}_X^{2\alpha}([0,T], L(V,W))$ be a controlled rough path. Then, for all $s, t \in [0, T]$, we have*
$$\left| \int_s^t Y_r \, d\mathbf{X}_r \right| \leq C \big(\|X\|_\alpha \|R^Y\|_{2\alpha} + \|\mathbb{X}\|_{2\alpha} \|Y'\|_\alpha \big) |t-s|^{3\alpha}$$
$$+ \|Y'\|_\infty \|\mathbb{X}\|_{2\alpha} |t-s|^{2\alpha} + \|Y\|_\infty \|X\|_\alpha |t-s|^\alpha,$$
with a constant $C > 0$ depending on α.

Proof. Let $s, t \in [0, T]$ be arbitrary. Then, we have

$$\left|\int_s^t Y_r \, d\mathbf{X}_r\right| \leq \left|\int_s^t Y_r \, d\mathbf{X}_r - Y_s X_{s,t} - Y'_s \mathbb{X}_{s,t}\right| + |Y_s X_{s,t}| + |Y'_s \mathbb{X}_{s,t}|.$$

Noting that

$$|Y_s X_{s,t}| \leq |Y_s| \, |X_{s,t}| \leq \|Y\|_\infty \|X\|_\alpha |t - s|^\alpha,$$
$$|Y'_s \mathbb{X}_{s,t}| \leq |Y'_s| \, |\mathbb{X}_{s,t}| \leq \|Y'\|_\infty \|\mathbb{X}\|_{2\alpha} |t - s|^{2\alpha},$$

the statement is a consequence of Theorem 2.1. □

For the following rough Fubini theorem, see Ref. [1, Exercise 4.10].

Proposition 2.2 (Rough Fubini). *Let* $(\Omega, \mathcal{F}, \mu)$ *be a measure space. Furthermore, let* $(Y, Y') : \Omega \to \mathscr{D}_X^{2\alpha}([0,T], L(V, W))$ *be a measurable mapping such that*

$$\Omega \to \mathbb{R}_+, \quad \omega \mapsto |Y(\omega)_0| + \|Y(\omega), Y(\omega)'\|_{X, 2\alpha}$$

belongs to $\mathcal{L}^1(\Omega, \mathcal{F}, \mu)$. *Then, the mapping*

$$\Omega \to W, \quad \omega \mapsto \int_0^T Y(\omega)_s \, d\mathbf{X}_s$$

belongs to $\mathcal{L}^1(\Omega, \mathcal{F}, \mu; W)$, *while the mapping*

$$s \mapsto \int_\Omega Y(\omega)_s \, \mu(d\omega)$$

belongs to $\mathscr{D}_X^{2\alpha}([0,T], L(V,W))$, *and we have the identity*

$$\int_\Omega \left(\int_0^T Y(\omega)_s \, d\mathbf{X}_s\right) \mu(d\omega) = \int_0^T \left(\int_\Omega Y(\omega)_s \, \mu(d\omega)\right) d\mathbf{X}_s.$$

2.4. Strongly Continuous Semigroups

In this section, we provide the required results concerning strongly continuous semigroups. Further details can be found, for example, in Ref. [29] or Ref. [30]. Let W be a Banach space. A family $(S_t)_{t \geq 0}$ of continuous linear operators $S_t \in L(W)$ is called a *strongly continuous semigroup* (or C_0-*semigroup*) on W if the following conditions are fulfilled:

(1) $S_0 = \mathrm{Id}$.
(2) $S_{s+t} = S_s S_t$ for all $s, t \geq 0$.
(3) $\lim_{t \to 0} S_t y = y$ for all $y \in W$.

Let $(S_t)_{t \geq 0}$ be a C_0-semigroup on W. It is well known that there are constants $M \geq 1$ and $\omega \in \mathbb{R}$ such that

$$|S_t| \leq M e^{\omega t} \quad \text{for all } t \geq 0. \tag{2.13}$$

If we can choose $M = 1$ and $\omega = 0$ in (2.13), that is, $|S_t| \leq 1$ for all $t \geq 0$, then we call the C_0-semigroup $(S_t)_{t \geq 0}$ a *semigroup of contractions*. Recall that the *infinitesimal generator* $A : W \supset D(A) \to W$ is the operator

$$A y := \lim_{h \to 0} \frac{S_h y - y}{h}$$

defined on the *domain*

$$D(A) := \left\{ y \in W : \lim_{h \to 0} \frac{S_h y - y}{h} \text{ exists} \right\}.$$

The following result is well known.

Lemma 2.9. *Let $(S_t)_{t \geq 0}$ be a C_0-semigroup with infinitesimal generator A. Then, the following statements are true:*

(a) *For every $y \in D(A)$, the mapping*

$$\mathbb{R}_+ \to W, \quad t \mapsto S_t y$$

belongs to the class $C^1(\mathbb{R}_+; W)$.

(b) *For all $t \geq 0$ and $y \in D(A)$, we have $S_t y \in D(A)$ and*

$$\frac{d}{dt} S_t y = A S_t y = S_t A y.$$

(c) *For all $t \geq 0$ and $y \in W$, we have $\int_0^t S_s y \, ds \in D(A)$ and*

$$A \left(\int_0^t S_s y \, ds \right) = S_t y - y.$$

(d) *For all $t \geq 0$ and $y \in D(A)$, we have*

$$\int_0^t S_s Ay\, ds = S_t y - y.$$

The domain $D(A)$ equipped with the graph norm

$$|y|_{D(A)} := |y| + |Ay| \quad \text{for all } y \in D(A)$$

is a Banach space. Inductively, for each $n \geq 2$, we define the *higher-order domain*

$$D(A^n) := \{y \in D(A^{n-1}) : A^{n-1}y \in D(A)\}.$$

Furthermore, we agree on the notation $D(A^0) := W$. Then, for each $n \in \mathbb{N}_0$, the space $D(A^n)$ equipped with the graph norm

$$|y|_{D(A^n)} := |y| + \sum_{j=1}^n |A^j y| \quad \text{for all } y \in D(A^n)$$

is a Banach space. Using Lemma 2.9, the upcoming auxiliary result is easy to verify.

Lemma 2.10. *For every $n \in \mathbb{N}$, the restriction $(S_t|_{D(A^n)})_{t \geq 0}$ is a C_0-semigroup on $(D(A^n), |\cdot|_{D(A^n)})$, with generator A on the domain $D(A^{n+1})$.*

The following auxiliary result is also straightforward to check.

Lemma 2.11. *For all $m, n \in \mathbb{N}_0$, the graph norm $|\cdot|_{D(A^{m+n})}$ and*

$$|y|_{D(A^m)} + \sum_{j=1}^n |A^j y|_{D(A^m)}, \quad y \in D(A^{m+n})$$

are equivalent norms on the space $D(A^{m+n})$.

For what follows, we agree on the notation

$$S_{s,t} := S_t - S_s \quad \text{for all } s, t \in \mathbb{R}_+.$$

Furthermore, we fix some $T \in \mathbb{R}_+$.

Proposition 2.3. *For all $y \in D(A)$ and all $s, t \in [0, T]$, we have*

$$|S_{s,t} y| \leq M e^{\omega T} |y|_{D(A)} |t - s|.$$

Proof. By Lemma 2.9 and the growth estimate (2.13), we obtain

$$|S_{s,t}y| = |S_t y - S_s y| = \left| \int_s^t S_u A y \, du \right| \leq \int_s^t |S_u A y| \, du$$

$$\leq \int_s^t |S_u| \, |Ay| \, du \leq M e^{\omega T} |y|_{D(A)} |t - s|,$$

completing the proof. □

Corollary 2.1. *For all $s, t \in [0, T]$, we have*

$$|S_{s,t}|_{L(D(A),W)} \leq M e^{\omega T} |t - s|.$$

Corollary 2.2. *Let V be another Banach space. Then, for all $Y \in L(V, D(A))$ and all $s, t \in [0, T]$, we have*

$$|S_{s,t} Y|_{L(V,W)} \leq M e^{\omega T} |Y|_{L(V,D(A))} |t - s|.$$

The following result will be crucial for the analysis of rough convolutions later on.

Proposition 2.4. *Let $y \in D(A^2)$ be arbitrary. Then, we have*

$$|S_{s-r,t-r} y - S_{s-q,t-q} y| \leq M e^{2\omega T} |y|_{D(A^2)} |t - s| \, |r - q| \quad (2.14)$$

for all $s, t \in [0, T]$, with $s \leq t$ and all $q, r \in [0, s]$.

Before we provide the proof of Proposition 2.4, let us state some consequences.

Corollary 2.3. *For all $s, t \in [0, T]$, with $s \leq t$ and all $q, r \in [0, s]$, we have*

$$|S_{s-r,t-r} - S_{s-q,t-q}|_{L(D(A^2),W)} \leq M e^{\omega T} |t - s| \, |r - q|.$$

Corollary 2.4. *Let V be another Banach space, and let $Y \in L(V, D(A^2))$ be arbitrary. Then, for all $s, t \in [0, T]$ with $s \leq t$ and all $q, r \in [0, s]$, we have*

$$|S_{s-r,t-r} Y - S_{s-q,t-q} Y|_{L(V,W)} \leq M e^{\omega T} |Y|_{L(V,D(A^2))} |t - s| \, |r - q|.$$

Now, we prepare some auxiliary results for the proof of Proposition 2.4.

Lemma 2.12. *Suppose that* $(S_t)_{t\in\mathbb{R}}$ *is a* C_0-*group satisfying*

$$|S_t| \leq e^{\omega|t|} \quad \text{for all } t \in \mathbb{R}.$$

Let $y \in D(A^2)$ *be arbitrary. Then, we have* (2.14) *for all* $s, t \in [0, T]$, *with* $s \leq t$ *and all* $q, r \in [0, s]$.

Proof. Using Proposition 2.3 and Lemma 2.10, we obtain

$$\begin{aligned}
|S_{s-r,t-r}y - S_{s-q,t-q}y| &= |(S_{t-r} - S_{s-r})y - (S_{t-q} - S_{s-q})y| \\
&= |(S_t - S_s)(S_{-r} - S_{-q})y| \\
&\leq e^{\omega T} |(S_{-r} - S_{-q})y|_{D(A)} |t - s| \\
&\leq e^{2\omega T} |y|_{D(A^2)} |t - s| |r - q|,
\end{aligned}$$

completing the proof. □

The following auxiliary result is easily verified.

Lemma 2.13. *Let* $(S_t)_{t\geq 0}$ *be a* C_0-*semigroup with generator* A *satisfying* (2.13). *Then, the family* $(T_t)_{t\geq 0}$ *given by* $T_t := e^{-\omega t} S_t$ *for all* $t \geq 0$ *is a* C_0-*semigroup with generator* $B := A - \omega$ *and*

$$|T_t| \leq M \quad \text{for all } t \geq 0. \tag{2.15}$$

The following auxiliary result is a consequence of Lemma I.5.1 and the proof of Theorem I.5.2 in Ref. [30].

Lemma 2.14. *Let* $(T_t)_{t\geq 0}$ *be a* C_0-*semigroup satisfying* (2.15). *Then, there exists an equivalent norm* $\|\cdot\|$ *on* W *such that*

$$|y| \leq \|y\| \leq M|y| \quad \text{for all } y \in W, \tag{2.16}$$

and $(T_t)_{t\geq 0}$ *is a semigroup of contractions under the norm* $\|\cdot\|$.

Let $(T_t)_{t\geq 0}$ be a semigroup of contractions with generator B. By the Hille–Yosida theorem, we have $(0, \infty) \subset \rho(B)$, where $\rho(B)$ denotes the resolvent set of B. For $\lambda > 0$, we define the *Yosida approximation* $B_\lambda \in L(W)$ as

$$B_\lambda := \lambda A R(\lambda, B) = \lambda^2 R(\lambda, B) - \lambda,$$

where $R(\lambda, B) \in L(W)$ denotes the resolvent given by

$$R(\lambda, B) := (\lambda - B)^{-1}.$$

Using Lemmas I.3.3, I.3.4, and equations (3.12) and (3.14) in Ref. [30], we obtain the following auxiliary result.

Lemma 2.15. *Let $(T_t)_{t \geq 0}$ be a C_0-semigroup of contractions with generator B. Then, the following statements are true:*

(1) *For each $\lambda > 0$, the family $(e^{tB_\lambda})_{t \in \mathbb{R}}$ is a uniformly continuous group of contractions.*
(2) *For all $t \geq 0$ and $y \in W$, we have*

$$\lim_{\lambda \to \infty} e^{tB_\lambda} y = T_t y.$$

(3) *For all $t \geq 0$ and $y \in D(B)$, we have*

$$\lim_{\lambda \to \infty} B_\lambda y = By.$$

Proof. Lemmas I.3.3, I.3.4, and equation (3.14) in Ref. [30]. □

Now, we are ready to provide the proof of Proposition 2.4.

Proof of Proposition 2.4. By Lemma 2.13, the family $(T_t)_{t \geq 0}$ given by $T_t := e^{-\omega t} S_t$ for all $t \geq 0$ is a C_0-semigroup with generator $B := A - \omega$, and we have (2.15). According to Lemma 2.14, there exists an equivalent norm $\|\cdot\|$ on W such that we have (2.16), and $(T_t)_{t \geq 0}$ is a semigroup of contractions under the norm $\|\cdot\|$. For $\lambda > 0$, we define the family $(T_t^\lambda)_{t \in \mathbb{R}}$ as $T_t^\lambda := e^{tB_\lambda}$ for all $t \in \mathbb{R}$. According to Lemma 2.15, the family $(T_t^\lambda)_{t \in \mathbb{R}}$ is a uniformly continuous group of contractions, and we have

$$\lim_{\lambda \to \infty} T_t^\lambda y = T_t y \quad \text{for all } t \geq 0 \text{ and } y \in W.$$

Now, for $\lambda > 0$, define the family $(S_t^\lambda)_{t \in \mathbb{R}}$ as $S_t^\lambda := e^{\omega t} T_t^\lambda$ for all $t \in \mathbb{R}$. Then, $(S_t^\lambda)_{t \in \mathbb{R}}$ is a uniformly continuous C_0-group satisfying

$$\|S_t^\lambda\| \leq e^{\omega |t|} \quad \text{for all } t \in \mathbb{R},$$

and we have
$$\lim_{\lambda\to\infty} S_t^\lambda y = S_t y \quad \text{for all } t \geq 0 \text{ and } y \in W. \tag{2.17}$$

Now, let $s, t \in [0, T]$ with $s \leq t$ and $q, r \in [0, s]$ be arbitrary. Using (2.16), (2.17), and Lemma 2.12, we obtain

$$|(S_{t-r} - S_{s-r})y - (S_{t-q} - S_{s-q})y|$$
$$\leq \|(S_{t-r} - S_{s-r})y - (S_{t-q} - S_{s-q})y\|$$
$$= \lim_{\lambda\to\infty} \|(S^\lambda_{t-r} - S^\lambda_{s-r})y - (S^\lambda_{t-q} - S^\lambda_{s-q})y\|$$
$$\leq \lim_{\lambda\to\infty} e^{2\omega T} \|y\|_{D(A_\lambda^2)} |t-s| \cdot |r-q|.$$

Furthermore, by Lemmas 2.15, 2.10, and (2.16), we have

$$\lim_{\lambda\to\infty} \|y\|_{D(A_\lambda^2)} = \lim_{\lambda\to\infty} \left(\|y\| + \|A_\lambda y\| + \|A_\lambda^2 y\|\right)$$
$$= \|y\| + \|Ay\| + \|A^2 y\|$$
$$\leq M\bigl(|y| + |Ay| + |A^2 y|\bigr) = M|y|_{D(A)},$$

completing the proof. $\square$

2.5. Compositions of Regular Paths with Functions

In this section, we briefly deal with compositions of regular paths with functions. Let us fix a time horizon $T \in \mathbb{R}_+$, and let $W, \bar{W}$ be Banach spaces.

Definition 2.5. We denote by $\text{Lip}([0,T] \times W, \bar{W})$ the space of all continuous functions $f_0 : [0,T] \times W \to \bar{W}$ such that for some constant $L > 0$, we have

$$|f_0(t, y) - f_0(t, z)| \leq L|y - z| \quad \text{for all } t \in [0, T] \text{ and } y, z \in W.$$

Note that for each $f_0 \in \text{Lip}([0,T] \times W, \bar{W})$, the seminorm

$$\|f_0\|_{\text{Lip}} := \sup_{t\in[0,T]} |f(t, 0)| + \sup_{t\in[0,T]} |f_0(t, \cdot)|_{\text{Lip}}$$

is finite.

Remark 2.2. Let $g_0 \in \mathrm{Lip}(W, \bar{W})$ be arbitrary, and define the function
$$f_0 : [0, T] \times W \to \bar{W}, \quad f_0(t, y) := g_0(y).$$
Then, we have $f_0 \in \mathrm{Lip}([0, T] \times W, \bar{W})$ with $\|f_0\|_{\mathrm{Lip}} = |g_0(0)| + |g_0|_{\mathrm{Lip}}$.

Lemma 2.16. *Let $f_0 \in \mathrm{Lip}([0, T] \times W, \bar{W})$ be arbitrary. Then, we have*
$$|f_0(t, y) - f_0(t, z)| \leq \|f_0\|_{\mathrm{Lip}} |y - z|, \quad y, z \in W \text{ and } t \in [0, T], \tag{2.18}$$
$$|f_0(t, y)| \leq \|f_0\|_{\mathrm{Lip}} (1 + |y|), \quad y \in W \text{ and } t \in [0, T]. \tag{2.19}$$

Proof. Setting $L := \sup_{t \in [0,T]} |f_0(t, \cdot)|_{\mathrm{Lip}}$, we have
$$|f_0(t, y) - f_0(t, z)| \leq L|y - z|, \quad y, z \in W, \text{ and } t \in [0, T],$$
showing (2.18). Now, for all $y \in W$ and $t \in [0, T]$, we obtain
$$|f_0(t, y)| \leq |f_0(t, y) - f_0(t, 0)| + |f_0(t, 0)|$$
$$\leq L|y| + \sup_{s \in [0,T]} |f_0(s, 0)| = \|f_0\|_{\mathrm{Lip}}(1 + |y|),$$
proving (2.19). □

For a path $Y : [0, T] \to W$ and a function $f_0 : [0, T] \times W \to \bar{W}$, we denote by $f_0(Y) : [0, T] \to \bar{W}$ the new path
$$f_0(Y)_t := f_0(t, Y_t), \quad t \in [0, T].$$
If the path Y is continuous and $f_0 \in \mathrm{Lip}([0, T] \times W, \bar{W})$, then the path $f_0(Y)$ is continuous and hence bounded.

2.6. Compositions of Controlled Rough Paths with Functions

In this section, we deal with compositions of controlled rough paths with functions. Let V, W, and $\bar{W}$ be Banach spaces. We fix an index $\alpha \in (\frac{1}{3}, \frac{1}{2}]$ and a time horizon $T \in \mathbb{R}_+$. Furthermore, let $\mathbf{X} = (X, \mathbb{X}) \in \mathscr{C}^\alpha([0, T], V)$ be a rough path.

2.6.1. Compositions with time-dependent smooth functions

In this section, we consider compositions of controlled rough paths with time-dependent smooth functions.

Definition 2.6. Let $\gamma \in (0,1]$ and $n \in \mathbb{N}$ be arbitrary. We denote by $C_b^{\gamma,n}([0,T] \times W, \bar{W})$ the space of all functions $f : [0,T] \times W \to \bar{W}$ such that we have:

(1) $f(t,\cdot) \in C_b^n(W, \bar{W})$ for all $t \in [0,T]$.
(2) $f(\cdot, y), \ldots, D_y^{n-1} f(\cdot, y) \in C^\gamma([0,T], \bar{W})$ for each $y \in W$.
(3)
$$\|f\|_{C_b^{\gamma,n}} := \sup_{t \in [0,T]} \|f(t,\cdot)\|_{C_b^n} + \sum_{k=0}^{n-1} \sup_{y \in W} \|D_y^k f(\cdot, y)\|_\gamma < \infty.$$

Remark 2.3. Let $g \in C_b^n(W, \bar{W})$ be arbitrary, and define the function
$$f : [0,T] \times W \to \bar{W}, \quad f(t,y) := g(y).$$
Then, we have $f \in C_b^{\gamma,n}([0,T] \times W, \bar{W})$, with $\|f\|_{C_b^{\gamma,n}} = \|g\|_{C_b^n}$.

Proposition 2.5. Let $\beta, \gamma \in (0,1]$ with $\beta \leq \gamma$ be arbitrary. Furthermore, let $Y \in C^\beta([0,T], W)$ and $f \in C_b^{\gamma,1}([0,T] \times W, \bar{W})$ be arbitrary. Then, we have $f(Y) \in C^\beta([0,T], \bar{W})$.

Proof. Let $s, t \in [0,T]$ be arbitrary. Then, we have
$$|f(t, Y_t) - f(s, Y_s)| \leq |f(t, Y_t) - f(t, Y_s)| + |f(t, Y_s) - f(s, Y_s)|$$
$$\leq \|f(t,\cdot)\|_{C_b^1} |Y_t - Y_s| + \|f(\cdot, Y_s)\|_\gamma |t-s|^\gamma$$
$$\leq \|f\|_{C_b^{\gamma,1}} \|Y\|_\beta |t-s|^\beta + \|f\|_{C_b^{\gamma,1}} |t-s|^\gamma$$
$$\leq \|f\|_{C_b^{\gamma,1}} (\|Y\|_\beta + T^{\gamma-\beta}) |t-s|^\beta,$$
showing that $f(Y) \in C^\beta([0,T], \bar{W})$. $\square$

For a controlled rough path $(Y, Y') \in \mathscr{D}_X^{2\alpha}([0,T], W)$ and a function $f \in C_b^{2\alpha,2}([0,T] \times W, \bar{W})$, we denote by $f(Y) : [0,T] \to \bar{W}$ the path
$$f(Y)_t := f(t, Y_t), \quad t \in [0, T],$$
and we denote by $f(Y)' : [0,T] \to L(V, \bar{W})$ the path
$$f(Y)'_t := D_y f(t, Y_t) Y'_t, \quad t \in [0, T].$$

Proposition 2.6. *Let* $(Y, Y') \in \mathscr{D}_X^{2\alpha}([0,T], W)$ *and* $f \in C_b^{2\alpha,2}([0,T] \times W, \bar{W})$ *be arbitrary. Then, we have*
$$(f(Y), f(Y)') \in \mathscr{D}_X^{2\alpha}([0,T], \bar{W}). \tag{2.20}$$

Furthermore, we have
$$\|f(Y)\|_\alpha \leq \|f\|_{C_b^{2\alpha,2}}(\|Y\|_\alpha + T^\alpha), \tag{2.21}$$
$$\|f(Y)'\|_\alpha \leq \|f\|_{C_b^{2\alpha,2}}(\|Y'\|_\alpha + \|Y\|_\alpha \|Y'\|_\infty + T^\alpha \|Y'\|_\infty), \tag{2.22}$$
$$\|R^{f(Y)}\|_{2\alpha} \leq \|f\|_{C_b^{2\alpha,2}}\left(1 + \frac{1}{2}\|Y\|_\alpha^2 + \|R^Y\|_{2\alpha}\right), \tag{2.23}$$

and for $T \leq 1$, *we have the estimates*
$$|f(Y), f(Y)'|_{X, 2\alpha} \leq C\big(1 + |Y, Y'|_{X, 2\alpha}^2\big), \tag{2.24}$$
$$\|\!|f(Y), f(Y)'|\!\|_{X, 2\alpha} \leq C\big(1 + \|\!|Y, Y'|\!\|_{X, 2\alpha}^2\big), \tag{2.25}$$

where the constant $C > 0$ *depends on* α, T, $\mathbf{X}$, *and* $\|f\|_{C_b^{2\alpha,2}}$. *Moreover, for* $T \leq 1$, *the constant* C *does not depend on* T.

Proof. Let $s, t \in [0, T]$ be arbitrary. Then, we have
$$|f(t, Y_t) - f(s, Y_s)| \leq |f(t, Y_t) - f(t, Y_s)| + |f(t, Y_s) - f(s, Y_s)|$$
$$\leq \|f(t, \cdot)\|_{C_b^2} |Y_t - Y_s| + \|f(\cdot, Y_s)\|_{2\alpha} |t - s|^{2\alpha}$$
$$\leq \|f\|_{C_b^{2\alpha,2}} \|Y\|_\alpha |t - s|^\alpha + \|f\|_{C_b^{2\alpha,2}} |t - s|^{2\alpha}$$
$$\leq \|f\|_{C_b^{2\alpha,2}} (\|Y\|_\alpha + T^\alpha) |t - s|^\alpha,$$

showing $f(Y) \in \mathcal{C}^\alpha([0,T], \bar{W})$ and the estimate (2.21). Furthermore, we have

$$\begin{aligned}
|D_y f(t, Y_t) Y'_t &- D_y f(s, Y_s) Y'_s| \\
&\leq |D_y f(t, Y_t) Y'_t - D_y f(t, Y_t) Y'_s| + |D_y f(t, Y_t) Y'_s - D_y f(t, Y_s) Y'_s| \\
&\quad + |D_y f(t, Y_s) Y'_s - D_y f(s, Y_s) Y'_s| \\
&\leq |D_y f(t, Y_s)| \, |Y'_t - Y'_s| + |D_y f(t, Y_t) - D_y f(t, Y_s)| \, |Y'_s| \\
&\quad + |D_y f(t, Y_s) - D_y f(s, Y_s)| \, |Y'_s| \\
&\leq \|f(t, \cdot)\|_{C_b^2} \|Y'\|_\alpha |t-s|^\alpha + \|f(t, \cdot)\|_{C_b^2} |Y_t - Y_s| \, \|Y'\|_\infty \\
&\quad + \|D_y f(\cdot, Y_s)\|_{2\alpha} |t-s|^{2\alpha} \|Y'\|_\infty \\
&\leq \|f\|_{C_b^{2\alpha,2}} \|Y'\|_\alpha |t-s|^\alpha + \|f\|_{C_b^{2\alpha,2}} \|Y\|_\alpha |t-s|^\alpha \|Y'\|_\infty \\
&\quad + \|f\|_{C_b^{2\alpha,2}} \|Y'\|_\infty |t-s|^{2\alpha} \\
&\leq \|f\|_{C_b^{2\alpha,2}} \bigl(\|Y'\|_\alpha + \|Y\|_\alpha \|Y'\|_\infty + T^\alpha \|Y'\|_\infty\bigr) |t-s|^\alpha,
\end{aligned}$$

showing $f(Y)' \in \mathcal{C}^\alpha([0,T], L(V, \bar{W}))$ and the estimate (2.22). Moreover, by Taylor's theorem, we obtain

$$\begin{aligned}
|R_{s,t}^{f(Y)}| &= |f(Y)_{s,t} - f(Y)'_s X_{s,t}| \\
&= |f(t, Y_t) - f(s, Y_s) - D_y f(s, Y_s) Y'_s X_{s,t}| \\
&\leq |f(t, Y_t) - f(s, Y_t)| + |f(s, Y_t) - f(s, Y_s) - D_y f(s, Y_s) Y_{s,t}| \\
&\quad + |D_y f(s, Y_s)(Y_{s,t} - Y'_s X_{s,t})| \\
&\leq \|f(\bullet, Y_t)\|_{2\alpha} |t-s|^{2\alpha} + \frac{1}{2} \|f(s, \bullet)\|_{C_b^2} |Y_t - Y_s|^2 \\
&\quad + \|f(s, \bullet)\|_{C_b^2} |R_{s,t}^Y| \\
&\leq \|f\|_{C_b^{2\alpha,2}} |t-s|^{2\alpha} + \frac{1}{2} \|f\|_{C_b^{2\alpha,2}} \|Y\|_\alpha^2 |t-s|^{2\alpha} \\
&\quad + \|f\|_{C_b^{2\alpha,2}} \|R^Y\|_{2\alpha} |t-s|^{2\alpha} \\
&\leq \|f\|_{C_b^{2\alpha,2}} \left(1 + \frac{1}{2} \|Y\|_\alpha^2 + \|R^Y\|_{2\alpha}\right) |t-s|^{2\alpha},
\end{aligned}$$

proving (2.20) and (2.23). Moreover, note that

$$|f(Y)_0| = |f(0, Y_0)| \leq \|f\|_{C_b^{2\alpha,2}},$$

$$|f(Y)'_0| = |D_y f(0, Y_0) Y'_0| \leq |D_y f(0, Y_0)| \, |Y'_0| \leq \|f\|_{C_b^{2\alpha,2}} |Y, Y'|_{X, 2\alpha}.$$

Together with Lemma 2.6, we obtain (2.24) and (2.25). □

2.6.2. Compositions with bilinear operators

In this section, we consider compositions of controlled rough paths with bilinear operators. First of all, let us consider compositions with linear operators. For a controlled rough path $(Y, Y') \in \mathscr{D}_X^{2\alpha}([0, T], W)$ and a continuous linear operator $\varphi \in L(W, \bar{W})$, we denote by $\varphi(Y) : [0, T] \to \bar{W}$ the path

$$\varphi(Y)_t := \varphi(Y_t), \quad t \in [0, T],$$

and we denote by $\varphi(Y)' : [0, T] \to L(V, \bar{W})$ the path

$$\varphi(Y)'_t := \varphi(Y'_t), \quad t \in [0, T].$$

Proposition 2.7. Let $(Y, Y') \in \mathscr{D}_X^{2\alpha}([0, T], W)$ and $\varphi \in L(W, \bar{W})$ be arbitrary. Then, we have

$$(\varphi(Y), \varphi(Y)') \in \mathscr{D}_X^{2\alpha}([0, T], \bar{W}) \tag{2.26}$$

and the estimates

$$\|\varphi(Y), \varphi(Y)'\|_{X, 2\alpha} \leq |\varphi| \, \|Y, Y'\|_{X, 2\alpha}, \tag{2.27}$$

$$|\varphi(Y), \varphi(Y)'|_{X, 2\alpha} \leq |\varphi| \, |Y, Y'|_{X, 2\alpha}, \tag{2.28}$$

$$\|\varphi(Y), \varphi(Y)'\|_{X, 2\alpha} \leq |\varphi| \, \|Y, Y'\|_{X, 2\alpha}. \tag{2.29}$$

Proof. Let $s, t \in [0, T]$ be arbitrary. Then, we have

$$|\varphi(Y'_t) - \varphi(Y'_s)| = |\varphi(Y'_t - Y'_s)| \leq |\varphi| \, |Y'_t - Y'_s| \leq |\varphi| \, \|Y'\|_\alpha |t - s|^\alpha,$$

showing $\|\varphi(Y')\|_\alpha \leq |\varphi| \, \|Y'\|_\alpha$. Furthermore, we have

$$R^{\varphi(Y)}_{s,t} = \varphi(Y_{s,t}) - \varphi(Y'_s) X_{s,t} = \varphi(Y_{s,t} - Y'_s X_{s,t}) = \varphi(R^Y_{s,t}),$$

and hence, we obtain

$$|R_{s,t}^{\varphi(Y)}| \leq |\varphi| |R_{s,t}^Y| \leq |\varphi| \|R^Y\|_{2\alpha} |t-s|^{2\alpha}.$$

This shows $\|R^{\varphi(Y)}\|_{2\alpha} \leq |\varphi| \|R^Y\|_{2\alpha}$, and we deduce (2.26). Noting that $|\varphi(Y_0)| \leq |\varphi| |Y_0|$ and $|\varphi(Y_0')| \leq |\varphi| |Y_0'|$, the desired estimates (2.27), (2.28), and (2.29) follow. $\square$

Now, let E be another Banach space. For controlled rough paths $(Y, Y') \in \mathscr{D}_X^{2\alpha}([0,T], W)$ and $(Z, Z') \in \mathscr{D}_X^{2\alpha}([0,T], E)$ and a continuous bilinear operator $B \in L^{(2)}(W \times E, \bar{W})$, we denote by $B(Y,Z) : [0,T] \to \bar{W}$ the path

$$B(Y,Z)_t := B(Y_t, Z_t), \quad t \in [0,T],$$

and we denote by $B(Y,Z)' : [0,T] \to L(V, \bar{W})$ the path

$$B(Y,Z)'_t := B(Y_t, Z_t') + B(Y_t', Z_t), \quad t \in [0,T].$$

Proposition 2.8. *Let* $(Y, Y') \in \mathscr{D}_X^{2\alpha}([0,T], W)$, $(Z, Z') \in \mathscr{D}_X^{2\alpha}([0,T], E)$ *and* $B \in L^{(2)}(W \times E, \bar{W})$ *be arbitrary. Then, we have*

$$(B(Y,Z), B(Y,Z)') \in \mathscr{D}_X^{2\alpha}([0,T], \bar{W}), \tag{2.30}$$

and the estimate

$$\||B(Y,Z), B(Y,Z)'\||_{X,2\alpha} \leq C \, \||Y,Y'\||_{X,2\alpha} \, \||Z,Z'\||_{X,2\alpha}, \tag{2.31}$$

where the constant $C > 0$ *depends on* α, T, $\mathbf{X}$ *and* $|B|$. *Moreover, for* $T \leq 1$, *the constant* C *does not depend on* T.

Proof. Let $s, t \in [0,T]$ be arbitrary. Then, we have

$$\begin{aligned}
|B(Y_t, Z_t) - B(Y_s, Z_s)| &\leq |B(Y_t, Z_t) - B(Y_t, Z_s)| \\
&\quad + |B(Y_t, Z_s) - B(Y_s, Z_s)| \\
&= |B(Y_t, Z_{s,t})| + |B(Y_{s,t}, Z_s)| \\
&\leq |B|(|Y_t| |Z_{s,t}| + |Y_{s,t}| |Z_s|) \\
&\leq |B|(\|Y\|_\infty \|Z\|_\alpha + \|Y\|_\alpha \|Z\|_\infty) |t-s|^\alpha,
\end{aligned}$$

showing that $B(Y, Z) \in C^\alpha([0,T], \bar{W})$. Furthermore, we have

$$\begin{aligned}
|B(Y_t, Z'_t) - B(Y_s, Z'_s)| &\leq |B(Y_t, Z'_t) - B(Y_t, Z'_s)| \\
&\quad + |B(Y_t, Z'_s) - B(Y_s, Z'_s)| \\
&= |B(Y_t, Z'_{s,t})| + |B(Y_{s,t}, Z'_s)| \\
&\leq |B|(|Y_t|\,|Z'_{s,t}| + |Y_{s,t}|\,|Z'_s|) \\
&\leq |B|(\|Y\|_\infty \|Z'\|_\alpha + \|Y\|_\alpha \|Z'\|_\infty)|t-s|^\alpha
\end{aligned}$$

as well as

$$\begin{aligned}
|B(Y'_t, Z_t) - B(Y'_s, Z_s)| &\leq |B(Y'_t, Z_t) - B(Y'_t, Z_s)| \\
&\quad + |B(Y'_t, Z_s) - B(Y'_s, Z_s)| \\
&= |B(Y'_t, Z_{s,t})| + |B(Y'_{s,t}, Z_s)| \\
&\leq |B|(|Y'_t|\,|Z_{s,t}| + |Y'_{s,t}|\,|Z_s|) \\
&\leq |B|(\|Y'\|_\infty \|Z\|_\alpha + \|Y'\|_\alpha \|Z\|_\infty)|t-s|^\alpha,
\end{aligned}$$

showing that $B(Y,Z)' \in C^\alpha([0,T], L(V, \bar{W}))$. Moreover, we have

$$\begin{aligned}
R^{B(Y,Z)}_{s,t} &= B(Y_t, Z_t) - B(Y_s, Z_s) - B(Y_s, Z_s)' X_{s,t} \\
&= B(Y_t, Z_{s,t}) + B(Y_{s,t}, Z_s) - \big(B(Y_s, Z'_s X_{s,t}) + B(Y'_s X_{s,t}, Z_s)\big) \\
&= B(Y_{s,t} - Y'_s X_{s,t}, Z_s) + B(Y_s, Z_{s,t} - Z'_s X_{s,t}) + B(Y_{s,t}, Z_{s,t}) \\
&= B(R^Y_{s,t}, Z_s) + B(Y_s, R^Z_{s,t}) + B(Y_{s,t}, Z_{s,t}),
\end{aligned}$$

and hence

$$\begin{aligned}
|R^{B(Y,Z)}_{s,t}| &\leq |B|(|R^Y_{s,t}|\,|Z_s| + |Y_s|\,|R^Z_{s,t}| + |Y_{s,t}|\,|Z_{s,t}|) \\
&\leq |B|(\|R^Y\|_{2\alpha}\|Z\|_\infty + \|Y\|_\infty \|R^Y\|_{2\alpha} + \|Y\|_\alpha \|Z\|_\alpha)|t-s|^{2\alpha},
\end{aligned}$$

which shows (2.30). Finally, noting that

$$|B(Y_0, Z_0)| \leq |B|\,|Y_0|\,|Z_0|,$$
$$|B(Y_0, Z'_0)| \leq |B|\,|Y_0|\,|Z'_0|,$$
$$|B(Y'_0, Z_0)| \leq |B|\,|Y'_0|\,|Z_0|$$

and applying Lemma 2.6 proves (2.31). □

Lemma 2.17. *Let E and F be two other Banach spaces. Furthermore, let $h : W \to L(W, F)$ be of class C^1, and let $B \in L(E, W)$ be arbitrary. Then, the mapping*

$$h_B : W \to L(E, F), \quad h_B(y) := h(y)B$$

is of class C^1, and we have

$$Dh_B(y)v = (Dh(y)v)B \quad \text{for all } y, v \in W.$$

Proof. We have $h_B = \ell \circ h$, where $\ell : L(W, F) \to L(E, F)$ is given by $\ell(z) = zB$. Note that ℓ is a linear operator. Moreover, we have

$$|\ell(z)| \leq |z| |B| \quad \text{for all } z \in L(W, F),$$

showing that ℓ is continuous. Therefore, the mapping h_B is of class C^1, and by the chain rule, we obtain

$$Dh_B(y)v = D(\ell \circ h)(y)v = D\ell(h(y)) \circ Dh(y)v$$
$$= \ell \circ Dh(y)v = (Dh(y)v)B,$$

completing the proof. □

Lemma 2.18. *Let $f \in C_b^{2\alpha,3}([0,T] \times W, \bar{W})$ be arbitrary. Then, there exists $g \in C_b^{2\alpha,2}([0,T] \times (W \times W), L(W, \bar{W}))$, with $\|g\|_{C_b^{2\alpha,2}} \leq \|f\|_{C_b^{2\alpha,3}}$, such that*

$$f(t, y_1) - f(t, y_2) = g(t, y)(y_1 - y_2) \tag{2.32}$$

for all $t \in [0, T]$ and all $y = (y_1, y_2) \in W \times W$.

Proof. We define $g : [0, T] \times W \times W \to L(W, \bar{W})$ as

$$g(t, y) := \int_0^1 D_2 f(t, \theta y_1 + (1 - \theta)y_2) \, d\theta.$$

Then, by Taylor's theorem, identity (2.32) is satisfied for all $t \in [0, T]$ and all $y = (y_1, y_2) \in W \times W$. For $\theta \in [0, 1]$, we consider the linear operator $B_\theta : W \times W \to W$ given by

$$B_\theta(y) := \theta y_1 + (1 - \theta)y_2, \quad y = (y_1, y_2) \in W \times W.$$

Then, for all $y = (y_1, y_2) \in W \times W$, we have

$$|B_\theta(y)| = |\theta y_1 + (1-\theta)y_2| \leq \theta|y_1| + (1-\theta)|y_2| \leq |y_1| + |y_2| = |y|.$$

Therefore, for each $\theta \in [0,1]$, we have $B_\theta \in L(W \times W, W)$, with

$$|B_\theta| \leq 1 \quad \text{for all } \theta \in [0,1]. \tag{2.33}$$

Now, we fix an arbitrary $t \in [0,T]$. Then, we have

$$g(t,y) = \int_0^1 D_2 f(t, B_\theta(y))\, d\theta, \quad y \in W \times W.$$

Therefore, we have $g(t, \cdot) \in C^2(W \times W, L(W, \bar{W}))$. Let $y \in W \times W$ be arbitrary. By the chain rule, we have

$$D_y g(t,y) = \int_0^1 D_y\big(D_2 f(t, B_\theta(y))\big)\, d\theta = \int_0^1 D_2^2 f(t, B_\theta(y)) B_\theta\, d\theta,$$

Moreover, by Lemma 2.17 and the chain rule for all $v \in W \times W$, we obtain

$$D_y^2 g(t,y) v = \int_0^1 D_y\big(D_2^2 f(t, B_\theta(y)) B_\theta\big) v\, d\theta$$

$$= \int_0^1 \big(D_2^3 f(t, B_\theta(y)) B_\theta(v)\big) B_\theta\, d\theta.$$

Therefore, noting (2.33), we have $g(t,\cdot) \in C_b^2(W \times W, L(W, \bar{W}))$ with

$$\|D_y^k g(t,\cdot)\|_\infty \leq \|D_y^{k+1} f(t,\cdot)\|_\infty, \quad k = 0,1,2. \tag{2.34}$$

Now, let $y \in W \times W$ be arbitrary. Furthermore, let $s, t \in [0,T]$ be arbitrary. Then, we have

$$|g(t,y) - g(s,y)| \leq \int_0^1 |D_2 f(t, B_\theta(y)) - D_2 f(s, B_\theta(y))|\, d\theta$$

$$\leq \int_0^1 \|D_2 f(\cdot, B_\theta(y))\|_{2\alpha} |t-s|^{2\alpha}\, d\theta.$$

Moreover, by (2.33), we obtain

$$|D_y g(t,y) - D_y g(s,y)| \leq \int_0^1 |D_2^2 f(t, B_\theta(y)) B_\theta - D_2^2 f(s, B_\theta(y)) B_\theta| \, d\theta$$

$$\leq \int_0^1 |D_2^2 f(t, B_\theta(y)) - D_2^2 f(s, B_\theta(y))| \, d\theta$$

$$\leq \int_0^1 \|D_2^2 f(\cdot, B_\theta(y))\|_{2\alpha} |t - s|^{2\alpha} \, d\theta.$$

Consequently, we have $g(\cdot, y), D_y g(\cdot, y) \in \mathcal{C}^{2\alpha}([0,T], W)$, with

$$\sup_{y \in W} \|D_y^k g(\cdot, y)\|_{2\alpha} \leq \sup_{y \in W} \|D_y^{k+1} f(\cdot, y)\|_{2\alpha}, \quad k = 0, 1. \tag{2.35}$$

From (2.34) and (2.35), we obtain $\|g\|_{C_b^{2\alpha,2}} \leq \|f\|_{C_b^{2\alpha,3}} < \infty$, which completes the proof. □

Proposition 2.9. Let $(Y, Y'), (Z, Z') \in \mathscr{D}_X^{2\alpha}([0,T], W)$ and $f \in C_b^{2\alpha,3}([0,T] \times W, \bar{W})$ be arbitrary. Then, we have

$$\||(f(Y), f(Y)') - (f(Z), f(Z)')\||_{X, 2\alpha}$$
$$\leq C\big(1 + |Y, Y'|_{X, 2\alpha}^2 + |Z, Z'|_{X, 2\alpha}^2\big) \||(Y, Y') - (Z, Z')\||_{X, 2\alpha},$$

where the constant $C > 0$ depends on α, T, $\mathbf{X}$, and $\|f\|_{C_b^{2\alpha,3}}$. Moreover, for $T \leq 1$, the constant C does not depend on T.

Proof. According to Lemma 2.18, there exists a mapping $g \in C_b^{2\alpha,2}([0,T] \times (W \times W), L(W, \bar{W}))$, with

$$\|g\|_{C_b^{2\alpha,2}} \leq \|f\|_{C_b^{2\alpha,3}}, \tag{2.36}$$

such that

$$f(t, y) - f(t, z) = g(t, y, z)(y - z), \quad t \in [0, T] \text{ and } y, z \in W.$$

We define the bilinear operator $B : L(W, \bar{W}) \times W \to \bar{W}$ as

$$B(T, y) := Ty.$$

Then, for all $(T, y) \in L(W, \bar{W}) \times W$, we have

$$|B(T, y)| \leq |T| \, |y|.$$

Therefore, we have $B \in L^{(2)}(L(W, \bar{W}) \times W, \bar{W})$ with $|B| \leq 1$. Hence, using Proposition 2.8, we obtain

$$\begin{aligned}
&\|\|(f(Y), f(Y)') - (f(Z), f(Z)')\|\|_{X,2\alpha} \\
&= \|\|g(Y,Z)(Y-Z), (g(Y,Z)(Y-Z))'\|\|_{X,2\alpha} \\
&= \|\|B(g(Y,Z), Y-Z), (B(g(Y,Z), Y-Z))'\|\|_{X,2\alpha} \\
&\leq C \|\|g(Y,Z), g(Y,Z)'\|\|_{X,2\alpha} \|\|(Y,Y') - (Z,Z')\|\|_{X,2\alpha},
\end{aligned}$$

where the constant $C > 0$ depends on α, T, and $\mathbf{X}$ and does not depend on $T \leq 1$. Furthermore, by Proposition 2.6 and Lemma 2.7, we have

$$\begin{aligned}
\|\|g(Y,Z), g(Y,Z)'\|\|_{X,2\alpha} &= |g(0, Y_0, Z_0)| + |g(Y,Z), g(Y,Z)'|_{X,2\alpha} \\
&\leq C(1 + |(Y,Z), (Y,Z)'|^2_{X,2\alpha}) \\
&\leq C(1 + |Y,Y'|^2_{X,2\alpha} + |Z,Z'|^2_{X,2\alpha}),
\end{aligned}$$

where the constant $C > 0$ depends on α, T, $\mathbf{X}$, and $\|g\|_{C_b^{2\alpha,2}}$ and does not depend on $T \leq 1$. Taking into account (2.36), this completes the proof. $\square$

2.6.3. Compositions with time-dependent linear operators

In this section, we consider compositions of controlled rough paths with time-dependent linear operators. For a controlled rough path $(Y, Y') \in \mathscr{D}_X^{2\alpha}([0,T], W)$ and a mapping $\varphi : [0,T] \times W \to \bar{W}$ such that $\varphi(t, \cdot) \in L(W, \bar{W})$ for each $t \in [0,T]$, we denote by $\varphi(Y) : [0,T] \to \bar{W}$ the path

$$\varphi(Y)_t := \varphi(t, Y_t), \quad t \in [0,T],$$

and we denote by $\varphi(Y)' : [0,T] \to L(V, \bar{W})$ the path

$$\varphi(Y)'_t := \varphi(t, Y'_t), \quad t \in [0,T].$$

Proposition 2.10. *Let $(Y, Y') \in \mathscr{D}_X^{2\alpha}([0,T], W)$ be arbitrary, and let $\varphi : [0,T] \times W \to \bar{W}$ be a function such that $\varphi(t, \cdot) \in L(W, \bar{W})$ for* each $t \in [0,T]$. We assume there are constants $K, L > 0$ such that

$$|\varphi(t,\cdot)| \leq K, \quad t \in [0,T],$$
$$|\varphi(t,y) - \varphi(s,y)| \leq L|y|\,|t-s|^{2\alpha}, \quad s,t \in [0,T] \text{ and } y \in W.$$

Then, we have

$$(\varphi(Y), \varphi(Y)') \in \mathscr{D}_X^{2\alpha}([0,T], \bar{W}). \tag{2.37}$$

Furthermore, we have

$$\|\varphi(Y)'\|_\alpha \leq LT^\alpha \|Y'\|_\infty + K\|Y'\|_\alpha, \tag{2.38}$$
$$\|R^{\varphi(Y)}\|_{2\alpha} \leq L\|Y\|_\infty + K\|R^Y\|_{2\alpha}, \tag{2.39}$$

and we have the estimate

$$\|\varphi(Y), \varphi(Y)'\|_{X,2\alpha} \leq C \, \|\!| Y, Y' \|\!|_{X,2\alpha}, \tag{2.40}$$

where the constant $C > 0$ depends on α, T, $\mathbf{X}$, K, and L. Moreover, for $T \leq 1$, the constant C does not depend on T.

Proof. Let $s, t \in [0,T]$ be arbitrary. Then, we have

$$|\varphi(t, Y_t) - \varphi(s, Y_s)| \leq |\varphi(t, Y_t) - \varphi(t, Y_s)| + |\varphi(t, Y_s) - \varphi(s, Y_s)|$$
$$\leq |\varphi(t,\cdot)|\,|Y_t - Y_s| + L|Y_s|\,|t-s|^{2\alpha}$$
$$\leq M\|Y\|_\alpha |t-s|^\alpha + L\|Y\|_\infty |t-s|^{2\alpha}$$
$$\leq (M\|Y\|_\alpha + LT^\alpha \|Y\|_\infty)|t-s|^\alpha,$$

and hence $\varphi(Y) \in C^\alpha([0,T], \bar{W})$. Furthermore, we have

$$|\varphi(t, Y_t') - \varphi(s, Y_s')| \leq |\varphi(t, Y_t') - \varphi(t, Y_s')| + |\varphi(t, Y_s') - \varphi(s, Y_s')|$$
$$\leq |\varphi(t,\cdot)|\,|Y_t' - Y_s'| + L|Y_s'|\,|t-s|^{2\alpha}$$
$$\leq M\|Y'\|_\alpha |t-s|^\alpha + L\|Y'\|_\infty |t-s|^{2\alpha}$$
$$\leq (M\|Y'\|_\alpha + LT^\alpha \|Y'\|_\infty)|t-s|^\alpha,$$

showing $\varphi(Y)' \in C^\alpha([0,T], L(V, \bar{W}))$ and the estimate (2.38). Moreover, we have

$$\begin{aligned}R_{s,t}^{\varphi(Y)} &= \varphi(Y_{s,t}) - \varphi(Y_s')X_{s,t} \\ &= \varphi(t, Y_t) - \varphi(s, Y_s) - \varphi(s, Y_s'X_{s,t}) \\ &= \varphi(t, Y_t) - \varphi(s, Y_t) + \varphi(s, Y_t) - \varphi(s, Y_s) - \varphi(s, Y_s'X_{s,t}) \\ &= \varphi(t, Y_t) - \varphi(s, Y_t) + \varphi(s, Y_{s,t} - Y_s'X_{s,t}).\end{aligned}$$

Therefore, we obtain

$$\begin{aligned}|R_{s,t}^{\varphi(Y)}| &\leq |\varphi(t, Y_t) - \varphi(s, Y_t)| + |\varphi(s, Y_{s,t} - Y_s'X_{s,t})| \\ &\leq L\|Y\|_\infty |t-s|^{2\alpha} + \|\varphi(s, \cdot)\| \, |R_{s,t}^Y| \\ &\leq L\|Y\|_\infty |t-s|^{2\alpha} + M\|R^Y\|_{2\alpha} |s-t|^{2\alpha} \\ &= (L\|Y\|_\infty + M\|R^Y\|_{2\alpha})|s-t|^{2\alpha},\end{aligned}$$

proving (2.37) and (2.39). Furthermore, note that

$$|\varphi(Y)_0| = |\varphi(0, Y_0)| \leq |\varphi(0, \cdot)| \, |Y_0| \leq M|Y_0|,$$
$$|\varphi(Y)_0'| = |\varphi(0, Y_0')| \leq |\varphi(0, \cdot)| \, |Y_0'| \leq M|Y_0'|.$$

Hence, the estimate (2.40) is a consequence of Lemma 2.6. □

For what follows, the upcoming auxiliary result will be useful.

Lemma 2.19. *Let E be another Banach space, and let $\varphi \in L(E, W)$ be arbitrary. Setting*

$$L(V, E) \to L(V, W), \quad S \mapsto \varphi S,$$

we may regard φ as an element from $L(L(V, E), L(V, W))$. Furthermore, we have

$$|\varphi|_{L(L(V,E),L(V,W))} \leq |\varphi|_{L(E,W)}.$$

Proof. We have

$$\begin{aligned}|\varphi|_{L(L(V,E),L(V,W))} &= \sup_{|S|\leq 1} |\varphi S|_{L(V,W)} \\ &\leq \sup_{|S|\leq 1} |\varphi|_{L(E,W)} |S|_{L(V,E)} \leq |\varphi|_{L(E,W)},\end{aligned}$$

completing the proof. □

Now, let A be the generator of a C_0-semigroup $(S_t)_{t\geq 0}$ on W. Then, there are constants $M \geq 1$ and $\omega \in \mathbb{R}$ such that the estimate (2.13) is satisfied.

Proposition 2.11. *Let $(Y, Y') \in \mathscr{D}_X^{2\alpha}([0,T], L(V, D(A)))$ be arbitrary, and let $t \in [0,T]$ be arbitrary. We define the paths $Z : [0,t] \to L(V, W)$ and $Z' : [0,t] \to L(V, L(V, W))$ as*

$$Z_s := S_{t-s}Y_s, \quad s \in [0,t],$$
$$Z'_s := S_{t-s}Y'_s, \quad s \in [0,t].$$

Then, we have

$$(Z, Z') \in \mathscr{D}_X^{2\alpha}([0,t], L(V, W)).$$

Furthermore, we have

$$\|Z, Z'\|_{X, 2\alpha} \leq C \, \| Y, Y' \|_{X, 2\alpha},$$

where the constant $C > 0$ depends on α, T, $\mathbf{X}$, M, and ω. Moreover, for $T \leq 1$, the constant C does not depend on T.

Proof. In view of Lemma 2.19, we can define

$$\varphi : [0,t] \times L(V, D(A)) \to L(V, W), \quad \varphi(s, y) := S_{t-s}y.$$

Let $s \in [0,t]$ be arbitrary. By Lemma 2.19, we have

$$\varphi(s, \cdot) \in L(L(V, D(A)), L(V, W)).$$

Furthermore, by (2.13), for each $y \in L(V, D(A))$, we have

$$|\varphi(s,y)|_{L(V,W)} = |S_{t-s}y|_{L(V,W)} \leq |S_{t-s}|_{L(D(A),W)} |y|_{L(V,D(A))}$$
$$\leq Me^{\omega T}|y|_{L(V,D(A))},$$

showing that

$$|\varphi(s,\cdot)| \leq Me^{\omega T}.$$

Now, let $s, t \in [0, T]$ and $y \in L(V, D(A))$ be arbitrary. Then, by Corollary 2.2, we have

$$|\varphi(s, y) - \varphi(r, y)|_{L(V,W)} = |S_{t-s}y - S_{t-r}y|_{L(V,W)} = |S_{t-r,t-s}y|_{L(V,W)}$$
$$\leq Me^{\omega T}|y|_{L(V,D(A))}|s - r|.$$

Hence, the statement follows from Proposition 2.10. □

Let $(Y, Y') \in \mathscr{D}_X^{2\alpha}([0, T], L(V, D(A)))$ be a controlled rough path. Then, Proposition 2.11 allows us to define the *rough convolution* $N : [0, T] \to W$ as follows. For each $t \in [0, T]$, let N_t be the Gubinelli integral

$$N_t := \int_0^t S_{t-s} Y_s \, d\mathbf{X}_s$$

according to Theorem 2.1. We investigate this rough convolution further in Section 2.8.

Lemma 2.20. *Let* $(Y, Y') \in \mathscr{D}_X^{2\alpha}([0, T], L(V, D(A^2)))$ *be arbitrary, and let* $s, t \in [0, T]$ *with* $s \leq t$ *be arbitrary. We define the paths* $Z : [s, t] \to L(V, W)$ *and* $Z' : [s, t] \to L(V, L(V, W))$ *as*

$$Z_r := (S_{t-r} - \mathrm{Id})Y_r, \quad r \in [s, t],$$
$$Z'_r := (S_{t-r} - \mathrm{Id})Y'_r, \quad t \in [s, t].$$

Then, we have

$$(Z, Z') \in \mathscr{D}_X^{2\alpha}([s, t], L(V, W)). \tag{2.41}$$

Furthermore, we have

$$\|Z'\|_\alpha \leq Me^{\omega T}(T^{1-\alpha}\|Y'\|_\infty + \|Y'\|_\alpha |t - s|), \tag{2.42}$$
$$\|R^Z\|_{2\alpha} \leq Me^{\omega T}(T^{1-2\alpha}\|Y\|_\infty + \|R^Y\|_{2\alpha}|t - s|), \tag{2.43}$$
$$\|Z\|_\infty \leq Me^{\omega T}\|Y\|_\infty |t - s|, \tag{2.44}$$
$$\|Z'\|_\infty \leq Me^{\omega T}\|Y'\|_\infty |t - s|. \tag{2.45}$$

Proof. We define $\varphi : [s,t] \times L(V, D(A^2)) \to L(V,W)$ as

$$\varphi(r,y) := (S_{t-r} - \text{Id})y.$$

Let $r \in [s,t]$ be arbitrary. Then, we have

$$\varphi(r, \cdot) \in L(L(V, D(A^2)), L(V,W)).$$

Furthermore, by Corollary 2.1, for each $y \in L(V, D(A^2))$, we have

$$\begin{aligned}|\varphi(r,y)|_{L(V,W)} &= |(S_{t-r} - \text{Id})y|_{L(V,W)} \\ &\leq |S_{t-r} - \text{Id}|_{L(D(A^2),W)} |y|_{L(V,D(A^2))} \\ &\leq |S_{0,t-r}|_{L(D(A),W)} |y|_{L(V,D(A^2))} \\ &\leq Me^{\omega T}|t-s| |y|_{L(V,D(A^2))},\end{aligned}$$

showing that

$$|\varphi(r, \cdot)| \leq Me^{\omega T}|t-s|.$$

Furthermore, we obtain

$$|Z_r|_{L(V,W)} = |\varphi(r, Y_r)|_{L(V,W)} \leq Me^{\omega T}|t-s| |Y_r|_{L(V,D(A^2))},$$

showing (2.44). Now, let $y \in L(V, D(A^2))$ and $r, q \in [s,t]$ be arbitrary. Then, by Corollary 2.2, we have

$$\begin{aligned}|\varphi(r,y) - \varphi(q,y)|_{L(V,W)} &= |S_{t-r}y - S_{t-q}y|_{L(V,W)} \\ &= |S_{t-q,t-r}y|_{L(V,W)} \\ &\leq Me^{\omega T}|y|_{L(V,D(A^2))}|r-q| \\ &\leq Me^{\omega T}|y|_{L(V,D(A^2))}T^{1-2\alpha}|r-q|^{2\alpha}.\end{aligned}$$

Hence, from Proposition 2.10, we obtain (2.41), (2.42), and (2.43). Now, we define $\Phi : [s,t] \times L(V, L(V, D(A^2))) \to L(V, L(V,W))$ as

$$\Phi(r, y') := (S_{t-r} - \text{Id})y'.$$

Let $r \in [s,t]$ be arbitrary. Then, by Lemma 2.19 and Corollary 2.1, for each $y' \in L(V, L(V, D(A^2)))$, we have

$$\begin{aligned}|\Phi(r,y')|_{L(V,L(V,W))} &= |(S_{t-r} - \mathrm{Id})y'|_{L(V,L(V,W))} \\ &\leq |S_{t-r} - \mathrm{Id}|_{L(L(V,D(A^2)),L(V,W))}|y'|_{L(V,L(V,D(A^2)))} \\ &\leq |S_{t-r} - \mathrm{Id}|_{L(D(A^2),W)}|y'|_{L(V,L(V,D(A^2)))} \\ &\leq |S_{0,t-r}|_{L(D(A),W)}|y'|_{L(V,L(V,D(A^2)))} \\ &\leq Me^{\omega T}|t-s|\,|y'|_{L(V,L(V,D(A^2)))}.\end{aligned}$$

Therefore, we obtain

$$\begin{aligned}|Z'_r|_{L(V,L(V,W))} &= |\Phi(r, Y'_r)|_{L(V,L(V,W))} \\ &\leq Me^{\omega T}|t-s|\,|Y'_r|_{L(V,L(V,D(A^2)))},\end{aligned}$$

showing (2.45). $\square$

Lemma 2.21. *Let $(Y, Y') \in \mathscr{D}_X^{2\alpha}([0,T], L(V, D(A^2)))$ be arbitrary, and let $s, t \in [0,T]$ with $s \leq t$ be arbitrary. We define the paths $Z : [0,s] \to L(V,W)$ and $Z' : [0,s] \to L(V, L(V,W))$ as*

$$\begin{aligned}Z_r &:= (S_{t-r} - S_{s-r})Y_r, \quad r \in [0,s], \\ Z'_r &:= (S_{t-r} - S_{s-r})Y'_r, \quad r \in [0,s].\end{aligned}$$

Then, we have

$$(Z, Z') \in \mathscr{D}_X^{2\alpha}([0,s], L(V,W)). \qquad (2.46)$$

Furthermore, we have

$$\|Z'\|_\alpha \leq Me^{\omega T}(T^{1-\alpha}e^{\omega T}\|Y'\|_\infty + \|Y'\|_\alpha)|t-s|, \qquad (2.47)$$

$$\|R^Z\|_{2\alpha} \leq Me^{\omega T}(T^{1-2\alpha}e^{\omega T}\|Y\|_\infty + \|R^Y\|_{2\alpha})|t-s|, \qquad (2.48)$$

$$\|Z\|_\infty \leq Me^{\omega T}\|Y\|_\infty|t-s|, \qquad (2.49)$$

$$\|Z'\|_\infty \leq Me^{\omega T}\|Y'\|_\infty|t-s|. \qquad (2.50)$$

Proof. We define $\varphi : [0,s] \times L(V, D(A^2)) \to L(V,W)$ as

$$\varphi(r, y) := S_{s-r, t-r}y.$$

Let $r \in [0, s]$ be arbitrary. Then, we have
$$\varphi(r, \cdot) \in L(L(V, D(A^2)), L(V, W)).$$
Furthermore, by Corollary 2.1, for each $y \in L(V, D(A^2))$, we have
$$\begin{aligned}|\varphi(r,y)|_{L(V,W)} &= |(S_{t-r} - S_{s-r})y|_{L(V,W)} \\ &\leq |S_{t-r} - S_{s-r}|_{L(D(A^2),W)} |y|_{L(V,D(A^2))} \\ &\leq |S_{s-r,t-r}|_{L(D(A),W)} |y|_{L(V,D(A^2))} \\ &\leq M e^{\omega T} |t-s| \, |y|_{L(V,D(A^2))},\end{aligned}$$
showing that
$$|\varphi(r, \cdot)| \leq M e^{\omega T} |t - s|.$$
Furthermore, we obtain
$$|Z_r|_{L(V,W)} = |\varphi(r, Y_r)|_{L(V,W)} \leq M e^{\omega T} |t-s| \, |Y_r|_{L(V,D(A^2))},$$
showing (2.49). Now, let $y \in L(V, D(A^2))$ and $q, r \in [0, s]$ be arbitrary. Then, by Corollary 2.4, we have
$$\begin{aligned}|\varphi(r,y) - \varphi(q,y)|_{L(V,W)} &= |S_{s-r,t-r} y - S_{s-q,t-q} y|_{L(V,W)} \\ &\leq M e^{2\omega T} |y|_{L(V,D(A^2))} |t-s| \, |r-q| \\ &\leq M e^{2\omega T} |y|_{L(V,D(A^2))} T^{1-2\alpha} |t-s| \, |r-q|^{2\alpha}.\end{aligned}$$
Hence, from Proposition 2.10, we obtain (2.46), (2.47), and (2.48). Now, we define $\Phi : [0, s] \times L(V, L(V, D(A^2))) \to L(V, L(V, W))$ as
$$\Phi(r, y') := S_{s-r, t-r} y'.$$
Let $r \in [0, s]$ be arbitrary. Then, by Lemma 2.19 and Corollary 2.1, for each $y' \in L(V, L(V, D(A^2)))$, we have
$$\begin{aligned}|\Phi(r,y')|_{L(V,L(V,W))} &= |S_{s-r,t-r} y'|_{L(V,L(V,W))} \\ &\leq |S_{s-r,t-r}|_{L(L(V,D(A^2)),L(V,W))} |y'|_{L(V,L(V,D(A^2)))} \\ &\leq |S_{s-r,t-r}|_{L(D(A^2),W)} |y'|_{L(V,L(V,D(A^2)))} \\ &\leq |S_{s-r,t-r}|_{L(D(A),W)} |y'|_{L(V,L(V,D(A^2)))} \\ &\leq M e^{\omega T} |t-s| \, |y'|_{L(V,L(V,D(A^2)))}.\end{aligned}$$

Therefore, we obtain

$$|Z'_r|_{L(V,L(V,W))} = |\Phi(r, Y'_r)|_{L(V,L(V,W))}$$
$$\leq Me^{\omega T}|t - s|\,|Y'_r|_{L(V,L(V,D(A^2)))},$$

showing (2.50). □

2.7. Regular Convolution Integrals

In this section, we consider regular convolution integrals. We fix $\alpha \in (0, \frac{1}{2}]$ and a time horizon $T \in \mathbb{R}_+$. Let $\mathbf{X} = (X, \mathbb{X}) \in \mathscr{C}^\alpha([0,T], V)$ be a rough path with values in a Banach space V. Furthermore, let A be the generator of a C_0-semigroup $(S_t)_{t \geq 0}$ on a Banach space W. Then, there are constants $M \geq 1$ and $\omega \in \mathbb{R}$ such that the estimate (2.13) is satisfied.

Proposition 2.12. *Let $\xi \in D(A)$ be arbitrary. We define the path $\Xi : [0, T] \to W$ as $\Xi_t := S_t \xi$ for all $t \in [0, T]$. Then, we have*

$$(\Xi, 0) \in \mathscr{D}_X^{2\alpha}([0, T], W).$$

Furthermore, we have

$$\|\Xi, 0\|_{X, 2\alpha} \leq Me^{\omega T}|\xi|_{D(A)} T^{1-2\alpha}.$$

Proof. Let $s, t \in [0, T]$ be arbitrary. Then, by Proposition 2.3, we have

$$|\Xi_{s,t}| = |S_t\xi - S_s\xi| \leq Me^{\omega T}|\xi|_{D(A)}|t - s|$$
$$\leq Me^{\omega T}|\xi|_{D(A)}T^{1-2\alpha}|t-s|^{2\alpha},$$

showing that $\Xi \in C^{2\alpha}([0,T], W)$. Together with Lemma 2.5, this completes the proof. □

Lemma 2.22. *Let $Y : [0, T] \to D(A)$ be measurable and bounded. We define the path $N : [0, T] \to W$ as*

$$N_t := \int_0^t S_{t-s} Y_s \, ds, \quad t \in [0, T].$$

Then, we have

$$|N_{s,t}| \leq (1+T)Me^{\omega T}\|Y\|_\infty |t - s| \quad \text{for all } s, t \in [0, T].$$

Proof. Let $s, t \in [0, T]$ with $s \leq t$ be arbitrary. Then, we have

$$N_{s,t} = \int_0^t S_{t-r} Y_r \, dr - \int_0^s S_{s-r} Y_r \, dr$$

$$= \int_s^t S_{t-r} Y_r \, dr + \int_0^s (S_{t-r} - S_{s-r}) Y_r \, dr.$$

Therefore, using the estimate (2.13) and Proposition 2.3, we obtain

$$|N_{s,t}| \leq \left| \int_s^t S_{t-r} Y_r \, dr \right| + \left| \int_0^s (S_{t-r} - S_{s-r}) Y_r \, dr \right|$$

$$\leq \int_s^t |S_{t-s} Y_r| \, dr + \int_0^s |S_{t-r} Y_r - S_{s-r} Y_r| \, dr$$

$$\leq M e^{\omega T} \int_s^t |Y_r| \, dr + M e^{\omega T} \int_0^s |Y_r|_{D(A)} |t - s| \, dr$$

$$\leq M e^{\omega T} \|Y\|_\infty |t - s| + M e^{\omega T} T \|Y\|_\infty |t - s|,$$

completing the proof. □

Proposition 2.13. *Let* $Y : [0, T] \to D(A)$ *be measurable and bounded. We define the path* $N : [0, T] \to W$ *as*

$$N_t := \int_0^t S_{t-s} Y_s \, ds, \quad t \in [0, T].$$

Then, N *is* $D(A)$*-valued, and we have* $(N, 0) \in \mathscr{D}_X^{2\alpha}([0, T], W)$ *and*

$$\|N, 0\|_{X, 2\alpha} \leq (1 + T) M e^{\omega T} \|Y\|_\infty T^{1-2\alpha}.$$

Proof. This is a consequence of Lemmas 2.5 and 2.22. □

Corollary 2.5. *Let* $Y, Z : [0, T] \to D(A)$ *be measurable and bounded. We define the paths* $N, P : [0, T] \to W$ *as*

$$N_t := \int_0^t S_{t-s} Y_s \, ds, \quad t \in [0, T],$$

$$P_t := \int_0^t S_{t-s} Z_s \, ds, \quad t \in [0, T].$$

Then, we have $(N,0), (P,0) \in \mathscr{D}_X^{2\alpha}([0,T], W)$ and
$$\|(N,0) - (P,0)\|_{X,2\alpha} \leq (1+T)Me^{\omega T}\|Y - Z\|_\infty T^{1-2\alpha}.$$

Proof. Noting that
$$N_t - P_t = \int_0^t S_{t-s}(Y_s - Z_s)\, ds, \quad t \in [0,T],$$
this is an immediate consequence of Proposition 2.13. □

2.8. Rough Convolution Integrals

In this section, we investigate rough convolution integrals. We fix $\alpha \in (\frac{1}{3}, \frac{1}{2}]$ and a time horizon $T \in \mathbb{R}_+$. Let $\mathbf{X} = (X, \mathbb{X}) \in \mathscr{C}^\alpha([0,T], V)$ be a rough path with values in a Banach space V. Furthermore, let A be the generator of a C_0-semigroup $(S_t)_{t\geq 0}$ on a Banach space W. Then, there are constants $M \geq 1$ and $\omega \in \mathbb{R}$ such that the estimate (2.13) is satisfied. For what follows, let a controlled rough path $(Y, Y') \in \mathscr{D}_X^{2\alpha}([0,T], L(V, D(A^2)))$ be given. According to Proposition 2.11, we can define the rough convolution $M : [0,T] \to W$ as
$$N_t := \int_0^t S_{t-s} Y_s\, d\mathbf{X}_s, \quad t \in [0,T].$$

Note that the path N is actually $D(A)$-valued. We also define the Gubinelli integral $I : [0,T] \to D(A^2)$ as
$$I_t := \int_0^t Y_s\, d\mathbf{X}_s, \quad t \in [0,T].$$

By Theorem 2.1, we have $(I, I') \in \mathscr{D}_X^{2\alpha}([0,T], D(A^2))$ with $I' = Y$. Furthermore, for all $s,t \in [0,T]$ with $s \leq t$, we have
$$N_{s,t} - I_{s,t} = \int_0^t S_{t-r} Y_r\, d\mathbf{X}_r - \int_0^s S_{s-r} Y_r\, d\mathbf{X}_r - \int_s^t Y_r\, d\mathbf{X}_r$$
$$= \int_s^t S_{t-r} Y_r\, d\mathbf{X}_r + \int_0^s (S_{t-r} - S_{s-r}) Y_r\, d\mathbf{X}_r - \int_s^t Y_r\, d\mathbf{X}_r$$
$$= \int_s^t (S_{t-r} - \mathrm{Id}) Y_r\, d\mathbf{X}_r + \int_0^s (S_{t-r} - S_{s-r}) Y_r\, d\mathbf{X}_r.$$
(2.51)

Lemma 2.23. *For all $s, t \in [0, T]$ with $s \leq t$, we have*

$$\left| \int_s^t (S_{t-r} - \mathrm{Id}) Y_r \, d\mathbf{X}_r \right|$$

$$\leq C \Big(\|X\|_\alpha (\|Y\|_\infty + \|R^Y\|_{2\alpha}) + \|\mathbb{X}\|_{2\alpha} (\|Y'\|_\infty + \|Y'\|_\alpha) \Big) |t - s|^{3\alpha},$$

where the constant $C > 0$ depends on α, T, M, and ω. Moreover, for $T \leq 1$, the constant C does not depend on T.

Proof. We define the paths $Z : [s,t] \to L(V,W)$ and $Z' : [s,t] \to L(V, L(V,W))$ as

$$Z_r := (S_{t-r} - \mathrm{Id}) Y_r, \quad r \in [s,t],$$
$$Z'_r := (S_{t-r} - \mathrm{Id}) Y'_r, \quad t \in [s,t].$$

By Lemma 2.20, we have

$$(Z, Z') \in \mathscr{D}_X^{2\alpha}([s,t], L(V,W)),$$

as well as

$$\|Z'\|_\alpha \leq M e^{\omega T} \big(T^{1-\alpha} \|Y'\|_\infty + \|Y'\|_\alpha |t-s| \big),$$
$$\|R^Z\|_{2\alpha} \leq M e^{\omega T} \big(T^{1-2\alpha} \|Y\|_\infty + \|R^Y\|_{2\alpha} |t-s| \big),$$
$$\|Z\|_\infty \leq M e^{\omega T} \|Y\|_\infty |t-s|,$$
$$\|Z'\|_\infty \leq M e^{\omega T} \|Y'\|_\infty |t-s|.$$

Furthermore, by Proposition 2.1, we have

$$\left| \int_s^t Z_r \, d\mathbf{X}_r \right| \leq C \big(\|X\|_\alpha \|R^Z\|_{2\alpha} + \|\mathbb{X}\|_{2\alpha} \|Z'\|_\alpha \big) |t-s|^{3\alpha}$$
$$+ \|Z'\|_\infty \|\mathbb{X}\|_{2\alpha} |t-s|^{2\alpha} + \|Z\|_\infty \|X\|_\alpha |t-s|^\alpha,$$

where the constant $C > 0$ depends on α. Moreover, note that $\alpha \leq \frac{1}{2}$ implies $3\alpha \leq 1 + \alpha$. This completes the proof. □

Corollary 2.6. *For all $s, t \in [0, T]$ with $s \leq t$, we have*

$$\left| \int_s^t (S_{t-r} - \mathrm{Id}) Y_r \, d\mathbf{X}_r \right| \leq C \, \||\mathbf{X}\||_\alpha \, \||Y, Y'\||_{X, 2\alpha} \, |t-s|^{3\alpha},$$

where the constant $C > 0$ depends on α, T, $\mathbf{X}$, M, and ω. Moreover, for $T \leq 1$, the constant C does not depend on T.

Proof. This is a consequence of Lemmas 2.6 and 2.23. □

Lemma 2.24. *For all $s, t \in [0, T]$ with $s \leq t$, we have*

$$\left| \int_0^s (S_{t-r} - S_{s-r}) Y_r \, d\mathbf{X}_r \right|$$
$$\leq C \Big(\|X\|_\alpha \big(\|Y\|_\infty + \|R^Y\|_{2\alpha} \big) + \|\mathbb{X}\|_{2\alpha} \big(\|Y'\|_\infty + \|Y'\|_\alpha \big) \Big) |t - s|,$$

where the constant $C > 0$ depends on α, T, M, and ω. Moreover, for $T \leq 1$, the constant C does not depend on T.

Proof. We define the paths $Z : [0, s] \to L(V, W)$ and $Z' : [0, s] \to L(V, L(V, W))$ as

$$Z_r := (S_{t-r} - S_{s-r}) Y_r, \quad r \in [0, s],$$
$$Z'_r := (S_{t-r} - S_{s-r}) Y'_r, \quad r \in [0, s].$$

By Lemma 2.21, we have

$$(Z, Z') \in \mathscr{D}_X^{2\alpha}([0, s], L(V, W))$$

and

$$\|Z'\|_\alpha \leq M e^{\omega T} \big(T^{1-\alpha} e^{\omega T} \|Y'\|_\infty + \|Y'\|_\alpha \big) |t - s|,$$
$$\|R^Z\|_{2\alpha} \leq M e^{\omega T} \big(T^{1-2\alpha} e^{\omega T} \|Y\|_\infty + \|R^Y\|_{2\alpha} \big) |t - s|,$$
$$\|Z\|_\infty \leq M e^{\omega T} \|Y\|_\infty |t - s|,$$
$$\|Z'\|_\infty \leq M e^{\omega T} \|Y'\|_\infty |t - s|.$$

Furthermore, by Proposition 2.1, we have

$$\left| \int_0^s Z_r \, d\mathbf{X}_r \right| \leq C \big(\|X\|_\alpha \|R^Z\|_{2\alpha} + \|\mathbb{X}\|_{2\alpha} \|Z'\|_\alpha \big) |s|^{3\alpha}$$
$$+ \|Z'\|_\infty \|\mathbb{X}\|_{2\alpha} |s|^{2\alpha} + \|Z\|_\infty \|X\|_\alpha |s|^\alpha,$$

where the constant $C > 0$ depends on α. This completes the proof.
□

Corollary 2.7. *For all $s, t \in [0, T]$ with $s \leq t$, we have*

$$\left| \int_0^s (S_{t-r} - S_{s-r}) Y_r \, d\mathbf{X}_r \right| \leq C \, ||| \mathbf{X} |||_\alpha \, |||Y, Y'|||_{X, 2\alpha} \, |t - s|,$$

where the constant $C > 0$ depends on α, T, $\mathbf{X}$, M, and ω. Moreover, for $T \leq 1$, the constant C does not depend on T.

Proof. This is a consequence of Lemmas 2.6 and 2.24. □

Corollary 2.8. *We have*

$$|N_{s,t} - I_{s,t}| \leq C \, ||| \mathbf{X} |||_\alpha \, |||Y, Y'|||_{X, 2\alpha} \, |t - s|^{3\alpha}, \quad s, t \in [0, T],$$

where the constant $C > 0$ depends on α, T, $\mathbf{X}$, M, and ω. Moreover, for $T \leq 1$, the constant C does not depend on T.

Proof. Note that $\alpha > \frac{1}{3}$ implies $1 < 3\alpha$. Therefore, taking into account equation (2.51), the assertion is an immediate consequence of Corollaries 2.6 and 2.7. □

Proposition 2.14. *Let $(Y, Y') \in \mathscr{D}_X^{2\alpha}([0, T], L(V, D(A^2)))$ be arbitrary. We define the paths $N : [0, T] \to W$ and $N' : [0, T] \to L(V, W)$ as*

$$N_t := \int_0^t S_{t-s} Y_s \, d\mathbf{X}_s, \quad t \in [0, T],$$

$$N'_t := Y_t, \quad t \in [0, T].$$

Then, N is $D(A)$-valued, and we have

$$(N, N') \in \mathscr{D}_X^{2\alpha}([0, T], W). \tag{2.52}$$

Furthermore, we have

$$\|N, N'\|_{X, 2\alpha} \leq C \, |||Y, Y'|||_{X, 2\alpha} \, (|||\mathbf{X}|||_\alpha + T^\alpha), \tag{2.53}$$

where the constant $C > 0$ depends on α, T, $\mathbf{X}$, M, and ω. Moreover, for $T \leq 1$, the constant C does not depend on T.

Proof. By Proposition 2.1, we have $I \in \mathcal{C}^\alpha([0,T], W)$, and by Corollary 2.8, we have $N - I \in \mathcal{C}^{3\alpha}([0,T], W)$. Therefore, it follows that $N \in \mathcal{C}^\alpha([0,T], W)$. Furthermore, we have $N' = Y \in \mathcal{C}^\alpha([0,T], L(V,W))$. Now, let $s, t \in [0, T]$ be arbitrary. Then, we have

$$R_{s,t}^N = N_{s,t} - N'_s X_{s,t} = N_{s,t} - Y_s X_{s,t}$$
$$= Y'_s \mathbb{X}_{s,t} + (N_{s,t} - I_{s,t}) + (I_{s,t} - Y_s X_{s,t} - Y'_s \mathbb{X}_{s,t}).$$

Furthermore, we have

$$|Y'_s \mathbb{X}_{s,t}| \leq \|Y'\|_\infty |\mathbb{X}_{s,t}| \leq \|Y'\|_\infty \|\mathbb{X}\|_{2\alpha} |t-s|^{2\alpha}.$$

Moreover, by Theorem 2.1, we have

$$|I_{s,t} - Y_s X_{s,t} - Y'_s \mathbb{X}_{s,t}| \leq C (\|X\|_\alpha \|R^Y\|_{2\alpha} + \|\mathbb{X}\|_{2\alpha} \|Y'\|_\alpha) |t-s|^{3\alpha},$$

where the constant $C > 0$ depends on α. Therefore, together with Corollary 2.8, we obtain $\|R^N\|_{2\alpha} < \infty$, proving (2.52). Moreover, by the previous estimates and Corollary 2.8, we have

$$\|N, N'\|_{X,2\alpha} = \|N'\|_\alpha + \|R^N\|_{2\alpha} = \|Y\|_\alpha + \|R^N\|_{2\alpha}$$
$$\leq \|Y\|_\alpha + \|Y'\|_\infty \|\mathbb{X}\|_{2\alpha} + C \|\mathbf{X}\|_\alpha \|Y, Y'\|_{X,2\alpha}$$
$$+ C(\|X\|_\alpha \|R^Y\|_{2\alpha} + \|\mathbb{X}\|_{2\alpha} \|Y'\|_\alpha),$$

where the constant $C > 0$ depends on α, T, $\mathbf{X}$, M, and ω and does not depend on $T \leq 1$. Hence, using Lemma 2.6 provides (2.53). □

Corollary 2.9. Let $(Y, Y'), (Z, Z') \in \mathscr{D}_X^{2\alpha}([0,T], L(V, D(A^2)))$ be arbitrary. We define the paths $N, P : [0, T] \to W$ as

$$N_t := \int_0^t S_{t-s} Y_s \, d\mathbf{X}_s, \quad t \in [0, T],$$

$$P_t := \int_0^t S_{t-s} Z_s \, d\mathbf{X}_s, \quad t \in [0, T].$$

Then, we have

$$(N, Y), (P, Z) \in \mathscr{D}_X^{2\alpha}([0,T], W).$$

Furthermore, for $T \leq 1$, we have

$$\|(N,Y) - (P,Z)\|_{X,2\alpha} \leq C \, \|\|(Y,Y') - (Z,Z')\|\|_{X,2\alpha} \, (\|\|\mathbf{X}\|\|_\alpha + T^\alpha),$$

where the constant $C > 0$ depends on α, $\mathbf{X}$, M, and ω.

Proof. Noting that

$$N_t - P_t = \int_0^t S_{t-s}(Y_s - Z_s) \, d\mathbf{X}_s, \quad t \in [0,T],$$

the result follows from Proposition 2.14. □

2.9. Rough Partial Differential Equations

In this section, we deal with RPDEs. Let V and W be Banach spaces, and let A be the generator of a C_0-semigroup $(S_t)_{t \geq 0}$ on W. Then, there are constants $M \geq 1$ and $\omega \in \mathbb{R}$ such that the estimate (2.13) is satisfied. We fix a time horizon $T \in \mathbb{R}_+$. Consider the RPDE

$$\begin{cases} dY_t = (AY_t + f_0(t, Y_t))dt + f(t, Y_t)d\mathbf{X}_t, \\ Y_0 = \xi \end{cases} \quad (2.54)$$

with a rough path $\mathbf{X} = (X, \mathbb{X}) \in \mathscr{C}^\alpha([0,T], V)$, for some index $\alpha \in (\frac{1}{3}, \frac{1}{2}]$ and appropriate mappings $f_0 : [0,T] \times W \to W$ and $f : [0,T] \times W \to L(V,W))$.

2.9.1. *Solution concepts*

In this section, we introduce the required solution concepts. Let $\mathbf{X} = (X, \mathbb{X}) \in \mathscr{C}^\alpha([0,T], V)$ be a rough path for some index $\alpha \in (\frac{1}{3}, \frac{1}{2}]$.

Definition 2.7. Suppose that $f_0 \in \mathrm{Lip}([0,T] \times W, W)$ and $f \in C_b^{2\alpha,2}([0,T] \times W, L(V,W))$. Furthermore, let $\xi \in D(A)$ be arbitrary. A path $(Y,Y') \in \mathscr{D}_X^{2\alpha}([0,T_0], W)$ for some $T_0 \in (0,T]$ is called a *local strong solution* to the RPDE (2.54) with $Y_0 = \xi$ if $Y \in D(A)$, the path $AY : [0,T] \to W$ is bounded, and we have $Y' = f(Y)$, as well as

$$Y_t = \xi + \int_0^t (AY_s + f_0(s, Y_s)) \, ds + \int_0^t f(s, Y_s) \, d\mathbf{X}_s, \quad t \in [0, T_0].$$

(2.55)

If we can choose $T_0 = T$, then we also call (Y, Y') a *(global) strong solution* to the RPDE (2.54) with $Y_0 = \xi$.

Remark 2.4. Note that the right-hand side of (2.55) is well defined. Indeed, let $(Y, Y') \in \mathscr{D}_X^{2\alpha}([0, T_0], W)$ for some $T_0 \in (0, T]$ be arbitrary. We define the paths $B(Y), \Phi(Y) : [0, T_0] \to W$ as

$$B(Y)_t := \int_0^t (AY_s + f_0(s, Y_s)) \, ds,$$

$$\Phi(Y)_t := \int_0^t f(s, Y_s) \, d\mathbf{X}_s.$$

By assumption, the path $AY : [0, T_0] \to W$ is bounded. Furthermore, the path $f_0(Y) : [0, T_0] \to W$ is continuous and hence bounded. Therefore, by Proposition 2.13 (applied with $S_t = \mathrm{Id}$ for all $t \geq 0$), it follows that

$$(B(Y), 0) \in \mathscr{D}_X^{2\alpha}([0, T_0], W).$$

Moreover, by Proposition 2.6, we have

$$(f(Y), f(Y)') \in \mathscr{D}_X^{2\alpha}([0, T_0], L(V, W)),$$

and hence by Proposition 2.14 (applied with $S_t = \mathrm{Id}$ for all $t \geq 0$), it follows that

$$(\Phi(Y), f(Y)) \in \mathscr{D}_X^{2\alpha}([0, T_0], W).$$

Consequently, the right-hand side of (2.55) is an element of $\mathcal{C}^\alpha([0, T_0], W)$.

Definition 2.8. Suppose that $f_0 \in \mathrm{Lip}([0, T] \times W, D(A))$ and $f \in C_b^{2\alpha,2}([0, T] \times W, L(V, D(A^2)))$. Furthermore, let $\xi \in D(A)$ be arbitrary. A path $(Y, Y') \in \mathscr{D}_X^{2\alpha}([0, T_0], W)$ for some $T_0 \in (0, T]$ is called a *local mild solution* to the RPDE (2.54) with $Y_0 = \xi$ if $Y' = f(Y)$ and

$$Y_t = S_t \xi + \int_0^t S_{t-s} f_0(s, Y_s) \, ds + \int_0^t S_{t-s} f(s, Y_s) \, d\mathbf{X}_s, \quad t \in [0, T_0]. \tag{2.56}$$

If we can choose $T_0 = T$, then we also call (Y, Y') a *(global) mild solution* to the RPDE (2.54) with $Y_0 = \xi$.

Remark 2.5. Note that the right-hand side of (2.56) is well defined. Indeed, let $(Y, Y') \in \mathscr{D}_X^{2\alpha}([0, T_0], W)$ for some $T_0 \in (0, T]$ be arbitrary. We define the paths $\Xi, \Gamma(Y), \Psi(Y) : [0, T_0] \to W$ as

$$\Xi_t := S_t \xi, \tag{2.57}$$

$$\Gamma(Y)_t := \int_0^t S_{t-s} f_0(s, Y_s) \, ds, \tag{2.58}$$

$$\Psi(Y)_t := \int_0^t S_{t-s} f(s, Y_s) \, d\mathbf{X}_s. \tag{2.59}$$

The path Ξ is $D(A)$-valued, and by Proposition 2.12, we have

$$(\Xi, 0) \in \mathscr{D}_X^{2\alpha}([0, T_0], W).$$

Furthermore, the path $f_0(Y) : [0, T_0] \to D(A)$ is continuous and hence bounded. Therefore, by Proposition 2.13, it follows that $\Gamma(Y)$ is $D(A)$-valued and

$$(\Gamma(Y), 0) \in \mathscr{D}_X^{2\alpha}([0, T_0], W).$$

Moreover, by Proposition 2.6, we have

$$(f(Y), f(Y)') \in \mathscr{D}_X^{2\alpha}([0, T_0], L(V, D(A^2))),$$

and hence by Proposition 2.14, it follows that $\Psi(Y)$ is $D(A)$-valued and

$$(\Psi(Y), f(Y)) \in \mathscr{D}_X^{2\alpha}([0, T_0], W).$$

Consequently, the right-hand side of (2.56) is an element of $C^\alpha([0, T_0], W)$, and we obtain that the path Y is $D(A)$-valued.

Remark 2.6. Suppose that $f_0 \in \mathrm{Lip}([0, T] \times W, D(A))$ and $f \in C_b^{2\alpha, 2}([0, T] \times W, L(V, D(A^2)))$. Furthermore, let $\xi \in D(A)$ be arbitrary, and let $(Y, Y') \in \mathscr{D}_X^{2\alpha}([0, T_0], W)$ be a local mild solution to the RPDE (2.54) with $Y_0 = \xi$ for some $T_0 \in (0, T]$. Then, we have $Y' = f(Y)$, and hence

$$f(Y)' = Df(Y)Y' = Df(Y)f(Y),$$

where $Df(Y)f(Y) : [0, T_0] \to L(V, L(V, D(A^2)))$ denotes the path

$$Df(Y)f(Y)_t := D_y f(t, Y_t) f(t, Y_t), \quad t \in [0, T_0].$$

In view of the upcoming result, recall that according to Remark 2.5, for every local mild solution (Y, Y') to the RPDE (2.54), we have $Y \in D(A)$.

Proposition 2.15. *Suppose that $f_0 \in \mathrm{Lip}([0,T] \times W, D(A))$ and $f \in C_b^{2\alpha,2}([0,T] \times W, L(V, D(A^2)))$. Let $\xi \in D(A)$ be arbitrary, and let $(Y, Y') \in \mathscr{D}_X^{2\alpha}([0,T_0], W)$ be a local mild solution to the RPDE (2.54) with $Y_0 = \xi$ for some $T_0 \in (0,T]$ such that the path $AY : [0,T_0] \to W$ is bounded. Then, (Y, Y') is also a local strong solution to the RPDE (2.54) with $Y_0 = \xi$.*

Proof. Let $t \in [0, T_0]$ be arbitrary. By Lemma 2.9, we have

$$S_t \xi - \xi = \int_0^t AS_s \xi \, ds.$$

Furthermore, by Lemma 2.9 and Fubini's theorem, we have

$$\int_0^t (S_{t-s} f_0(s, Y_s) - f_0(s, Y_s)) ds = \int_0^t \left(\int_0^{t-s} AS_u f_0(s, Y_s) du \right) ds$$

$$= \int_0^t \left(\int_u^t AS_{s-u} f_0(u, Y_u) ds \right) du = \int_0^t \left(\int_0^s AS_{s-u} f_0(u, Y_u) du \right) ds$$

$$= \int_0^t A \left(\int_0^s S_{s-u} f_0(u, Y_u) du \right) ds.$$

For the upcoming calculation, we would like to use the rough Fubini (Proposition 2.2). In order to verify that the required conditions are fulfilled, let us define the mapping $Z : [0,t] \to \mathscr{D}_X^{2\alpha}([0,t], W)$ as follows. We fix $s \in [0,t]$ and define $Z(s) : [0,t] \to W$ as

$$Z(s)_u := AS_{s-u} f(u, Y_u) \mathbf{1}_{[0,s]}(u), \quad u \in [0,t].$$

Restricting the controlled rough path to the interval $[0,s]$, we have

$$(Y, Y') \in \mathscr{D}_X^{2\alpha}([0,s], W).$$

By Proposition 2.6, we have

$$(f(Y), f(Y)') \in \mathscr{D}_X^{2\alpha}([0,s], L(V, D(A^2)))$$

and the estimate
$$\||f(Y), f(Y)'\||_{X,2\alpha} \leq C\bigl(1 + \||Y, Y'\||^2_{X,2\alpha}\bigr),$$
where the constant $C > 0$ depends on α, T, $\mathbf{X}$ and $\|f\|_{C_b^{2\alpha,2}}$. Consider the mapping $\varphi^s : [0, s] \times L(V, D(A^2)) \to L(V, D(A))$ given by
$$\varphi^s(u, y) = S_{s-u} y.$$
Taking into account Lemmas 2.10 and 2.11, by Proposition 2.11, we have
$$(\varphi^s(f(Y)), \varphi^s(f(Y))') \in \mathscr{D}_X^{2\alpha}([0, s], L(V, D(A)))$$
and the estimate
$$\||\varphi^s(f(Y)), \varphi^s(f(Y))'\||_{X,2\alpha} \leq C \||f(Y), f(Y)'\||_{X,2\alpha},$$
where the constant $C > 0$ depends on α, T, $\mathbf{X}$, M, and ω. Moreover, taking into account Lemma 2.19, by Proposition 2.7, we have
$$(Z, Z') \in \mathscr{D}_X^{2\alpha}([0, s], L(V, W))$$
and the estimate
$$\||Z, Z'\||_{X,2\alpha} \leq |A| \, \||\varphi^s(f(Y)), \varphi^s(f(Y))'\||_{X,2\alpha},$$
where A is regarded as a continuous linear operator $A \in L(D(A), W)$. Consequently, the mapping
$$[0, t] \to \mathbb{R}_+, \quad u \mapsto |Z(s)_0| + \|Z(s), Z(s)'\|_{X,2\alpha}$$
is bounded. Therefore, we may apply the rough Fubini (Proposition 2.2), which, together with Lemma 2.9, gives us
$$\int_0^t (S_{t-s} f(s, Y_s) - f(s, Y_s)) d\mathbf{X}_s = \int_0^t \left(\int_0^{t-s} A S_u f(s, Y_s) du \right) d\mathbf{X}_s$$
$$= \int_0^t \left(\int_u^t A S_{s-u} f(u, Y_u) ds \right) d\mathbf{X}_u = \int_0^t \left(\int_0^s A S_{s-u} f(u, X_u) d\mathbf{X}_u \right) ds$$
$$= \int_0^t A \left(\int_0^s S_{s-u} f(u, X_u) d\mathbf{X}_u \right) ds.$$

Since Y is a local mild solution to the RPDE (2.54) with $Y_0 = \xi$, we have

$$Y_t = S_t \xi + \int_0^t S_{t-s} f_0(s, Y_s) ds + \int_0^t S_{t-s} f(s, Y_s) d\mathbf{X}_s$$

$$= \xi + \int_0^t f_0(s, Y_s) ds + \int_0^t f(s, Y_s) d\mathbf{X}_s$$

$$+ (S_t \xi - \xi) + \int_0^t (S_{t-s} f_0(s, Y_s) - f_0(s, Y_s)) ds$$

$$+ \int_0^t (S_{t-s} f(s, Y_s) - f(s, Y_s)) d\mathbf{X}_s,$$

and hence, combining the latter identities, we obtain

$$Y_t = \xi + \int_0^t f_0(s, Y_s) ds + \int_0^t f(s, Y_s) d\mathbf{X}_s$$

$$+ \int_0^t A S_s \xi \, ds + \int_0^t A \left(\int_0^s S_{s-u} f_0(u, Y_u) du \right) ds$$

$$+ \int_0^t A \left(\int_0^s S_{s-u} f(u, Y_u) d\mathbf{X}_u \right) ds,$$

which implies

$$Y_t = \xi + \int_0^t f_0(s, Y_s) ds + \int_0^t f(s, Y_s) d\mathbf{X}_s$$

$$+ \int_0^t A \underbrace{\left(S_s \xi + \int_0^s S_{s-u} f_0(u, Y_u) du + \int_0^s S_{s-u} f(u, Y_u) d\mathbf{X}_u \right)}_{=Y_s} ds$$

$$= \xi + \int_0^t (AY_s + f_0(s, Y_s)) ds + \int_0^t f(s, Y_s) d\mathbf{X}_s.$$

This proves that Y is also a local strong solution to the RPDE (2.54) with $Y_0 = \xi$. □

2.9.2. The space for the fixed point problem

Note that equation (2.56) may be regarded as a fixed point problem. In this section, we analyze the space for this fixed point problem.

Let $\mathbf{X} = (X, \mathbb{X}) \in \mathscr{C}^\alpha([0,T], V)$ be a rough path for some index $\alpha \in (\frac{1}{3}, \frac{1}{2}]$. Furthermore, let $f_0 \in \mathrm{Lip}([0,T] \times W, D(A))$ and $f \in C_b^{2\alpha,3}([0,T] \times W, L(V, D(A^2)))$ be arbitrary. We also fix an initial condition $\xi \in D(A)$. For every $t \in [0,T]$, we define the subset $\mathbb{B}_t \subset \mathscr{D}_X^{2\alpha}([0,t], W)$ as

$$\mathbb{B}_t := \{(Y, Y') \in \mathscr{D}_X^{2\alpha}([0,t], W) : Y_0 = \xi, Y'_0 = f(0, \xi),$$
$$\|Y, Y'\|_{X, 2\alpha; [0,t]} \leq 1\}.$$

Lemma 2.25. *For each $t \in [0,T]$, the set $\mathbb{B}_t$ equipped with the metric*

$$d((Y, Y'), (Z, Z')) := \|(Y, Y') - (Z, Z')\|_{X, 2\alpha; [0,t]},$$
$$(Y, Y'), (Z, Z') \in \mathbb{B}_t,$$

is a complete metric space.

Proof. The space $\mathscr{D}_X^{2\alpha}([0,t], W)$ equipped with the norm $\|\|\cdot\|\|_{X, 2\alpha; [0,t]}$ is a Banach space and hence a complete metric space. Furthermore, for all $(Y, Y'), (Z, Z') \in \mathbb{B}_t$, we have

$$d((Y, Y'), (Z, Z')) = \|Y - Z, Y' - Z'\|_{X, 2\alpha; [0,t]}$$
$$= |Y_0 - Z_0| + |Y'_0 - Z'_0| + \|Y - Z, Y' - Z'\|_{X, 2\alpha; [0,t]}$$
$$= \|\|Y - Z, Y' - Z'\|\|_{X, 2\alpha; [0,t]}.$$

Moreover, the set $\mathbb{B}_t$ is a closed subset of $\mathscr{D}_X^{2\alpha}([0,t], W)$, completing the proof. $\square$

For $t \in [0,T]$, we define the mapping $\Phi_t : \mathscr{D}_X^{2\alpha}([0,t], W) \to \mathscr{D}_X^{2\alpha}([0,t], W)$ as

$$\Phi_t(Y, Y') := \Xi + \Gamma(Y) + \Psi(Y), \tag{2.60}$$

where the paths $\Xi, \Gamma(Y), \Psi(Y) : [0,t] \to W$ are defined according to (2.57)–(2.59). Note that the mapping Φ_t is well defined due to Remark 2.5.

Proposition 2.16. *For all $t \in [0,T]$ with $t \leq 1$ and all $(Y, Y') \in \mathbb{B}_t$, we have*

$$\|\Phi_t(Y, Y')\|_{X, 2\alpha; [0,t]} \leq C(1 + |\xi|_{D(A)}) t^{1-2\alpha} + C(\|\|\mathbf{X}\|\|_{\alpha; [0,t]} + t^\alpha),$$

where the constant $C > 0$ depends on α, $\mathbf{X}$, $\|f_0\|_{\text{Lip}}$, $\|f\|_{C_b^{2\alpha,2}}$, M, and ω.

Proof. For convenience of notation, we skip the subscript $[0,t]$ in the following calculations. Note that

$$\|\Phi_t(Y,Y')\|_{X,2\alpha} \leq \|\Xi,0\|_{X,2\alpha} + \|\Gamma(Y),0\|_{X,2\alpha} + \|\Psi(Y),f(Y)\|_{X,2\alpha}.$$

By Proposition 2.12, we have

$$\|\Xi,0\|_{X,2\alpha} \leq Me^\omega |\xi|_{D(A)} t^{1-2\alpha}.$$

Moreover, by Proposition 2.13, we have

$$\|\Gamma(Y),0\|_{X,2\alpha} \leq 2Me^\omega \|f_0(Y)\|_\infty t^{1-2\alpha}.$$

Noting that $Y_0 = \xi$, $Y_0' = f(0,\xi)$, and $\|Y,Y'\|_{X,2\alpha} \leq 1$, by Lemmas 2.6 and 2.16, we obtain

$$\begin{aligned}
\|f_0(Y)\|_\infty &\leq \|f_0\|_{\text{Lip}}(1 + \|Y\|_\infty) \\
&\leq \|f_0\|_{\text{Lip}}(1 + \|Y,Y'\|_{X,2\alpha}(\|\mathbf{X}\|_\alpha + 2)) \\
&\leq \|f_0\|_{\text{Lip}}(1 + (|\xi| + |f(0,\xi)| + 1)(\|\mathbf{X}\|_\alpha + 2)) \\
&\leq \|f_0\|_{\text{Lip}}(1 + (|\xi| + \|f\|_{C_b^{2\alpha,2}} + 1)(\|\mathbf{X}\|_\alpha + 2)).
\end{aligned}$$

Moreover, noting that $Y_0 = \xi$, $Y_0' = f(0,\xi)$, and $\|Y,Y'\|_{X,2\alpha} \leq 1$, by Propositions 2.6 and 2.14, we have

$$\begin{aligned}
\|\Psi(Y),f(Y)\|_{X,2\alpha} &\leq C \|f(Y),f(Y)'\|_{X,2\alpha}(\|\mathbf{X}\|_\alpha + t^\alpha) \\
&= C(|f(0,\xi)| + |f(Y),f(Y)'|_{X,2\alpha})(\|\mathbf{X}\|_\alpha + t^\alpha) \\
&\lesssim C(1 + |Y,Y'|_{X,2\alpha}^2)(\|\mathbf{X}\|_\alpha + t^\alpha) \\
&\lesssim C(1 + (|f(0,\xi)| + 1)^2)(\|\mathbf{X}\|_{\alpha;[0,t]} + t^\alpha) \\
&\lesssim C(\|\mathbf{X}\|_\alpha + t^\alpha),
\end{aligned}$$

where the constant $C > 0$, which changes from line to line, depends on α, $\mathbf{X}$, $\|f\|_{C_b^{2\alpha,2}}$, M, and ω. This completes the proof. $\square$

Proposition 2.17. *For all $t \in [0,T]$ with $t \le 1$ and all (Y,Y'), $(Z,Z') \in \mathbb{B}_t$, we have*

$$\|\Phi_t(Y,Y') - \Phi_t(Z,Z')\|_{X,2\alpha;[0,t]} \le C\|(Y,Y') - (Z,Z')\|_{X,2\alpha;[0,t]}$$
$$\times \left(\|\mathbf{X}\|_{\alpha;[0,t]} + t^\alpha + t^{1-2\alpha} \right),$$

where the constant $C > 0$ depends on α, $\mathbf{X}$, $\|f_0\|_{\mathrm{Lip}}$, $\|f\|_{C_b^{2\alpha,3}}$, M, and ω.

Proof. For convenience of notation, we skip the subscript $[0,t]$ in the following calculations. Note that

$$\|\Phi(Y,Y') - \Phi(Z,Z')\|_{X,2\alpha} \le \|(\Gamma(Y),0) - (\Gamma(Z),0)\|_{X,2\alpha}$$
$$+ \|(\Psi(Y), f(Y)) - (\Psi(Z), f(Z))\|_{X,2\alpha}.$$

By Corollary 2.5, we have

$$\|(\Gamma(Y),0) - (\Gamma(Z),0)\|_{X,2\alpha} \le 2Me^\omega \|f_0(Y) - f_0(Z)\|_\infty t^{1-2\alpha}.$$

Furthermore, noting that $Y_0 = Z_0$ and $Y_0' = Z_0'$, by Lemmas 2.6 and 2.16, we obtain

$$\|f_0(Y) - f_0(Z)\|_\infty \le \|f_0\|_{\mathrm{Lip}} \|Y - Z\|_\infty$$
$$\le \|f_0\|_{\mathrm{Lip}} \|(Y,Y') - (Z,Z')\|_{X,2\alpha}(\|\mathbf{X}\|_\alpha + 2).$$

Moreover, by Corollary 2.9 and Proposition 2.9, we have

$$\|(\Psi(Y), f(Y)) - (\Psi(Z), f(Z))\|_{X,2\alpha}$$
$$\le C \| (f(Y), f(Y)') - (f(Z), f(Z)') \|_{X,2\alpha} (\|\mathbf{X}\|_\alpha + t^\alpha)$$
$$\lesssim C(1 + |Y,Y'|_{X,2\alpha}^2 + |Z,Z'|_{X,2\alpha}^2)$$
$$\times \|(Y,Y') - (Z,Z')\|_{X,2\alpha} (\|\mathbf{X}\|_\alpha + t^\alpha),$$

where the constant $C > 0$, which changes from line to line, depends on α, $\mathbf{X}$, $\|f\|_{C_b^{2\alpha,3}}$, M, and ω. Now, noting that $Y_0 = Z_0 = \xi$, $Y_0' = Z_0' = f(0,\xi)$, and $\|Y,Y'\|_{X,2\alpha}, \|Z,Z'\|_{X,2\alpha} \le 1$, we obtain

$$|Y,Y'|_{X,2\alpha} = |f(0,\xi)| + \|Y,Y'\|_{X,2\alpha} \le \|f\|_{C_b^{2\alpha,3}} + 1,$$
$$|Z,Z'|_{X,2\alpha} = |f(0,\xi)| + \|Z,Z'\|_{X,2\alpha} \le \|f\|_{C_b^{2\alpha,3}} + 1.$$

Consequently, noting again that $Y_0 = Z_0$ and $Y_0' = Z_0'$, we have

$$\|(Y, Y') - (Z, Z')\|_{X, 2\alpha} = \|(Y, Y') - (Z, Z')\|_{X, 2\alpha},$$

concluding the proof. □

2.9.3. Auxiliary results

In this section, we derive auxiliary results which are required to establish the existence and uniqueness of local mild solutions for the RPDE (2.54). Let $f_0 \in \text{Lip}([0, T] \times W, D(A))$ and $f \in C_b^{2\alpha, 2}([0, T] \times W, L(V, D(A^2)))$ be arbitrary.

Lemma 2.26. *Let* $\mathbf{X} = (X, \mathbb{X}) \in \mathscr{C}^\alpha([0, T], V)$ *be a rough path for some index* $\alpha \in (\frac{1}{3}, \frac{1}{2}]$. *Furthermore, let* $\xi \in D(A)$ *be arbitrary, and let* $(Y, Y') \in \mathscr{D}_X^{2\alpha}([0, T_0], W)$ *be a local mild solution to the RPDE* (2.54) *with* $Y_0 = \xi$ *for some* $T_0 \in (0, T]$. *Then, we have*

$$\|f(Y)'\|_\alpha \leq C(1 + \|Y\|_\alpha),$$

where the constant $C > 0$ *depends on* α, T, *and* $\|f\|_{C_b^{2\alpha, 2}}$.

Proof. By Proposition 2.6, we have

$$\|f(Y)\|_\alpha \leq \|f\|_{C_b^{2\alpha, 2}}(\|Y\|_\alpha + T^\alpha).$$

Therefore, by Proposition 2.6 and Remark 2.6, we obtain

$$\|f(Y)'\|_\alpha \leq \|f\|_{C_b^{2\alpha, 2}} \Big(\|f(Y)\|_\alpha + \|Y\|_\alpha \|f(Y)\|_\infty + T^\alpha \|f(Y)\|_\infty \Big)$$

$$\leq \|f\|_{C_b^{2\alpha, 2}} \Big(\|f\|_{C_b^{2\alpha, 2}}(\|Y\|_\alpha + T^\alpha)$$

$$+ \|Y\|_\alpha \|f\|_{C_b^{2\alpha, 2}} + T^\alpha \|f\|_{C_b^{2\alpha, 2}} \Big),$$

completing the proof. □

Lemma 2.27. *Let* $\mathbf{X} = (X, \mathbb{X}) \in \mathscr{C}^\alpha([0, T], V)$ *be a rough path for some index* $\alpha \in (\frac{1}{3}, \frac{1}{2}]$. *Furthermore, let* $\xi \in D(A)$ *be arbitrary, and* let $(Y, Y') \in \mathscr{D}_X^{2\alpha}([0, T_0], W)$ be a local mild solution to the RPDE (2.54) with $Y_0 = \xi$ for some $T_0 \in (0, T]$. Let $s, t \in [0, T_0]$ with $s \leq t$ be arbitrary, and define the interval $I := [s, t]$. Then, we have

$$|R_{s,t}^Y| \leq C\Big(1 + \|Y\|_{\alpha;I} + \|X\|_{\alpha;I}(1 + \|R^{f(Y)}\|_{2\alpha;I})$$
$$+ \|\mathbb{X}\|_{2\alpha;I}(1 + \|Y\|_{\alpha;I})\Big)|t - s| + C|\mathbb{X}_{s,t}|,$$

where the constant $C > 0$ depends on ξ, α, T, $\|f_0\|_{\text{Lip}}$, $\|f\|_{C_b^{2\alpha,2}}$, M, and ω.

Proof. For convenience of notation, we skip the subscript I in the following calculations. By equation (2.51), we have

$$Y_{s,t} = S_{s,t}\xi + \Gamma(Y)_{s,t} + \int_s^t (S_{t-r} - \text{Id})f(Y_r)\, d\mathbf{X}_r$$
$$+ \int_0^s (S_{t-r} - S_{s-r})f(Y_r)\, d\mathbf{X}_r + \int_s^t f(Y_r)\, d\mathbf{X}_r,$$

where the path $\Gamma(Y) : [0, T_0] \to W$ is defined according to (2.58). Therefore, by Remark 2.6, we have

$$|R_{s,t}^Y| = |Y_{s,t} - Y_s'X_{s,t}| = |Y_{s,t} - f(Y_s)X_{s,t}|$$
$$\leq |S_{s,t}\xi| + |\Gamma(Y)_{s,t}| + \left|\int_s^t (S_{t-r} - \text{Id})f(Y_r)\, d\mathbf{X}_r\right|$$
$$+ \left|\int_0^s (S_{t-r} - S_{s-r})f(Y_r)\, d\mathbf{X}_r\right|$$
$$+ \left|\int_s^t f(Y_r)\, d\mathbf{X}_r - f(Y_s)X_{s,t} - Df(Y_s)f(Y_s)\mathbb{X}_{s,t}\right|$$
$$+ |Df(Y_s)f(Y_s)\mathbb{X}_{s,t}|.$$

By Proposition 2.3, we have

$$|S_{s,t}\xi| \leq Me^{\omega T}|\xi|_{D(A)}|t - s|,$$

and by Lemmas 2.16 and 2.22, we have

$$\begin{aligned}
|\Gamma(Y)_{s,t}| &\leq (1+T)Me^{\omega T}\|f_0(Y)\|_\infty |t-s| \\
&\leq (1+T)Me^{\omega T}\|f_0\|_{\mathrm{Lip}}(1+\|Y\|_\infty)|t-s| \\
&\leq (1+T)Me^{\omega T}\|f_0\|_{\mathrm{Lip}}(1+|Y_0|+\|Y\|_\alpha T^\alpha)|t-s| \\
&\leq (1+T)Me^{\omega T}\|f_0\|_{\mathrm{Lip}}(1+|\xi|+\|Y\|_\alpha T^\alpha)|t-s|.
\end{aligned}$$

Furthermore, by Lemma 2.23, we have

$$\left| \int_s^t (S_{t-r} - \mathrm{Id}) f(Y_r)\, d\mathbf{X}_r \right|$$
$$\leq C\Big(\|X\|_\alpha \big(\|f(Y)\|_\infty + \|R^{f(Y)}\|_{2\alpha}\big) + \|\mathbb{X}\|_{2\alpha}\big(\|f(Y)'\|_\infty + \|f(Y)'\|_\alpha\big)\Big)|t-s|^{3\alpha},$$

where the constant $C > 0$ depends on α, T, M, and ω. Similarly, by Lemma 2.24, we have

$$\left| \int_0^s (S_{t-r} - S_{s-r}) f(Y_r)\, d\mathbf{X}_r \right|$$
$$\leq C\Big(\|X\|_\alpha \big(\|f(Y)\|_\infty + \|R^{f(Y)}\|_{2\alpha}\big) + \|\mathbb{X}\|_{2\alpha}\big(\|f(Y)'\|_\infty + \|f(Y)'\|_\alpha\big)\Big)|t-s|,$$

where the constant $C > 0$ depends on α, T, M, and ω. Furthermore, by Theorem 2.1 and Remark 2.6, we have

$$\left| \int_s^t f(Y_r)\, d\mathbf{X}_r - f(Y_s) X_{s,t} - Df(Y_s) f(Y_s) \mathbb{X}_{s,t} \right|$$
$$\leq C(\|X\|_\alpha \|R^{f(Y)}\|_{2\alpha} + \|\mathbb{X}\|_{2\alpha} \|f(Y)'\|_\alpha)|t-s|^{3\alpha},$$

where the constant $C > 0$ depends on α. Moreover, we have

$$|Df(Y_s) f(Y_s) \mathbb{X}_{s,t}| \leq |Df(Y_s)|\, |f(Y_s)|\, |\mathbb{X}_{s,t}| \leq \|f\|_{C_b^{2\alpha,2}}^2 |\mathbb{X}_{s,t}|.$$

Noting that $\alpha > \frac{1}{3}$, which implies $3\alpha > 1$, using Lemma 2.26, the proof is complete. □

Lemma 2.28. *Let $\beta \in (\frac{1}{3}, \frac{1}{2}]$ be arbitrary. We choose $\alpha \in (\frac{1}{3}, \beta)$ such that $\beta \leq \frac{3}{2}\alpha$, and let*
$$\mathbf{X} = (X, \mathbb{X}) \in \mathscr{C}^{\beta}([0,T], V) \subset \mathscr{C}^{\alpha}([0,T], V)$$
be a rough path. Furthermore, let $\xi \in D(A)$ be arbitrary, and let $(Y, Y') \in \mathscr{D}_X^{2\alpha}([0, T_0], W)$ be a local mild solution to the RPDE (2.54) with $Y_0 = \xi$ for some $T_0 \in (0, T]$. Then, we even have $(Y, Y') \in \mathscr{D}_X^{2\beta}([0, T_0], W)$.

Proof. Let $s, t \in [0, T_0]$ be arbitrary. Since $\beta \leq \frac{3}{2}\alpha < 2\alpha$, we have
$$|Y_{s,t}| = |Y'_s X_{s,t} + R^Y_{s,t}| \leq \|Y'\|_\infty |X_{s,t}| + |R^Y_{s,t}|$$
$$\leq \|Y'\|_\infty \|X\|_\beta |t-s|^\beta + \|R^Y\|_{2\alpha} |t-s|^{2\alpha}$$
$$\leq \|Y'\|_\infty \|X\|_\beta |t-s|^\beta + \|R^Y\|_{2\alpha} |t-s|^\beta,$$

showing that $Y \in \mathcal{C}^\beta([0, T_0], W)$. Furthermore, from Remark 2.6 and Proposition 2.5, we obtain
$$Y' = f(Y) \in \mathcal{C}^\beta([0, T_0], L(V, D(A^2))) \subset \mathcal{C}^\beta([0, T_0], L(V, W)).$$
Moreover, noting that $2\beta \leq 1$, by Lemma 2.27, we have
$$|R^Y_{s,t}| \leq C\Big(1 + \|Y\|_{\alpha;I} + \|X\|_{\alpha;I}(1 + \|R^{f(Y)}\|_{2\alpha;I})$$
$$+ \|\mathbb{X}\|_{2\alpha;I}(1 + \|Y\|_{\alpha;I})\Big)|t-s| + C|\mathbb{X}_{s,t}|$$
$$\leq C\Big(1 + \|Y\|_{\alpha;I} + \|X\|_{\alpha;I}(1 + \|R^{f(Y)}\|_{2\alpha;I})$$
$$+ \|\mathbb{X}\|_{2\alpha;I}(1 + \|Y\|_{\alpha;I})\Big) T^{1-2\beta}|t-s|^{2\beta} + C\|\mathbb{X}\|_{2\beta}|t-s|^{2\beta},$$

where the constant $C > 0$ depends on ξ, α, T, $\|f_0\|$, $\|f\|_{C_b^{2\alpha,2}}$, M, and ω. This shows $\|R^Y\|_{2\beta} < \infty$, completing the proof. □

2.9.4. *Existence and uniqueness of local mild solutions*

In this section, we present a result concerning the existence and uniqueness of local mild solutions to the RPDE (2.54).

Theorem 2.2. *Let $\mathbf{X} = (X, \mathbb{X}) \in \mathscr{C}^\beta([0,T], V)$ be a rough path for some index $\beta \in (\frac{1}{3}, \frac{1}{2}]$, and let $f_0 \in \text{Lip}([0,T] \times W, D(A))$ and*

$f \in C_b^{2\beta,3}([0,T] \times W, L(V, D(A^2)))$ be mappings. Then, for every $\xi \in D(A)$, there exist $T_0 \in (0,T]$ and a unique local mild solution $(Y, Y') \in \mathscr{D}_X^{2\beta}([0, T_0], W)$ to the RPDE (2.54) with $Y_0 = \xi$.

Proof. We choose $\alpha \in (\frac{1}{3}, \beta)$ such that $\beta \leq \frac{3}{2}\alpha$. Since $\alpha < \beta$, we also have $\mathbf{X} \in \mathscr{C}^\alpha([0,T], V)$. For $t \in [0,T]$, we define the mapping $\Phi_t : \mathscr{D}_X^{2\alpha}([0,t], W) \to \mathscr{D}_X^{2\alpha}([0,t], W)$ according to (2.60). By Propositions 2.16 and 2.17 and Lemma 2.1, for every $t \in [0,T]$ with $t \leq 1$ and all $(Y, Y'), (Z, Z') \in \mathbb{B}_t$, we have

$$\|\Phi_t(Y, Y')\|_{X, 2\alpha; [0,t]} \leq C(1 + |\xi|_{D(A)}) t^{1-2\alpha}$$
$$+ C\big(\|X\|_{\beta, [0,t]} t^{\beta-\alpha} + \|\mathbb{X}\|_{2\beta, [0,t]} t^{2(\beta-\alpha)} + t^\alpha\big),$$
$$\|\Phi(Y, Y') - \Phi(Z, Z')\|_{X, 2\alpha; [0,t]} \leq C \|\|(Y, Y') - (Z, Z')\|\|_{X, 2\alpha, [0,t]}$$
$$\big(\|X\|_{\beta, [0,t]} t^{\beta-\alpha} + \|\mathbb{X}\|_{2\beta, [0,t]} t^{2(\beta-\alpha)} + t^\alpha + t^{1-2\alpha}\big),$$

where the constant $C > 0$ depends on α, $\mathbf{X}$, $\|f_0\|_{\text{Lip}}$, $\|f\|_{C_b^{2\alpha,3}}$, M, and ω. We choose $T_0 \in (0,T]$ with $T_0 \leq 1$ small enough such that

$$C(1 + |\xi|_{D(A)}) T_0^{1-2\alpha}$$
$$+ C\big(\|X\|_{\beta, [0,T_0]} T_0^{\beta-\alpha} + \|\mathbb{X}\|_{2\beta, [0,T_0]} T_0^{2(\beta-\alpha)} + T_0^\alpha\big) \leq 1,$$
$$C\big(\|X\|_{\beta, [0,T_0]} T_0^{\beta-\alpha} + \|\mathbb{X}\|_{2\beta, [0,t]} T_0^{2(\beta-\alpha)} + T_0^\alpha + T_0^{1-2\alpha}\big) \leq \frac{1}{2}.$$

Then, Φ_{T_0} is a contraction, and by the Banach fixed point theorem, there is a unique mild solution $(Y, Y') \in \mathscr{D}_X^{2\alpha}([0, T_0], W)$ to the RPDE (2.54) with $Y_0 = \xi$. Now, applying Lemma 2.28 completes the proof. $\square$

2.9.5. *Further auxiliary results*

In this section, we derive further auxiliary results which are required to establish the existence and uniqueness of global mild solutions for the RPDE (2.54).

Proposition 2.18. *Let $\mathbf{X} = (X, \mathbb{X}) \in \mathscr{C}^\alpha([0,T], V)$ be a rough path for some index $\alpha \in (\frac{1}{3}, \frac{1}{2})$, let $f_0 \in \text{Lip}([0,T] \times W, D(A))$ and $f \in C_b^{2\alpha,2}([0,T] \times W, L(V, D(A^2)))$ be mappings, and let $\xi \in D(A)$ be arbitrary. Then, there exists a constant $K > 0$, depending on ξ, α,*

T, $\mathbf{X}$, $\|f_0\|_{\mathrm{Lip}}$, $\|f\|_{C_b^{2\alpha,2}}$, M, and ω, such that for every mild solution $(Y, Y') \in \mathscr{D}_X^{2\alpha}([0,T], W)$ to the RPDE (2.54) with $Y_0 = \xi$, we have

$$|Y_t| \leq K \quad \text{for all } t \in [0,T].$$

Proof. Let $s, t \in [0,T]$ with $s \leq t$ be arbitrary, and define the interval $I := [s,t]$. By Lemma 2.27, we have

$$|R^Y_{s,t}| \leq C\Big(1 + \|Y\|_{\alpha;I} + \|X\|_{\alpha;I}(1 + \|R^{f(Y)}\|_{2\alpha;I}) $$
$$+ \|\mathbb{X}\|_{2\alpha;I}(1 + \|Y\|_{\alpha;I})\Big)|t-s| + C\|X\|_{\alpha;I}|t-s|^{2\alpha},$$

where the constant $C > 0$ depends on ξ, α, T, $\|f_0\|_{\mathrm{Lip}}$, $\|f\|_{C_b^{2\alpha,2}}$, M, and ω. We set $\beta := 1 - 2\alpha$. Since $\alpha < \frac{1}{2}$, we have $\beta > 0$. Furthermore, let $h > 0$ be arbitrary. Then, we have

$$\|R^Y\|_{2\alpha;h} \leq C\|X\|_{\alpha;h} + C\Big(1 + \|Y\|_{\alpha;h} + \|X\|_{\alpha;h}(1 + \|R^{f(Y)}\|_{2\alpha;h})$$
$$+ \|\mathbb{X}\|_{2\alpha;h}(1 + \|Y\|_{\alpha;h})\Big)h^\beta. \qquad (2.61)$$

By Proposition 2.6, we have

$$\|R^{f(Y)}\|_{2\alpha;I} \leq \|f\|_{C_b^{2\alpha,2}}\Big(1 + \frac{1}{2}\|Y\|_{\alpha;I}^2 + \|R^Y\|_{2\alpha;I}\Big),$$

and hence

$$\|R^{f(Y)}\|_{2\alpha;h} \leq \|f\|_{C_b^{2\alpha,2}}\big(1 + \|Y\|_{\alpha;h}^2 + \|R^Y\|_{2\alpha;h}\big). \qquad (2.62)$$

Therefore, combining (2.61) and (2.62), there is a constant $c_1 > 0$ only depending on α, T, and $\|f\|_{C_b^{2\alpha,2}}$, such that

$$\|R^Y\|_{2\alpha;h} \leq c_1\|X\|_{\alpha;h}$$
$$+ c_1\Big(1 + \|Y\|_{\alpha;h} + \|X\|_{\alpha;h}(1 + \|Y\|_{\alpha;h}^2 + \|R^Y\|_{2\alpha;h})$$
$$+ \|\mathbb{X}\|_{2\alpha;h}(1 + \|Y\|_{\alpha;h})\Big)h^\beta.$$

Thus, there is a constant $c_2 > 0$ only depending on α, T, and $\|f\|_{C_b^{2\alpha,2}}$, such that for all $h \in (0,1]$, we obtain

$$\|R^Y\|_{2\alpha;h} \leq c_2(\|X\|_{\alpha;h} + \|\mathbb{X}\|_{2\alpha;h} + 1) + c_2\|X\|_{\alpha;h}h^\beta\|Y\|_{\alpha;h}^2$$
$$+ c_2\|X\|_{\alpha;h}h^\beta\|R^Y\|_{2\alpha;h} + c_2(\|\mathbb{X}\|_{2\alpha;h} + 1)h^\beta\|Y\|_{\alpha;h}. \quad (2.63)$$

Now, we consider $h \in (0,1]$ to be so small that

$$c_2\|X\|_\alpha h^\beta \leq \frac{1}{2} \quad \text{and} \quad c_2(\|X\|_\alpha + \|\mathbb{X}\|_{2\alpha} + 1)^{1/2}h^\beta \leq \frac{1}{2}. \quad (2.64)$$

Then, by (2.63) and (2.64), we have

$$\|R^Y\|_{2\alpha;h} \leq c_2(\|X\|_{\alpha;h} + \|\mathbb{X}\|_{2\alpha;h} + 1) + \frac{1}{2}\|Y\|_{\alpha;h}^2 + \frac{1}{2}\|R^Y\|_{2\alpha;h}$$
$$+ \frac{1}{2}(\|\mathbb{X}\|_{2\alpha;h} + 1)^{1/2}\|Y\|_{\alpha;h}.$$

Therefore, using the elementary inequality $xy \leq \frac{x^2}{2} + \frac{y^2}{2}$, $x, y \in \mathbb{R}$, we obtain

$$\|R^Y\|_{2\alpha;h} \leq 2c_2(\|X\|_{\alpha;h} + \|\mathbb{X}\|_{2\alpha;h} + 1) + \|Y\|_{\alpha;h}^2$$
$$+ (\|\mathbb{X}\|_{2\alpha;h} + 1)^{1/2}\|Y\|_{\alpha;h}$$
$$\leq c_3(\|X\|_{\alpha;h} + \|\mathbb{X}\|_{2\alpha;h} + 1) + \frac{3}{2}\|Y\|_{\alpha;h}^2, \quad (2.65)$$

with a constant $c_3 > 0$ only depending on α, T, and $\|f\|_{C_b^{2\alpha,2}}$. On the other hand, since $Y_{s,t} = f(Y_s)X_{s,t} + R^Y_{s,t}$, we have

$$|Y_t - Y_s| \leq |f(Y_s)X_{s,t}| + |R^Y_{s,t}| \leq \|f\|_{C_b^{2\alpha,2}}\|X\|_\alpha|t-s|^\alpha$$
$$+ \|R^Y\|_{2\alpha}|t-s|^{2\alpha},$$

and hence

$$\|Y\|_{\alpha;h} \leq C(\|X\|_{\alpha;h} + \|R^Y\|_{2\alpha;h}h^\alpha),$$

where the constant $C > 0$ depends on $\|f\|_{C_b^{2\alpha,2}}$. Note that $\beta < \alpha$. Indeed, since $3\alpha > 1$, we have $\beta = 1 - 2\alpha < 3\alpha - 2\alpha = \alpha$. Therefore,

we have
$$\|Y\|_{\alpha;h} \leq C(\|X\|_{\alpha;h} + \|R^Y\|_{2\alpha;h}h^\beta).$$
Thus, using (2.65) and (2.64), we obtain
$$\|Y\|_{\alpha;h} \leq c_4\|X\|_{\alpha;h} + c_4(\|X\|_{\alpha;h} + \|\mathbb{X}\|_{2\alpha;h} + 1)h^\beta + c_4\|Y\|_{\alpha,h}^2 h^\beta$$
$$\leq c_4\|X\|_{\alpha;h} + c_5(\|X\|_{\alpha;h} + \|\mathbb{X}\|_{2\alpha;h} + 1)^{1/2} + c_4\|Y\|_{\alpha,h}^2 h^\beta,$$
with constants $c_4, c_5 > 0$ only depending on α, T, and $\|f\|_{C_b^{2\alpha,2}}$. Hence, multiplying this inequality with $c_4 h^\beta$, we have
$$c_4\|Y\|_{\alpha;h}h^\beta \leq c_6(\||\mathbf{X}\||_\alpha + 1)h^\beta + (c_4\|Y\|_{\alpha,h}h^\beta)^2,$$
where $c_6 := c_4 + c_5$. Now, we set
$$\psi_h := c_4\|Y\|_{\alpha;h}h^\beta \quad \text{and} \quad \lambda_h := c_6(\||\mathbf{X}\||_\alpha + 1)h^\beta.$$
Then, we have
$$\psi_h \leq \lambda_h + \psi_h^2.$$

Now, exactly the same argumentation as in the proof given in Ref. [1, Proposition 8.2] shows the existence of a constant $N > 0$, depending on α, T, $\mathbf{X}$, $\|f_0\|_{\text{Lip}}$, $\|f\|_{C_b^{2\alpha,2}}$, M, and ω, such that $\|Y\|_\alpha \leq N$. Consequently, setting $K := |\xi| + NT^\alpha$, we obtain $\|Y\|_\infty \leq |Y_0| + \|Y\|_\alpha T^\alpha \leq K$, completing the proof. $\square$

The following auxiliary result shows how two local mild solutions of the RPDE (2.54) can be concatenated.

Lemma 2.29. *Let* $\mathbf{X} = (X, \mathbb{X}) \in \mathscr{C}^\alpha([0, T], V)$ *be a rough path for some index* $\alpha \in (\frac{1}{3}, \frac{1}{2}]$, *and let* $f_0 \in \text{Lip}([0, T] \times W, D(A))$ *and* $f \in C_b^{2\alpha,2}([0, T] \times W, L(V, D(A^2)))$ *be mappings. Moreover, let* $0 \leq q \leq r \leq s \leq T$ *and* $\xi \in D(A)$ *be arbitrary. Let* $(Y(q, r, \xi), Y(q, r, \xi)') \in \mathscr{D}_X^{2\alpha}([q, r], W)$ *be a solution to the equation* $Y(q, r, \xi)' = f(Y(q, r, \xi))$, *and*
$$Y(q, r, \xi)_t = S_{t-q}\xi + \int_q^t S_{t-u}f_0(u, Y(q, r, \xi)_u) \, du$$
$$+ \int_q^t S_{t-u}f(u, Y(q, r, \xi)_u) \, d\mathbf{X}_u \quad t \in [q, r]. \quad (2.66)$$

Furthermore, we set $\eta := Y(q,r,\xi)_r$, and let $(Y(r,s,\eta), Y(r,s,\eta)') \in \mathscr{D}_X^{2\alpha}([r,s], W)$ be a solution to the equation $Y(r,s,\eta)' = f(Y(r,s,\eta))$, and

$$Y(r,s,\eta)_t = S_{t-r}\eta + \int_r^t S_{t-u} f_0(u, Y(r,s,\eta)_u)\, du$$

$$+ \int_r^t S_{t-u} f(u, Y(r,s,\eta)_u)\, d\mathbf{X}_u, \quad t \in [r,s]. \quad (2.67)$$

We define the concatenated path $Y(q,s,\xi) : [q,s] \to W$ as

$$Y(q,s,\xi)_t := \begin{cases} Y(q,r,\xi)_t, & t \in [q,r], \\ Y(r,s,\eta)_t, & t \in [r,s], \end{cases}$$

and we define the concatenated path $Y(q,s,\xi)' : [q,s] \to L(V,W)$ as

$$Y(q,s,\xi)'_t := \begin{cases} Y(q,r,\xi)'_t, & t \in [q,r], \\ Y(r,s,\eta)'_t, & t \in [r,s]. \end{cases}$$

Then, we have $(Y(q,s,\xi), Y(q,s,\xi)') \in \mathscr{D}_X^{2\alpha}([q,r], W)$, and this path is a solution to $Y(q,s,\xi)' = f(Y(q,s,\xi))$ and

$$Y(q,s,\xi)_t = S_{t-q}\xi + \int_q^t S_{t-u} f_0(u, Y(q,s,\xi)_u)\, du$$

$$+ \int_q^t S_{t-u} f(u, Y(q,s,\xi)_u)\, d\mathbf{X}_u, \quad t \in [q,s]. \quad (2.68)$$

Proof. It is easy to check that $(Y(q,s,\xi), Y(q,s,\xi)') \in \mathscr{D}_X^{2\alpha}([q,r], W)$ and $Y(q,s,\xi)' = f(Y(q,s,\xi))$. Furthermore, taking into account (2.66), it is evident that equation (2.68) is satisfied for all $t \in [q,r]$. Moreover, by (2.67), for each $t \in [r,s]$, we obtain

$$Y(q,s,\xi)_t = S_{t-r} Y(q,r,\xi)_r + \int_r^t S_{t-u} f_0(u, Y(r,s,Y(q,r,\xi)_r)_u)\, du$$

$$+ \int_r^t S_{t-u} f(u, Y(r,s,Y(q,r,\xi)_r)_u)\, d\mathbf{X}_u$$

$$= S_{t-r}\left(S_{r-q}\xi + \int_q^r S_{r-u}f_0(u,Y(q,r,\xi)_u)du\right.$$

$$\left. + \int_q^r S_{r-u}f(u,Y(q,r,\xi)_u)d\mathbf{X}_u\right)$$

$$+ \int_r^t S_{t-u}f_0(u,Y(r,s,Y(q,r,\xi)_r)_u)du$$

$$+ \int_r^t S_{t-u}f(u,Y(r,s,Y(q,r,\xi)_r)_u)d\mathbf{X}_u$$

$$= S_{t-q}\xi + \int_q^r S_{t-u}f_0(u,Y(q,s,\xi)_u)du$$

$$+ \int_q^r S_{t-u}f(u,Y(q,s,\xi)_u)d\mathbf{X}_u$$

$$+ \int_r^t S_{t-u}f_0(u,Y(r,s,Y(q,r,\xi)_r)_u)du$$

$$+ \int_r^t S_{t-u}f(u,Y(r,s,Y(q,r,\xi)_r)_u)d\mathbf{X}_u,$$

completing the proof. $\square$

2.9.6. Existence and uniqueness of global mild solutions

In this section, we present a result concerning the existence and uniqueness of global mild solutions to the RPDE (2.54).

Theorem 2.3. *Let* $\mathbf{X} = (X,\mathbb{X}) \in \mathscr{C}^\beta([0,T],V)$ *be a rough path for some index* $\beta \in (\frac{1}{3},\frac{1}{2}]$, *and let* $f_0 \in \mathrm{Lip}([0,T] \times W, D(A))$ *and* $f \in C_b^{2\beta,3}([0,T] \times W, L(V,D(A^2)))$ *be mappings such that*

$$f_0|_{[0,T] \times D(A)} \in \mathrm{Lip}([0,T] \times D(A), D(A^2)), \tag{2.69}$$

$$f|_{[0,T] \times D(A)} \in C_b^{2\beta,3}([0,T] \times D(A), L(V,D(A^3))). \tag{2.70}$$

Then, for every $\xi \in D(A^2)$, *there exists a unique mild solution* $(Y,Y') \in \mathscr{D}_X^{2\beta}([0,T],W)$ *to the RPDE (2.54) with* $Y_0 = \xi$, *which is also a strong solution.*

Proof. We choose $\alpha \in (\frac{1}{3}, \beta)$ such that $\beta \leq \frac{3}{2}\alpha$. Since $\alpha < \beta$, we have $\mathbf{X} \in \mathscr{C}^\alpha([0,T], V)$. Taking into account (2.69), (2.70), and Lemmas 2.8, 2.10, and 2.11, by Proposition 2.18, for every mild solution $(Y, Y') \in \mathscr{D}_X^{2\alpha}([0,T], W)$ to the RPDE (2.54) with $Y_0 = \xi$, we have

$$|Y_t|_{D(A)} \leq K \quad \text{for all } t \in [0,T], \qquad (2.71)$$

where the constant $K > 0$ depends on ξ, α, T, $\mathbf{X}$, $\|f_0\|_{\text{Lip}}$, $\|f\|_{C_b^{2\alpha,3}}$, M, and ω. For $t \in [0,T]$, we define the mapping $\Phi_t : \mathscr{D}_X^{2\alpha}([0,t], W) \to \mathscr{D}_X^{2\alpha}([0,t], W)$ according to (2.60). By Propositions 2.16 and 2.17 and Lemma 2.1, for every $t \in [0,T]$ with $t \leq 1$ and all $(Y,Y'), (Z,Z') \in \mathbb{B}_t$, we have

$$\|\Phi_t(Y,Y')\|_{X,2\alpha;[0,t]} \leq C(1 + |\xi|_{D(A)})t^{1-2\alpha}$$
$$+ C\big(\|X\|_{\beta,[0,t]}t^{\beta-\alpha} + \|\mathbb{X}\|_{2\beta,[0,t]}t^{2(\beta-\alpha)} + t^\alpha\big),$$
$$\|\Phi(Y,Y') - \Phi(Z,Z')\|_{X,2\alpha;[0,t]} \leq C \, \|\!|(Y,Y') - (Z,Z')|\!\|_{X,2\alpha;[0,t]}$$
$$\big(\|X\|_{\beta,[0,t]}t^{\beta-\alpha} + \|\mathbb{X}\|_{2\beta,[0,t]}t^{2(\beta-\alpha)} + t^\alpha + t^{1-2\alpha}\big),$$

where the constant $C > 0$ depends on α, $\mathbf{X}$, $\|f_0\|_{\text{Lip}}$, $\|f\|_{C_b^{2\alpha,3}}$, M, and ω. We choose $t \in (0,1]$ such that $T = nt$ for some $n \in \mathbb{N}$ and

$$C(1+K)t^{1-2\alpha} + C\big(\|X\|_{\beta,[0,t]}t^{\beta-\alpha} + \|\mathbb{X}\|_{2\beta,[0,t]}t^{2(\beta-\alpha)} + t^\alpha\big) \leq 1,$$
$$C\big(\|X\|_{\beta,[0,t]}t^{\beta-\alpha} + \|\mathbb{X}\|_{2\beta,[0,t]}t^{2(\beta-\alpha)} + t^\alpha + t^{1-2\alpha}\big) \leq \frac{1}{2}.$$

Then, Φ_t is a contraction, and by the Banach fixed point theorem, there is a unique mild solution $(Y(1), Y(1)') \in \mathscr{D}_X^{2\alpha}([0,t], W)$ to the RPDE (2.54) with $Y(1)_0 = \xi$. Taking into account (2.71), we can inductively apply this argument to the intervals $[(k-1)t, kt]$ for all $k = 2, \ldots, n$ to obtain a unique mild solution $(Y(k), Y(k)') \in \mathscr{D}_X^{2\alpha}([(k-1)t, kt])$ to the RPDE (2.54) with $Y(k)_{(k-1)t} = Y(k-1)_{(k-1)t}$. Using Lemma 2.29, we can concatenate these solutions and deduce the existence of a unique mild solution $(Y, Y') \in \mathscr{D}_X^{2\alpha}([0,T], W)$. By Lemma 2.28, we even have $(Y, Y') \in \mathscr{D}_X^{2\beta}([0,T])$, showing that (Y, Y') is the unique mild solution to the RPDE (2.54) with $Y_0 = \xi$. By (2.71), the path $AY : [0,T] \to W$ is bounded. Hence, applying Proposition 2.15 concludes the proof. $\square$

Under the conditions of Theorem 2.3, we can consider the Itô–Lyons map

$$(\xi, \mathbf{X}) \mapsto (Y, Y'). \tag{2.72}$$

With certain adaptations, the following result regarding the continuity of the Itô–Lyons map can be proven similarly as for rough ordinary differential equations (RODEs); see, for example, Ref. [1, Section 8.6].

Proposition 2.19. *Let* $\beta \in (\frac{1}{3}, \frac{1}{2}]$ *be an index, and let* $f_0 \in \mathrm{Lip}([0,T] \times W, D(A))$ *and* $f \in C_b^{2\beta,3}([0,T] \times W, L(V, D(A^2)))$ *be mappings such that* (2.69) *and* (2.70) *are fulfilled. Then, the Itô–Lyons map*

$$D(A^2) \times \mathscr{C}^{\beta}([0,T], V) \to \mathcal{C}^{\beta}([0,T], W) \oplus \mathcal{C}^{2\beta}([0,T]^2, W \otimes W)$$

given by (2.72) *is locally Lipschitz continuous.*

Now, consider an RODE of the form

$$\begin{cases} dY_t = f_0(t, Y_t)dt + f(t, Y_t)d\mathbf{X}_t, \\ Y_0 = \xi. \end{cases} \tag{2.73}$$

Corollary 2.10. *Let* $\mathbf{X} = (X, \mathbb{X}) \in \mathscr{C}^{\beta}([0,T], V)$ *be a rough path for some index* $\beta \in (\frac{1}{3}, \frac{1}{2}]$, *and let* $f_0 \in \mathrm{Lip}([0,T] \times W, W)$ *and* $f \in C_b^{2\beta,3}([0,T] \times W, L(V, W))$ *be mappings. Then, for every* $\xi \in W$, *there exists a unique solution* $(Y, Y') \in \mathscr{D}_X^{2\beta}([0,T], W)$ *to the RODE* (2.73) *with* $Y_0 = \xi$.

In particular, the previous result applies to time-homogeneous RODEs of the form

$$\begin{cases} dY_t = f_0(Y_t)dt + f(Y_t)d\mathbf{X}_t, \\ Y_0 = \xi. \end{cases} \tag{2.74}$$

Corollary 2.11. *Let* $\mathbf{X} = (X, \mathbb{X}) \in \mathscr{C}^{\beta}([0,T], V)$ *be a rough path for some index* $\beta \in (\frac{1}{3}, \frac{1}{2}]$, *and let* $f_0 : W \to W$ *be Lipschitz continuous and* $f \in C_b^3(W, L(V,W))$. *Then, for every* $\xi \in W$, *there exists a unique solution* $(Y, Y') \in \mathscr{D}_X^{2\beta}([0,T], W)$ *to the time-homogeneous RODE* (2.74) *with* $Y_0 = \xi$.

2.10. Stochastic Partial Differential Equations Driven by Infinite-Dimensional Wiener Processes

In this section, we apply our existence and uniqueness result for RPDEs (see Theorem 2.3) to SPDEs driven by infinite-dimensional Wiener processes. For this purpose, we start with some preparations. In the sequel, $(\Omega, \mathcal{F}, (\mathcal{F}_t)_{t \in \mathbb{R}_+}, \mathbb{P})$ denotes a filtered probability space satisfying the usual conditions.

2.10.1. *Infinite-dimensional Wiener process as a rough path*

In this section, we demonstrate how typical sample paths of an infinite-dimensional Wiener process can be realized as rough paths. Let U be a separable Hilbert space, and let X be an U-valued Q-Wiener process for some nuclear, self-adjoint, positive definite linear operator $Q \in L_1^{++}(U)$; see Ref. [18, Def. 4.2]. There exist an orthonormal basis $\{e_k\}_{k \in \mathbb{N}}$ of U and a sequence $(\lambda_k)_{k \in \mathbb{N}} \subset (0, \infty)$ with $\sum_{k=1}^{\infty} \lambda_k < \infty$ such that

$$Q e_k = \lambda_k e_k \quad \text{for all } k \in \mathbb{N}.$$

The space $U_0 := Q^{1/2}(U)$, equipped with the inner product

$$\langle u, v \rangle_{U_0} := \langle Q^{-1/2} u, Q^{-1/2} v \rangle_U, \quad u, v \in U_0,$$

is another separable Hilbert space. Note that

$$Q^{1/2} : (U, \|\cdot\|_U) \to (U_0, \|\cdot\|_{U_0})$$

is an isometric isomorphism and that $\{\sqrt{\lambda_k} e_k\}_{k \in \mathbb{N}}$ is an orthonormal basis of U_0. Now, let H be another separable Hilbert space. We denote by $L_2(U_0, H)$ the space of all Hilbert–Schmidt operators from U_0 into H. Furthermore, we define

$$L(U, H)_0 := \{\Phi|_{U_0} : \Phi \in L(U, H)\}.$$

The following result shows that $L(U, H)$ is continuously embedded in $L_2(U_0, H)$.

Lemma 2.30. *We have* $L(U,H)_0 \subset L_2(U_0, H)$ *and*
$$|\Phi|_{U_0}|_{L_2(U_0,H)} \leq \sqrt{\operatorname{tr}(Q)} |\Phi|_{L(U,H)} \quad \text{for all } \Phi \in L(U,H).$$

Proof. Recalling that $\{\sqrt{\lambda_k} e_k\}_{k \in \mathbb{N}}$ is an orthonormal basis of U_0, we have
$$|\Phi|^2_{L_2(U_0,H)} = \sum_{k=1}^{\infty} |\Phi(\sqrt{\lambda_k} e_k)|^2 = \sum_{k=1}^{\infty} \lambda_k |\Phi(e_k)|^2 \leq \operatorname{tr}(Q) |\Phi|^2_{L(U,H)},$$
completing the proof. □

For what follows, we fix a time horizon $T \in \mathbb{R}_+$. The H-valued Itô integral $\int_0^T \Phi_s dX_s$ can be defined for every predictable $L_2(U_0, H)$-valued process Φ such that
$$\mathbb{E}\left[\int_0^T |\Phi_s|^2_{L_2(U_0,H)} ds\right] < \infty, \tag{2.75}$$
and for each such process Φ, we have the *Itô isometry*
$$\mathbb{E}\left[\left|\int_0^T \Phi_s dX_s\right|^2\right] = \mathbb{E}\left[\int_0^T |\Phi_s|^2_{L_2(U_0,H)} ds\right].$$

We refer to Refs. [17], [18], or [19] for further details.

Remark 2.7. Actually, the Itô integral $\int_0^T \Phi_s dX_s$ can even be defined for every predictable $L_2(U_0, H)$-valued process Φ satisfying
$$\mathbb{P}\left(\int_0^T |\Phi_s|^2_{L_2(U_0,H)} ds < \infty\right) = 1,$$
but this is not required here.

The following Burkholder-type inequality is a consequence of Ref. [17, Lemma 3.3.b].

Proposition 2.20. *For each $p \geq 2$, there exists a constant $c_p > 0$ such that for every predictable $L_2(U_0, H)$-valued process Φ satisfying*
$$\mathbb{E}\left[\left(\int_0^T |\Phi_s|^2_{L_2(U_0,H)} ds\right)^{p/2}\right] < \infty,$$

we have the estimate

$$\mathbb{E}\left[\sup_{t\in[0,T]}\left|\int_0^t \Phi_s dX_s\right|^p\right] \leq c_p \mathbb{E}\left[\left(\int_0^T |\Phi_s|^2_{L_2(U_0,H)}ds\right)^{p/2}\right].$$

According to Ref. [18, Prop. 4.3], the sequence $(\beta^k)_{k\in\mathbb{N}}$ defined as

$$\beta^k := \frac{1}{\sqrt{\lambda_k}}\langle X, e_k\rangle_U, \quad k \in \mathbb{N},$$

is a sequence of independent real-valued standard Wiener processes, and according to Ref. [19, Prop. 2.1.10], the Q-Wiener process X admits the series representation

$$X_t = \sum_{k=1}^{\infty} \sqrt{\lambda_k}\beta_t^k e_k, \quad t \in [0,T]. \tag{2.76}$$

More generally, we have the following consequence of Ref. [19, Prop. 2.4.5].

Proposition 2.21. *For every predictable $L_2(U_0, H)$-valued process Φ satisfying (2.75), we have the series representation*

$$\int_0^t \Phi_s\, dX_s = \sum_{k=1}^{\infty} \sqrt{\lambda_k} \int_0^t \Phi_s(e_k)\, d\beta_s^k, \quad t \in [0,T].$$

Moreover, the series representation (2.76) shows that

$$|X_{s,t}|^2 = \sum_{k=1}^{\infty} \lambda_k |\beta_{s,t}^k|^2, \quad s,t \in [0,T]. \tag{2.77}$$

Lemma 2.31. *For each $p \geq 2$, there is a constant $C_p > 0$ such that*

$$\mathbb{E}\big[|X_{s,t}|^p\big] \leq C_p|t-s|^{p/2}, \quad s,t \in [0,T].$$

Proof. Let $s,t \in [0,T]$ with $s < t$ be arbitrary. We define the predictable $L(U)$-valued process $\Phi = \mathrm{Id} \cdot \mathbf{1}_{(s,t]}$. Then, we have

$$X_{s,t} = \int_0^T \Phi_r|_{U_0} dX_r.$$

By Proposition 2.20 and Lemma 2.30, we obtain

$$\mathbb{E}[|X_{s,t}|^p] = \mathbb{E}\left[\left|\int_0^T \Phi_r|_{U_0} dX_r\right|^p\right]$$

$$\leq c_p \mathbb{E}\left[\left(\int_0^T |\Phi_r|_{U_0}|^2_{L_2(U_0,U)} dr\right)^{p/2}\right]$$

$$\leq c_p \operatorname{tr}(Q)^{p/2} \mathbb{E}\left[\left(\int_0^T |\Phi_r|^2_{L(U)} dr\right)^{p/2}\right] = c_p|t-s|^{p/2},$$

where the constant $c_p > 0$ stems from Proposition 2.20. $\square$

Since U is a separable Hilbert space, we have $U \otimes U \simeq L_2(U)$. We define the predictable $L(U, L_2(U))$-valued process Ψ as

$$\Psi_{s,r}(e_j)e_k = \sqrt{\lambda_k}\beta^k_{s,r} e_j, \quad s,r \in [0,T] \text{ and } j,k \in \mathbb{N}. \tag{2.78}$$

According to Lemma 2.30, the process $\Psi|_{U_0}$ is $L_2(U_0, L_2(U))$-valued.

Lemma 2.32. *For all $s, r \in [0,T]$ we have*

$$|\Psi_{s,r}|_{U_0}|_{L_2(U_0, L_2(U))} = \sqrt{\operatorname{tr}(Q)}\, |X_{s,r}|.$$

Proof. Recalling that $\{\sqrt{\lambda_j}e_j\}_{j\in\mathbb{N}}$ is an orthonormal basis of U_0 and that $\{e_k\}_{k\in\mathbb{N}}$ is an orthonormal basis of U, by (2.77), we have

$$|\Psi_{s,r}|_{U_0}|^2_{L_2(U_0,L_2(U))} = \sum_{j=1}^\infty |\Psi_{s,r}(\sqrt{\lambda_j}e_j)|^2_{L_2(U)}$$

$$= \sum_{j=1}^\infty \lambda_j \sum_{k=1}^\infty |\Psi_{s,r}(e_j)e_k|^2_U$$

$$= \sum_{j=1}^\infty \lambda_j \sum_{k=1}^\infty \lambda_k |\beta^k_{s,r}|^2 = \operatorname{tr}(Q)\,|X_{s,r}|^2,$$

completing the proof. $\square$

Consequently, we can define the $L_2(U)$-valued second-order process $\mathbb{X}$ as the Itô integral

$$\mathbb{X}_{s,t} := \int_s^t \Psi_{s,r}|_{U_0} dX_r, \quad s,t \in [0,T].$$

Lemma 2.33. *For all $s,t \in [0,T]$, we have*

$$\mathbb{X}_{s,t} = \sum_{j,k=1}^{\infty} \sqrt{\lambda_j \lambda_k} \left(\int_s^t \beta_{s,r}^j d\beta_r^k \right) (e_j \otimes e_k).$$

Proof. By Proposition 2.21, we have

$$\mathbb{X}_{s,t} = \sum_{k=1}^{\infty} \sqrt{\lambda_k} \int_s^t \Phi_{s,r}(e_k) d\beta_r^k.$$

Thus, taking into account (2.78), for each $j \in \mathbb{N}$, we obtain

$$\mathbb{X}_{s,t} e_j = \sum_{k=1}^{\infty} \sqrt{\lambda_k} \int_s^t \Phi_{s,r}(e_k) e_j \, d\beta_r^k = \sum_{k=1}^{\infty} \sqrt{\lambda_k} \int_s^t \sqrt{\lambda_j} \beta_{s,r}^j e_k \, d\beta_r^k,$$

completing the proof. $\square$

Lemma 2.34. *For each $p \geq 2$, there is a constant $C_p > 0$ such that*

$$\mathbb{E}\big[|\mathbb{X}_{s,t}|^p\big] \leq C_p |t-s|^p, \quad s,t \in [0,T].$$

Proof. Applying Proposition 2.20 twice as well as Hölder's inequality and Lemma 2.32, we obtain

$$\mathbb{E}\big[|\mathbb{X}_{s,t}|^p\big] = \mathbb{E}\left[\left| \int_s^t \Psi_{s,r}|_{U_0} dX_r \right|^p \right]$$

$$\leq c_p \mathbb{E}\left[\left| \int_s^t |\Psi_{s,r}|_{U_0}|_{L_2(U_0,L_2(U))}^2 dr \right|^{p/2} \right]$$

$$\leq c_p \mathbb{E}\left[\left| \int_s^t |\Psi_{s,r}|_{U_0}|_{L_2(U_0,L_2(U))}^p dr \right| \right] |t-s|^{\frac{p}{2}-1}$$

$$= c_p \mathbb{E}\left[\left| \int_s^t \big(|\Psi_{s,r}|_{U_0}|_{L_2(U_0,L_2(U))}^2\big)^{p/2} dr \right| \right] |t-s|^{\frac{p}{2}-1}$$

$$\leq c_p \mathrm{tr}(Q)^{p/2}\, \mathbb{E}\left[\left|\int_s^t |X_{s,r}|^p dr\right|\right] |t-s|^{\frac{p}{2}-1}$$

$$\leq c_p \mathrm{tr}(Q)^{p/2}\, \mathbb{E}\left[\sup_{r\in[s,t]} |X_{s,r}|^p\right] |t-s|^{\frac{p}{2}} \leq c_p^2 |t-s|^p,$$

where the constant $c_p > 0$ stems from Proposition 2.20. □

Now, we define the Itô-enhanced Q-Wiener process $\mathbf{X} := (X, \mathbb{X})$.

Proposition 2.22. *For each $\alpha \in (\frac{1}{3}, \frac{1}{2})$, we have $\mathbb{P}$-almost surely $\mathbf{X} \in \mathscr{C}^\alpha([0,T], U)$.*

Proof. Using Lemma 2.33, it follows that the process $\mathbf{X}$ satisfies Chen's relation (2.4). Let $\alpha \in (\frac{1}{3}, \frac{1}{2})$ be arbitrary. We choose $\beta = \frac{1}{2}$ and $q > 6$ arbitrarily. Then, we have $\beta - \frac{1}{q} > \frac{1}{3}$. By Lemmas 2.31 and 2.34, we have

$$|X_{s,t}|_{\mathcal{L}^q} \leq C_q |t-s|^{1/2} = C_q |t-s|^\beta, \quad s,t \in [0,T],$$

$$|\mathbb{X}_{s,t}|_{\mathcal{L}^{q/2}} \leq C_q |t-s| = C_q |t-s|^{2\beta}, \quad s,t \in [0,T],$$

with a constant $C_q > 0$. Consequently, applying Ref. [1, Theorem 3.1] completes the proof. □

2.10.2. Coincidence of the two integrals

In this section, we show that the rough integral and the Itô integral coincide. Let $\mathbf{X} = (X, \mathbb{X})$ be the Itô-enhanced Q-Wiener process, as introduced in Section 2.10.1. We start with two auxiliary results.

Lemma 2.35. *Let Y be a continuous, adapted, and bounded $L(U, H)$-valued process. Furthermore, let $(\Pi_n)_{n\in\mathbb{N}}$ be a sequence of partitions of the interval $[0,T]$ such that $|\Pi_n| \to 0$. Then, we have*

$$\sum_{[u,v]\in\Pi_n} Y_u X_{u,v} \xrightarrow{L^2} \int_0^T Y_s|_{U_0}\, dX_s \quad \text{as } n \to \infty.$$

Proof. Let $\Pi = \{0 = t_1 < \cdots < t_{K+1} = T\}$ be an arbitrary partition of the interval $[0, T]$. We define the elementary $L(U, H)$-valued process as

$$Y^{\Pi} := \sum_{i=1}^{K} Y_{t_i} \mathbf{1}_{(t_i, t_{i+1}]}.$$

Note that Y^{Π} is predictable because Y is adapted. By the definition of the elementary Itô integral, we have

$$\sum_{[u,v] \in \Pi} Y_u X_{u,v} = \sum_{i=1}^{K} Y_{t_i} X_{t_i, t_{i+1}} = \int_0^T Y^{\Pi} dX_s. \tag{2.79}$$

Now, we set $Y^n := Y^{\Pi_n}$ for each $n \in \mathbb{N}$. By the continuity of Y, we have $Y^n \to Y$ pointwise. Therefore, since Y is bounded, by Lemma 2.30 and Lebesgue's dominated convergence theorem, we obtain

$$\mathbb{E}\left[\int_0^T |Y_s|_{U_0} - Y_s^n|_{U_0}|^2_{L_2(U_0, H)} ds\right]$$

$$\leq \operatorname{tr}(Q) \mathbb{E}\left[\int_0^T |Y_s - Y_s^n|^2_{L(U,H)} ds\right] \to 0.$$

Consequently, it follows that

$$\int_0^T Y_s^n \, dX_s \xrightarrow{L^2} \int_0^T Y_s|_{U_0} \, dX_s \quad \text{as } n \to \infty.$$

In view of (2.79), this completes the proof. $\square$

Lemma 2.36. Let $(\mathcal{G}_n)_{n=0,\ldots,K}$ be a time-discrete filtration for some $K \in \mathbb{N}$, and let $(Y_n)_{n=1,\ldots,K}$ be a predictable, bounded $L(U, H)$-valued process. Furthermore, let $(X_n)_{n=1,\ldots,K}$ be an adapted, square-integrable U-valued process such that X_n is independent of $\mathcal{G}_{n-1}$ and $\mathbb{E}[X_n] = 0$ for all $n = 1, \ldots, K$. We define the H-valued process $S = (S_n)_{n=0,\ldots,K}$ as $S_0 := 0$ and

$$S_n := \sum_{j=1}^{n} Y_j X_j, \quad n = 1, \ldots, K.$$

Then, the following statements are true:
(a) S is a square-integrable $(\mathcal{G}_n)_{n=0,\ldots,K}$-martingale.
(b) For all $n = 1,\ldots, K$, we have

$$\mathbb{E}\left[\left|\sum_{j=1}^{n}(S_j - S_{j-1})\right|^2\right] = \mathbb{E}\left[\sum_{j=1}^{n}|S_j - S_{j-1}|^2\right].$$

Proof. It is clear that the process S is square integrable. Let $n = 1,\ldots, K$ be arbitrary. Then, we have

$$\mathbb{E}[S_n - S_{n-1}|\mathcal{G}_{n-1}] = \mathbb{E}[Y_n X_n|\mathcal{G}_{n-1}] = Y_n \mathbb{E}[X_n|\mathcal{G}_{n-1}] = Y_n \mathbb{E}[X_n] = 0,$$

proving that S is a $(\mathcal{G}_n)_{n=0,\ldots,K}$-martingale. Furthermore, we obtain

$$\mathbb{E}\left[\left|\sum_{j=1}^{n}(S_j - S_{j-1})\right|^2\right]$$

$$= \mathbb{E}\left[\sum_{j=1}^{n}|S_j - S_{j-1}|^2 + 2\sum_{\substack{j,k=1 \\ j<k}}^{n}\langle S_j - S_{j-1}, S_k - S_{k-1}\rangle\right].$$

Moreover, by the martingale property of S, we have

$$\mathbb{E}\left[\sum_{\substack{j,k=1 \\ j<k}}^{n}\langle S_j - S_{j-1}, S_k - S_{k-1}\rangle\right]$$

$$= \sum_{\substack{j,k=1 \\ j<k}}^{n}\mathbb{E}\big[\mathbb{E}[\langle S_j - S_{j-1}, S_k - S_{k-1}\rangle\,|\,\mathcal{F}_{k-1}]\big]$$

$$= \sum_{\substack{j,k=1 \\ j<k}}^{n}\mathbb{E}\big[\langle S_j - S_{j-1}, \underbrace{\mathbb{E}[S_k - S_{k-1}\,|\,\mathcal{F}_{k-1}]}_{=0}\rangle\big] = 0,$$

completing the proof. □

For the following result, we recall that $U \otimes U \simeq L_2(U)$ and $L(U, L(U, H)) \hookrightarrow L(L_2(U), H)$. Furthermore, recall that $L(U, H)$ is continuously embedded in $L_2(U_0, H)$, as stated in Lemma 2.30. Let $\alpha \in (\frac{1}{3}, \frac{1}{2})$ be an arbitrary index. According to Proposition 2.22, there is a $\mathbb{P}$-nullset N_1 such that $\mathbf{X} \in \mathscr{C}^\alpha([0,T], U)$ on N_1^c.

Proposition 2.23. Let Y be a continuous $L(U, H)$-valued process, and let Y' be a continuous $L(U, L(U, H))$-valued processes such that $(Y, Y') \in \mathscr{D}_X^{2\alpha}([0, T], L(U, H))$ on N_1^c. Then, the following statements are true:

(a) The H-valued rough integral

$$\int_0^T Y_s \, d\mathbf{X}_s = \lim_{|\Pi| \to 0} \sum_{[u,v] \in \Pi} (Y_u X_{u,v} + Y'_u \mathbb{X}_{u,v}) \qquad (2.80)$$

exists on N_1^c.

(b) If Y and Y' are adapted and bounded, then there is a $\mathbb{P}$-nullset N with $N_1 \subset N$ such that

$$\int_0^T Y_s \, d\mathbf{X}_s = \int_0^T Y_s|_{U_0} \, dX_s \quad \text{on } N^c.$$

Proof. The first statement is a consequence of Theorem 2.1. We proceed with the proof of the second statement. By hypothesis, we have $|Y|, |Y'| \leq M$ for some constant $M > 0$. Let $(\Pi_n)_{n \in \mathbb{N}}$ be a sequence of partitions of the interval $[0, T]$ with $|\Pi_n| \to 0$. By Lemma 2.35, we have

$$\sum_{[u,v] \in \Pi_n} Y_u X_{u,v} \xrightarrow{\mathbb{P}} \int_0^t Y_s|_{U_0} \, dX_s \quad \text{as } n \to \infty.$$

Passing to a subsequence, if necessary, we obtain

$$\sum_{[u,v] \in \Pi_n} Y_u X_{u,v} \xrightarrow{\text{a.s.}} \int_0^t Y_s \, dX_s \quad \text{as } n \to \infty. \qquad (2.81)$$

Hence, there is a $\mathbb{P}$-nullset N_2 such that the above convergence (2.81) holds true on N_2^c. Together with (2.80), we obtain

$$\lim_{n \to \infty} \sum_{[u,v] \in \Pi_n} Y'_u \mathbb{X}_{u,v} = \lim_{n \to \infty} \sum_{[u,v] \in \Pi_n} (Y_u X_{u,v} + Y'_u \mathbb{X}_{u,v})$$

$$- \lim_{n \to \infty} \sum_{[u,v] \in \Pi_n} Y_u X_{u,v}$$

$$= \int_0^t Y_s \, d\mathbf{X}_s - \int_0^t Y_s|_{U_0} \, dX_s$$

on $(N_1 \cup N_2)^c$. Let $\Pi = \{0 = t_1 < \cdots < t_{K+1} = T\}$ be an arbitrary partition of the interval $[0,T]$. By Lemma 2.36 (with U replaced by $L_2(U)$), the H-valued process $S = (S_n)_{n=0,\ldots,K}$ given by $S_0 := 0$ and

$$S_n := \sum_{j=1}^{n} Y'_{t_j} \mathbb{X}_{t_j, t_{j+1}}, \quad n = 1, \ldots, K,$$

is a time-discrete square-integrable $(\mathcal{F}_{t_{n+1}})_{n=0,\ldots,K}$-martingale. Furthermore, by Lemmas 2.36 and 2.34, we have

$$\mathbb{E}\left[\left|\sum_{[u,v]\in\Pi} Y'_u \mathbb{X}_{u,v}\right|^2\right] = \mathbb{E}[|S_K|^2] = \mathbb{E}\left[\left|\sum_{j=1}^{K}(S_j - S_{j-1})\right|^2\right]$$

$$= \mathbb{E}\left[\sum_{j=1}^{K}|S_j - S_{j-1}|^2\right] \le M^2 \sum_{j=1}^{K} \mathbb{E}\left[|\mathbb{X}_{t_j, t_{j+1}}|^2\right]$$

$$\le M^2 C \sum_{j=1}^{K} |t_{j+1} - t_j|^2 \le M^2 CT|\Pi|,$$

where the constant $C > 0$ stems from Lemma 2.34. Therefore, we have

$$\sum_{[u,v]\in\Pi_n} Y'_u \mathbb{X}_{u,v} \xrightarrow{L^2} 0 \quad \text{as } n \to \infty.$$

Thus, passing to a subsequence, if necessary, we obtain

$$\sum_{[u,v]\in\Pi_n} Y'_u \mathbb{X}_{u,v} \xrightarrow{\text{a.s.}} 0 \quad \text{as } n \to \infty. \tag{2.82}$$

Hence, there is a $\mathbb{P}$-nullset N_3 such that the above convergence (2.82) holds true on N_3^c. Consequently, setting $N := N_1 \cup N_2 \cup N_3$ completes the proof. $\square$

2.10.3. *Stochastic partial differential equations*

Now, we are ready for our study of SPDEs. Let $\mathbf{X} = (X, \mathbb{X})$ be the Itô-enhanced Q-Wiener process, as introduced in Section 2.10.1.

Furthermore, let W be a Banach space, and let A be the generator of a C_0-semigroup $(S_t)_{t \geq 0}$ on W. Consider the random RPDE

$$\begin{cases} dY_t = (AY_t + f_0(t, Y_t))dt + f(t, Y_t)d\mathbf{X}_t, \\ Y_0 = \xi \end{cases} \quad (2.83)$$

with mappings $f_0 : [0,T] \times W \to W$ and $f : [0,T] \times W \to L(U,W))$. Let $\beta \in (\frac{1}{3}, \frac{1}{2})$ be an index. By Proposition 2.22, there is a $\mathbb{P}$-nullset N such that $\mathbf{X} \in \mathscr{C}^\beta([0,T], U)$ on N^c.

Theorem 2.4. *Let* $f_0 \in \mathrm{Lip}([0,T] \times W, D(A))$ *and* $f \in C_b^{2\beta,3}([0,T] \times H, L(U, D(A^2)))$ *be mappings such that*

$$f_0|_{[0,T] \times D(A)} \in \mathrm{Lip}([0,T] \times D(A), D(A^2)),$$

$$f|_{[0,T] \times D(A)} \in C_b^{2\beta,3}([0,T] \times D(A), L(U, D(A^3))).$$

Then, for every $\xi \in D(A^2)$, *the following statements are true:*

(a) *There exists a unique mild solution* $(Y, Y') \in \mathscr{D}_X^{2\beta}([0,T], W)$ *to the random RPDE* (2.83) *with* $Y_0 = \xi$ *on* N^c, *which is also a strong solution.*
(b) *If* W *is a separable Hilbert space* H, *then there is a* $\mathbb{P}$-*nullset* N_1 *with* $N \subset N_1$ *such that the associated stochastic process* Y *restricted to* N_1^c *is also a mild and strong solution to the Itô SPDE*

$$\begin{cases} dY_t = (AY_t + f_0(t, Y_t))dt + f(t, Y_t)|_{U_0} dX_t, \\ Y_0 = \xi. \end{cases}$$

Proof. The first statement is an immediate consequence of Theorem 2.3. For the proof of the second statement, note that $\mathcal{F}_t^X = \mathcal{F}_t^{\mathbf{X}} \subset \mathcal{F}_t$ for all $t \in [0,T]$. By the continuity of the Itô–Lyons maps (see Proposition 2.19), it follows that the stochastic process (Y, Y') is adapted. Therefore, applying Proposition 2.23 concludes the proof. $\square$

Remark 2.8. Compared to other SPDE results in the literature (see, for example, Ref. [17] or Ref. [18]), we have imposed stronger conditions on the coefficients; in particular, usually only Lipschitz-type assumptions on f are required. On the other hand, we immediately obtain the *flow* of solutions $Y^\xi(\omega)$ for $\xi \in D(A^2)$ and $\omega \in N^c$.

2.11. Stochastic Partial Differential Equations Driven by Infinite-Dimensional Fractional Brownian Motion

In this section, we apply our existence and uniqueness result for RPDEs (see Theorem 2.3) to SPDEs driven by infinite-dimensional fractional Brownian motion. Fractional Brownian motion in Hilbert spaces has been studied, for example, in Refs. [26, 27].

Let $(\Omega, \mathcal{F}, (\mathcal{F}_t)_{t \in \mathbb{R}_+}, \mathbb{P})$ be a filtered probability space satisfying the usual conditions. Furthermore, let U be a separable Hilbert space. Recall that a U-valued process X is called a *Gaussian process* if, for all $n \in \mathbb{N}$ and all $t_1, \ldots, t_n \in \mathbb{R}_+$ with $t_1 < \cdots < t_n$, the U^n-valued random variable $(X_{t_1}, \ldots, X_{t_n})$ is a Gaussian random variable. An U-valued Gaussian process X is called *centered* if $\mathbb{E}[X_t] = 0$ for all $t \in \mathbb{R}_+$. For a U-valued Gaussian process X, we introduce the *covariance function*

$$R : \mathbb{R}_+^2 \to L_2(U), \quad (s, t) \mapsto \mathbb{E}[X_s \otimes X_t].$$

Let $Q \in L_1^{++}(U)$ be a nuclear, self-adjoint, positive definite linear operator.

Definition 2.9. A centered Gaussian process X is called a *Q-fractional Brownian motion* with the Hurst index $H \in (\frac{1}{3}, \frac{1}{2}]$ if its covariance function is given by

$$R(s, t) = \frac{1}{2}\left(s^{2H} + t^{2H} - |t - s|^{2H}\right)Q, \quad s, t \in \mathbb{R}_+.$$

For what follows, let X be a Q-fractional Brownian motion with the Hurst index $H \in (\frac{1}{3}, \frac{1}{2}]$. We also fix a time horizon $T \in \mathbb{R}_+$. There exist an orthonormal basis $\{e_k\}_{k \in \mathbb{N}}$ of U and a sequence $(\lambda_k)_{k \in \mathbb{N}} \subset (0, \infty)$ with $\sum_{k=1}^{\infty} \lambda_k < \infty$ such that

$$Qe_k = \lambda_k e_k \quad \text{for all } k \in \mathbb{N}.$$

The Q-fractional Brownian motion X admits the series representation

$$X_t = \sum_{k=1}^{\infty} \sqrt{\lambda_k} \beta_t^k e_k, \quad t \in [0, T],$$

with a sequence of independent real-valued fractional Brownian motions $(\beta^k)_{k \in \mathbb{N}}$ with the Hurst index H; see, for example, Refs. [20], [22], or [27].

Proposition 2.24. *There exists a Lévy area $\mathbb{X}$ such that $\mathbb{P}$-almost surely $\mathbf{X} := (X, \mathbb{X}) \in \mathscr{C}_g^\alpha([0,T], U)$ for each $\alpha \in (\frac{1}{3}, H)$.*

Proof. Let $\alpha \in (\frac{1}{3}, H)$ be arbitrary. According to Ref. [11, Lemma 2.4], there exist a Lévy area $\mathbb{X}$ and a sequence $(X^n)_{n \in \mathbb{N}}$ of continuous, piecewise linear functions $X^n : [0,T] \to U$ starting at zero such that $\mathbb{P}$-almost surely $|||\mathbf{X}^n - \mathbf{X}|||_\alpha \to 0$, where we have set $\mathbf{X}^n := (X^n, \mathbb{X}^n)$, $n \in \mathbb{N}$, and the second-order processes are given by

$$\mathbb{X}^n_{s,t} = \int_s^t X^n_{s,r} \otimes dX^n_r, \quad s,t \in [0,T]. \tag{2.84}$$

By Lemma 2.3, we have $\mathbf{X}^n \in \mathscr{C}_g^\alpha([0,T], U)$ for each $n \in \mathbb{N}$. Therefore, by Lemma 2.4, we deduce that $\mathbb{P}$-almost surely $\mathbf{X} \in \mathscr{C}_g^\alpha([0,T], U)$. □

Remark 2.9. If $H = \frac{1}{2}$, then X is a Q-Wiener process (see Section 2.10.1), and $\mathbf{X} = (X, \mathbb{X})$, according to Proposition 2.24, is a Stratonovich-enhanced Q-Wiener process.

Remark 2.10. For each $n \in \mathbb{N}$, the Lévy area $\mathbb{X}^n$ given by (2.84) can be expressed as

$$\mathbb{X}^n_{s,t} = \sum_{j,k=1}^\infty \sqrt{\lambda_j \lambda_k} \left(\int_s^t \beta^{j,n}_{s,r} d\beta^{k,n}_r \right) (e_j \otimes e_k),$$

where $\beta^{j,n}$ denotes the corresponding nth linear interpolation of β^j.

Now, let W be a Banach space, and let A be the generator of a C_0-semigroup $(S_t)_{t \geq 0}$ on W. Consider the random RPDE

$$\begin{cases} dY_t = (AY_t + f_0(t, Y_t))dt + f(t, Y_t)d\mathbf{X}_t, \\ Y_0 = \xi \end{cases} \tag{2.85}$$

with mappings $f_0 : [0,T] \times W \to W$ and $f : [0,T] \times W \to L(U, W))$. Let $\beta \in (\frac{1}{3}, H)$ be an index. By Proposition 2.24, there is a $\mathbb{P}$-nullset N such that $\mathbf{X} \in \mathscr{C}_g^\beta([0,T], U)$ on N^c. As an immediate consequence of Theorem 2.3, we obtain the following result.

Theorem 2.5. Let $f_0 \in \text{Lip}([0,T] \times W, D(A))$ and $f \in C_b^{2\beta,3}([0,T] \times H, L(U, D(A^2)))$ be mappings such that

$$f_0|_{[0,T] \times D(A)} \in \text{Lip}([0,T] \times D(A), D(A^2)),$$

$$f|_{[0,T] \times D(A)} \in C_b^{2\beta,3}([0,T] \times D(A), L(U, D(A^3))).$$

Then, for every $\xi \in D(A^2)$, there exists a unique mild solution $(Y, Y') \in \mathscr{D}_X^{2\beta}([0,T], W)$ to the random RPDE (2.85) with $Y_0 = \xi$ on N^c, which is also a strong solution.

Remark 2.11. As for random RPDEs driven by infinite-dimensional Wiener processes, we obtain the flow of solutions $Y^\xi(\omega)$ for $\xi \in D(A^2)$ and $\omega \in N^c$.

Remark 2.12. If W is a separable Hilbert space H, then there exists a theory of stochastic integration (see Ref. [26]), and one may check that the associated stochastic process Y from Theorem 2.5 is also a mild and strong solution to the fractional SPDE

$$\begin{cases} dY_t = (AY_t + f_0(t, Y_t))dt + f(t, Y_t)dX_t, \\ Y_0 = \xi. \end{cases}$$

Funding Statement

The author gratefully acknowledges financial support from the Deutsche Forschungsgemeinschaft (DFG, German Research Foundation) – project number 444121509.

References

[1] P. K. Friz and M. Hairer, *A Course on Rough Paths*, 2nd edn. Springer, Cham (2020).
[2] A. L. Allan, Rough Path Theory. Lecture Notes, ETH Zürich (2021). https://metaphor.ethz.ch/x/2021/fs/401-4611-21L/notes/RP_lecture_notes_Allan.pdf.
[3] M. Gubinelli and S. Tindel, Rough evolution equations. *Annals of Probability*. **38**(1), 1–75 (2010).
[4] M. Hairer, Rough stochastic PDEs. *Communications on Pure and Applied Mathematics*. **64**(11), 1547–1585 (2011).

[5] P. K. Friz and H. Oberhauser, On the splitting-up method for rough (partial) differential equations. *Journal of Differential Equations.* **251**(2), 316–338 (2011).

[6] J. Teichmann, Another approach to some rough and stochastic partial differential equations. *Stochastics and Dynamics.* **11**(2-3), 535–550 (2011).

[7] P. K. Friz and H. Oberhauser, Rough path stability of (semi-)linear SPDEs. *Probability Theory and Related Fields.* **158**(1-2), 401–434 (2014).

[8] A. Gerasimovičs and M. Hairer, Hörmander's theorem for semilinear SPDEs. *Electronic Journal of Probability.* **24**(132), 1–56 (2019).

[9] A. Gerasimovičs, A. Hocquet, and T. Nilssen, Non-autonomous rough semilinear PDEs and the multiplicative sewing lemma. *Journal of Functional Analysis.* **281**(10), 109200 (2021).

[10] M. J. Garrido-Atienza, K. Lu, and B. Schmalfuß, Local pathwise solutions to stochastic evolution equations driven by fractional Brownian motions with Hurst parameters $H \in (1/3, 1/2]$. *Discrete and Continuous Dynamical Systems — Series B.* **20**(8), 2553–2581 (2015).

[11] R. Hesse and A. Neamţu, Local mild solutions for rough stochastic partial differential equations. *Journal of Differential Equations.* **267**(11), 6480–6538 (2019).

[12] R. Hesse and A. Neamţu, Global solutions for semilinear rough partial differential equations. *Stochastics and Dynamics.* **22**(2), 2240011 (2022).

[13] M. J. Garrido-Atienza, K. Lu, and B. Schmalfuß, Random dynamical systems for stochastic evolution equations driven by multiplicative fractional Brownian noise with Hurst parameters $H \in (1/3, 1/2]$. *SIAM Journal on Applied Dynamical Systems.* **15**(1), 625–654 (2016).

[14] R. Hesse and A. Neamţu, Global solutions and random dynamical systems for rough evolution equations. *Discrete and Continuous Dynamical Systems — Series B.* **25**(7), 2723–2748 (2020).

[15] C. Kuehn and A. Neamţu, Center manifolds for rough partial differential equations. *Electronic Journal of Probability.* **28**(48), 1–31 (2023).

[16] M. Ghani Varzaneh and S. Riedel, An integrable bound for rough stochastic partial differential equations with applications to invariant manifolds and stability (2023). Preprint arXiv:2307.01679v2.

[17] L. Gawarecki and V. Mandrekar, *Stochastic Differential Equations in Infinite Dimensions with Applications to SPDEs.* Springer, Berlin (2011).

[18] G. Da Prato and J. Zabczyk, *Stochastic Equations in Infinite Dimensions*, 2nd edn. Cambridge University Press, Cambridge (2014).

[19] W. Liu and M. Röckner, *Stochastic Partial Differential Equations: An Introduction.* Springer, Heidelberg (2015).
[20] W. Grecksch and V. V. Anh, A parabolic stochastic differential equation with fractional Brownian motion input. *Statistics & Probability Letters.* **41**(4), 337–346 (1999).
[21] B. Maslowski and D. Nualart, Evolution equations driven by a fractional Brownian motion. *Journal of Functional Analysis.* **202**(1), 277–305 (2003).
[22] T. E. Duncan, B. Maslowski, and B. Pasic-Duncan, Fractional Brownian motion and stochastic equations in a Hilbert space. *Stochastics and Dynamics.* **2**(2), 225–250 (2002).
[23] T. E. Duncan, B. Maslowski, and B. Pasic-Duncan, Stochastic equations in Hilbert space with a multiplicative fractional Gaussian noise. *Stochastic Processes and their Applications.* **115**(8), 1357–1383 (2005).
[24] T. E. Duncan, B. Maslowski, and B. Pasic-Duncan, Semilinear stochastic equations in a Hilbert space with a fractional Brownian motion. *SIAM Journal on Mathematical Analysis.* **40**(6), 2286–2315 (2009).
[25] B. Maslowski and J. van Neerven, Equivalence of laws and null controllability for SPDEs driven by a fractional Brownian motion. *Nonlinear Differential Equations and Applications.* **20**, 1473–1498 (2013).
[26] T. E. Duncan, J. Jakubowski, and B. Pasic-Duncan, Stochastic integration for fractional Brownian motion in a Hilbert space. *Stochastics and Dynamics.* **6**(1), 53–75 (2006).
[27] W. Grecksch, C. Roth, and V. V. Anh, Q-fractional Brownian motion in infinite dimensions with application to fractional Black-Scholes market. *Stochastic Analysis and Applications.* **27**(1), 149–175 (2009).
[28] M. J. Garrido-Atienza, K. Lu, and B. Schmalfuß, Lévy-areas of Ornstein-Uhlenbeck processes in Hilbert-spaces. In V. Sadovnichiy and M. Zgurovsky (eds.), *Continuous and Distributed Systems II.* Studies in Systems, Decision and Control, Vol. 30. Springer (2015), pp. 167–188.
[29] K.-J. Engel and R. Nagel, *One-Parameter Semigroups for Linear Evolution Equations.* Springer, New York (2010).
[30] A. Pazy, *Semigroups of Linear Operators and Applications to Partial Differential Equations.* Springer, New York (1983).

© 2025 World Scientific Publishing Company
https://doi.org/10.1142/9789819802104_0003

Chapter 3

Fractional Noise-Perturbed Nonlinear Schrödinger Equations: Stochastic Minimization Problems

Wilfried Grecksch[*,‡] **and Hannelore Lisei**[†,§]

[*]*Faculty of Natural Sciences II, Institute of Mathematics,*
Martin Luther University Halle-Wittenberg,
D-06099 Halle (Saale), Germany

[†]*Faculty of Mathematics and Computer Science,*
Babeş-Bolyai University,
Kogălniceanu Street 1, 400084 Cluj-Napoca, Romania
[‡]*wilfried.grecksch@mathematik.uni-halle.de*
[§]*hannelore.lisei@math.ubbcluj.ro*

An optimal control problem for a class of stochastic Schrödinger equations with power-type nonlinearity driven by a multiplicative fractional Brownian motion with the Hurst index $H \in (0,1)$ is discussed. The state equation is defined in a variational sense. A separation approach is used, and the solution is given by the product of the solution of a controlled pathwise problem and the solution of a stochastic differential equation. A general cost function for the optimal control problem is introduced. Finite-dimensional Galerkin approximations and a linearization method are presented and used to derive ε-optimal solutions.

3.1. Introduction

A change in the quantum state of a physical system is described by the Schrödinger equation. The concrete equation depends on the physical background. Nonlinear Schrödinger equations appear, for example, in the study of laser beams [20], in the theory of solids [1], and crystals [5]. Stochastic Schrödinger equations are part of modern research in applied stochastics, where the noise terms are modeled by stochastic integrals. As in the deterministic case, a distinction is made between Lipschitz nonlinearities (for example, Refs. [15] and [16]) and power-type (Kerr) nonlinearities and their generalizations (for example, Refs. [2], [3], [11], [19], and [21]). In the case of a multiplicative noise term, the stochastic Schrödinger equation can be transformed into a pathwise partial differential equation (PDE). This method is used in Refs. [2] and [19], where the stochastic integrals are defined with respect to Brownian motions. If the noise process is a fractional Brownian motion (fBm), then the effects of long-range or short-range dependence can be described. The above transformation method is used in the case of fBm; for example, see Refs. [17] and [18].

There are several methods to approximate the solutions of stochastic Schrödinger equations. A successive linearization method is used in Ref. [7] to approximate a stochastic Ginzburg–Landau equation. A similar method is used in Ref. [8] in the case of a Lipschitz nonlinearity.

Optimal control problems for stochastic Schrödinger-type equations arise in various fields of modern physics. The admissible controls are time- or space-dependent potentials or external forces. In Ref. [8], in the case of feedback controls and cylindrical Wiener processes, ε-optimal controls are determined through successive linearization and using the Galerkin method. The case of fBm is discussed in Ref. [18], where the admissible controls are complex-valued time-dependent potentials and the power-type nonlinearity has a complex coefficient, while in this chapter, we consider space- and time-dependent external influences and the power-type nonlinearity has a real-valued, negative coefficient.

In this chapter, we consider the complex Sobolev spaces

$$\mathbb{H} = \mathcal{L}^2(0, L) \quad \text{and} \quad \mathbb{V} = H_0^1(0, L)$$

and the operator

$$A = -\Delta : D_A \to \mathbb{H}, \quad \text{where } D_A = \{v \in \mathbb{V} : Av \in \mathbb{H}\}.$$

We investigate the following nonlinear stochastic partial differential equation of Schrödinger type:

$$dX(t) = \bigl(-\mathrm{i}AX(t) - |X(t)|^2 X(t) + \alpha(t)X(t) + g(t)\bigr)dt$$
$$+ \mathrm{i}\beta X(t)dB^H(t), \qquad (3.1)$$

where $t \in [0,T]$, $\beta \in \mathbb{R}$, $\alpha \in C(0,T;\mathbb{C})$. We assume that the admissible open loop controls g belong to

$$\mathcal{G} = \left\{ g: (0,T) \times (0,L) \to \mathbb{C} \Big|\ \underset{t\in(0,T)}{\mathrm{ess\,sup}} \|g(t)\|_{D_A} \leq \rho \right\}, \qquad (3.2)$$

where $\rho > 0$ is fixed. We show that $X = y \cdot Z$ is a variational solution of (3.1), where

$$y(t) = \exp\left\{ \int_0^t \alpha(s)ds + \frac{1}{2}\beta^2 t^{2H} + \mathrm{i}\beta B^H(t) \right\}, \quad t \in [0,T],$$

and Z is the variational solution of the nonlinear pathwise PDE

$$dZ(t) = \bigl(\mathrm{i}\Delta Z(t) - |y(t)|^2|Z(t)|^2 Z(t) + \frac{1}{y(t)}g(t)\bigr)dt, \quad t \in [0,T]. \qquad (3.3)$$

The following optimal control problem is investigated:

$$(\mathcal{P}) \quad \begin{cases} J(g,X) \to \min, \\ g \in \mathcal{G}, \end{cases}$$

where the cost function is given by

$$J(g,X) = \mathbb{E}\int_0^T J_1(t,X(t))dt + \mathbb{E}J_2(X(T)) + \int_0^T J_3(t,g(t))dt.$$

The assumptions are formulated in Section 3.6, including quadratic functionals as a special case. Because of the external controls g, which depend on space and time, we need more regularity assumptions in our investigations than in the case of controls which are only time-dependent (as studied in Ref. [18]). Also, the initial condition is assumed to belong to D_A. By these assumptions, we can include in J_1, J_2, J_3 (in the cost functional J) the norm in $\mathbb{V}$ (see Example 3.1).

This chapter is organized as follows. Section 3.2 contains some preliminaries from operator theory. The exact formulation of (3.1) and the finite-dimensional Galerkin equations are given in Section 3.3. In Section 3.4, *a priori* estimates are derived for the Galerkin approximations corresponding to (3.3) (see Theorem 3.3). Then, it is shown that these approximations are convergent to the unique solution of (3.3) (Theorems 3.2 and 3.4). The variational solution of the original equation and its Galerkin approximations are given by a product approach (as shown in Theorem 3.5). In the case of weak* convergence of the controls, the strong convergence of the corresponding solutions is studied for (3.3) and its finite-dimensional counterpart (see Theorems 3.6, 3.7, and 3.8). Then, an approximation of the Galerkin equations through successive linearization is discussed in Section 3.5. In Section 3.6, it is proved that the optimal control problem ($\mathcal{P}$) and each of the three corresponding finite-dimensional control problems, ($\mathcal{P}_n$), ($\hat{\mathcal{P}}_{n,k}$), and ($\hat{\mathcal{P}}_n$), admit at least one solution (Theorem 3.14). The presented finite-dimensional approximations (obtained through the Galerkin method and the linearization method) are used to derive ε-optimal solutions for the original problem (Theorems 3.15, 3.16, and 3.17).

Notations: For simplicity, we usually omit writing the dependence on $x \in [0, L]$ of the functions which are integrated with respect to x. In the following notations, $\mathbb{B}$ denotes a Banach space:

i	imaginary unit in $\mathbb{C}$
$\|z\|$	modulus of $z \in \mathbb{C}$
$\bar{v}$	the complex conjugate of $v \in \mathbb{C}$
A	$-\Delta$ with respect to Dirichlet boundary conditions
$(\mu_k)_k$	the sequence of eigenvalues of A
$(\varphi_k)_k$	the sequence of eigenfunctions of A
$\hookrightarrow$	continuous embedding
$\stackrel{c}{\hookrightarrow}$	compact embedding
$\overline{\mathbb{V}}^*$	the antidual space of $\mathbb{V}$
$\mathbb{I}$	the identity operator on $\mathbb{H}$
$\langle \cdot, \cdot \rangle$	antiduality pairing

$\rightharpoonup$	weak convergence
$\stackrel{*}{\rightharpoonup}$	weak* convergence
$C(0,T;\mathbb{B})$	space of all continuous functions $u:[0,T]\to\mathbb{B}$
$\mathcal{L}^p(0,T;\mathbb{B})$	Bochner space of all functions $u:(0,T)\to\mathbb{B}$ such that $\int_0^T \|u(s)\|_{\mathbb{B}}^p ds < \infty$, $p \geq 1$
$\mathcal{L}^\infty(0,T;\mathbb{B})$	Bochner space of all functions $u:(0,T)\to\mathbb{B}$ such that $\operatorname*{ess\,sup}_{t\in(0,T)} \|u(t)\|_{\mathbb{B}} < \infty$

3.2. Preliminaries

Consider the complex Sobolev spaces over the bounded interval $(0,L)$:

$$\mathbb{H} = \mathcal{L}^2(0,L), \quad \mathbb{V} = H_0^1(0,L), \quad \text{and} \quad \mathbb{U} = \mathcal{L}^4(0,L).$$

The inner product in $\mathbb{H}$ is given by

$$(u,v) = \int_0^L u \cdot \bar{v}\, dx, \quad \forall\, u,v \in \mathbb{H},$$

while the inner product in $\mathbb{V}$ is

$$(u,v)_{\mathbb{V}} = \int_0^L \frac{du}{dx} \cdot \frac{d\bar{v}}{dx}\, dx, \quad \forall\, u,v \in \mathbb{V}.$$

The norms in $\mathbb{H}$ and $\mathbb{V}$ are denoted by $\|\cdot\|$ and $\|\cdot\|_{\mathbb{V}}$, respectively. The norm in $\mathbb{U}$ is

$$\|u\|_{\mathbb{U}} = \left(\int_0^L |u|^4 dx\right)^{\frac{1}{4}}, \quad \text{for all } u \in \mathbb{U}.$$

We consider the operator $A:\mathbb{V}\to\overline{\mathbb{V}}^*$ given by

$$\langle Au, v\rangle = (u,v)_{\mathbb{V}}, \quad \text{for all } u,v \in \mathbb{V}.$$

Let $(\mu_k)_k$ be the increasing sequence of eigenvalues, and let $(\varphi_k)_k$ be the corresponding eigenfunctions of A with respect to Dirichlet boundary conditions. The eigenfunctions $(\varphi_k)_k$ form an orthonormal system in $\mathbb{H}$, and they are orthogonal in $\mathbb{V}$. Recall that

$$\operatorname{Im}\langle Av, v\rangle = 0, \quad \forall\, v \in \mathbb{V}, \tag{3.4}$$

and $\mathbb{V} \overset{c}{\hookrightarrow} \mathbb{H}$ with

$$\|v\|^2 \leq \frac{L^2}{\pi^2}\|v\|_{\mathbb{V}}^2, \quad \forall\, v \in \mathbb{V}. \tag{3.5}$$

Note that the best embedding constant is the first eigenvalue of A.

The operator A is in fact the extension of $A : D_A \subset \mathbb{H} \to \mathbb{H}$, where

$$D_A = D(A) = \{v \in \mathbb{V} : Av \in \mathbb{H}\} = H^2(0, L) \cap H_0^1(0, L),$$

and it holds that

$$\langle Au, v\rangle = (Au, v), \quad \forall\, u \in D_A, v \in \mathbb{V}. \tag{3.6}$$

D_A is dense in $\mathbb{V}$ (with respect to $\|\cdot\|_{\mathbb{V}}$) and in $\mathbb{H}$ (with respect to $\|\cdot\|$); for details, see Ref. [10, Chapter VI, & 3. Linear Operators in Hilbert Spaces]. We consider

$$(u, v)_{D_A} = (Au, Av), \quad \forall u, v \in D_A,$$

$$\|u\|_{D_A}^2 = \|Au\|^2 = \int_0^L \left|\frac{d^2 u}{dx^2}\right|^2 dx, \quad \forall\, u \in D_A. \tag{3.7}$$

For each $u \in D_A$, we write

$$\|u\|_{\mathbb{V}}^2 = (Au, u) \leq \|Au\|\|u\| \leq \frac{L}{\pi}\|u\|_{D_A}\|u\|_{\mathbb{V}}.$$

Therefore,

$$\|u\|_{\mathbb{V}}^2 \leq \frac{L^2}{\pi^2}\|u\|_{D_A}^2, \quad \forall\, u \in D_A. \tag{3.8}$$

Hence, $D_A \hookrightarrow \mathbb{V}$. Moreover, $D_A \overset{c}{\hookrightarrow} \mathbb{V}$ since $H^2(0, L) \overset{c}{\hookrightarrow} H^1(0, L)$ by Theorem 4.18 in Ref. [13]. Since $H_0^1(0, L) = \{u \in H^1(0, L) : u(0) = u(L) = 0\}$ (see Ref. [25, Appendix (48b)]), it holds that

$$D_A = \{u \in H^2(0, L) : u(0) = u(L) = 0\}.$$

We have

$$(u,v) = \sum_{j=1}^{\infty} (u,\varphi_j)\overline{(v,\varphi_j)}, \quad \forall\, u,v \in \mathbb{H},$$

$$(u,v)_{\mathbb{V}} = \langle Au, v\rangle = \sum_{j=1}^{\infty} \mu_j (u,\varphi_j)\overline{(v,\varphi_j)}, \quad \forall\, u,v \in \mathbb{V},$$

$$(u,v)_{D_A} = (Au, Av) = \sum_{j=1}^{\infty} \mu_j^2 (u,\varphi_j)\overline{(v,\varphi_j)}, \quad \forall\, u,v \in D_A. \quad (3.9)$$

$(D_A, (\cdot,\cdot)_{D_A})$, $(\mathbb{V},(\cdot,\cdot)_{\mathbb{V}})$, and $(\mathbb{H},(\cdot,\cdot))$ are separable Hilbert spaces.

Similarly to Ref. [14, p. 285, Lemma 1.1] (where the case $L=1$ is treated), one can prove that for each $u \in H^1(0,L)$, it holds that

$$\sup_{x \in [0,L]} |u(x)|^2 \le \|u\| \left(\frac{1}{L}\|u\| + 2\left\|\frac{du}{dx}\right\|\right) \le K_L \|u\|_{\mathbb{V}}^2, \quad (3.10)$$

where $K_L = \frac{L}{\pi^2} + \frac{2L}{\pi}$. Since $\frac{du}{dx} \in H^1(0,L)$ for each $u \in D_A$, we have, by (3.10),

$$\sup_{x \in [0,L]} \left|\frac{du}{dx}(x)\right|^2 \le K_L \|u\|_{D_A}^2. \quad (3.11)$$

Lemma 3.1. *The following inequalities hold:*

$$\||u|^2 u - |v|^2 v\| \le \frac{3}{2} K_L (\|u\|_{\mathbb{V}}^2 + \|v\|_{\mathbb{V}}^2)\|u - v\|, \quad \forall\, u,v \in \mathbb{V}, \quad (3.12)$$

$$\||u|^2 u\|^2 \le K_L^2 \|u\|_{\mathbb{V}}^4 \|u\|^2 \le \frac{L^2 K_L^2}{\pi^2}\|u\|_{\mathbb{V}}^6, \quad \forall\, u \in \mathbb{V}, \quad (3.13)$$

$$\||u|^2 u\|_{\mathbb{V}}^2 \le 9 K_L^2 \|u\|_{\mathbb{V}}^6, \quad \forall\, u \in \mathbb{V},$$

$$-\operatorname{Re}(|u|^2 u, u)_{\mathbb{V}} \le 0, \quad \forall\, u \in \mathbb{V}, \quad (3.14)$$

$$-\operatorname{Re}(|u|^2 u, u)_{D_A} \le 5 K_L \|u\|_{\mathbb{V}}^2 \|u\|_{D_A}^2, \quad \forall\, u \in D_A.$$

Proof. The inequalities (3.10) and (A.2) imply that (3.12) holds. By using (3.10) and (3.5), we have, for each $u \in \mathbb{V}$,

$$\||u|^2 u\|^2 \leq \sup_{x \in [0,L]} |u(x)|^4 \int_0^L |u|^2 dx \leq K_L^2 \|u\|_\mathbb{V}^4 \|u\|^2 \leq \frac{L^2 K_L^2}{\pi^2} \|u\|_\mathbb{V}^6$$

and

$$\||u|^2 u\|_\mathbb{V}^2 = \int_0^L \left|\frac{d(|u|^2 u)}{dx}\right|^2 dx \leq 9 \int_0^L |u|^4 \left|\frac{du}{dx}\right|^2 dx \leq 9 K_L^2 \|u\|_\mathbb{V}^6.$$

We compute

$$\operatorname{Re}(|u|^2 u, u)_\mathbb{V} = \operatorname{Re} \int_0^L \frac{d|u|^2 u}{dx} \cdot \frac{d\bar{u}}{dx} dx$$

$$= \int_0^L \left(\frac{1}{2}\left(\frac{d|u|^2}{dx}\right)^2 + |u|^2 \left|\frac{du}{dx}\right|^2\right) dx \geq 0.$$

By using (3.10) and (3.11), we write

$$-\operatorname{Re}(|u|^2 u, u)_{D_A} = -\operatorname{Re} \int_0^L \frac{d^2 |u|^2 u}{dx^2} \cdot \frac{d^2 \bar{u}}{dx^2} dx$$

$$\leq \int_0^L \left(|u|^2 \left|\frac{d^2 u}{dx^2}\right|^2 + 4|u| \left|\frac{du}{dx}\right|^2 \left|\frac{d^2 u}{dx^2}\right|\right) dx$$

$$\leq \int_0^L \left(3|u|^2 \left|\frac{d^2 u}{dx^2}\right|^2 + 2\left|\frac{du}{dx}\right|^4\right) dx$$

$$\leq 3 \sup_{x \in [0,L]} |u(x)|^2 \|u\|_{D_A}^2 + 2 \sup_{x \in [0,L]} \left|\frac{du}{dx}(x)\right|^2 \|u\|_\mathbb{V}^2 \leq 5 K_L \|u\|_\mathbb{V}^2 \|u\|_{D_A}^2. \quad \square$$

For each $n \in \mathbb{N}^*$, let $\mathcal{H}_n = \operatorname{sp}\{h_1, h_2, \ldots, h_n\}$. We denote the finite-dimensional Hilbert spaces as follows:

$$(\mathbb{H}_n, \|\cdot\|) = (\mathcal{H}_n, \|\cdot\|),$$
$$(\mathbb{V}_n, \|\cdot\|_\mathbb{V}) = (\mathcal{H}_n, \|\cdot\|_\mathbb{V}),$$
$$(\mathbb{D}_n, \|\cdot\|_{D_A}) = (\mathcal{H}_n, \|\cdot\|_{D_A}).$$

Note that for fixed $n \in \mathbb{N}^*$, these three norms are equivalent on $\mathcal{H}_n$, but we choose to use the specific norm in each of these spaces since in our investigations, we also take $n \to \infty$ and leave the finite-dimensional spaces.

For each $\varphi \in \mathcal{H}_n$, it holds that

$$\varphi = \sum_{j=1}^{n} (\varphi, \varphi_j) \varphi_j \quad \text{and} \quad \|\varphi\|^2 = \sum_{j=1}^{n} |(\varphi, \varphi_j)|^2, \tag{3.15}$$

$$A\varphi = \sum_{j=1}^{n} \mu_j (\varphi, \varphi_j) \varphi_j \quad \text{and}$$

$$(A\varphi, \varphi) = (\varphi, A\varphi) = \|\varphi\|_{\mathbb{V}}^2 = \sum_{j=1}^{n} \mu_j |(\varphi, \varphi_j)|^2. \tag{3.16}$$

We define the orthogonal projection

$$P_n : \mathbb{H} \to \mathcal{H}_n \text{ given by } P_n u = \sum_{j=1}^{n} (u, \varphi_j) \varphi_j, \quad \forall\, u \in \mathbb{H}.$$

Denote $\gamma_{0n} = P_n \gamma_0$ and $\mathcal{G}_n = \{P_n g : g \in \mathcal{G}\}$ (the set of admissible controls for the finite-dimensional control problems).

Observe that

$$\|P_n u\|^2 \leq \|u\|^2 \text{ and } (P_n u, \varphi) = (u, \varphi), \quad \forall\, u \in \mathbb{H}, \varphi \in \mathcal{H}_n, \tag{3.17}$$

$$(P_n u, P_n \varphi) = (P_n u, \varphi), \quad \forall\, u, \varphi \in \mathbb{H}. \tag{3.18}$$

If $u \in \mathbb{H}$, then

$$\|u - P_n u\| = \|(\mathbb{I} - P_n)u\| = \sum_{j=n+1}^{\infty} |(u, \varphi_j)|^2 \to 0, \quad \text{as } n \to \infty, \tag{3.19}$$

and if $v \in \mathbb{V}$, then

$$\|v - P_n v\|_{\mathbb{V}} = \sum_{j=n+1}^{\infty} \mu_j |(v, \varphi_j)|^2 \to 0, \quad \text{as } n \to \infty, \tag{3.20}$$

since the reminder of a convergent series tends to zero. Moreover,

$$|(P_n u, A\varphi)| \leq \|P_n u\|_{\mathbb{V}} \|\varphi\|_{\mathbb{V}} \leq \|u\|_{\mathbb{V}} \|\varphi\|_{\mathbb{V}}, \quad \forall\, u \in \mathbb{V}, \varphi \in \mathcal{H}_n, \tag{3.21}$$

and

$$\|P_n u\|_{D_A} \leq \|u\|_{D_A}, \quad \forall\, u \in D_A. \tag{3.22}$$

3.3. Formulation of the Problem

Let $(\Omega, \mathcal{F}, \mathbb{P})$ be a complete probability space. Denote by $\mathbb{E}$ the expectation with respect to $\mathbb{P}$. Consider $(B^H(t))_{t \in [0,T]}$ to be a real-valued fBm with the Hurst index $H \in (0,1)$, which generates an increasing family of σ-algebras $(\mathcal{F}_t)_{t \in [0,T]}$.

We investigate the following stochastic nonlinear Schrödinger equation:

$$dX(t) = \bigl(-\mathrm{i}AX(t) - |X(t)|^2 X(t) + \alpha(t)X(t) + g(t)\bigr)dt$$
$$+ \mathrm{i}\beta X(t) dB^H(t), \tag{3.23}$$

$t \in [0,T]$, with initial condition $X(0) = \gamma_0 \in D_A$ and with fixed $T > 0$, $\beta \in \mathbb{R}$, $\alpha \in C(0,T;\mathbb{C})$, $g \in \mathcal{G}$ (see (3.2)). All stochastic integrals in this chapter are Wick–Itô–Skorohod integrals.

A process $X \in \mathcal{L}^\infty(\Omega; C(0,T;\mathbb{V}) \cap L^\infty(0,T;D_A))$ is called a *variational solution* of the stochastic nonlinear Schrödinger equation (3.23) if

$$(X(t), v) = (\gamma_0, v) - \mathrm{i}\int_0^t \langle AX(s), v\rangle\, ds - \int_0^t (|X(s)|^2 X(s), v) ds$$
$$+ \int_0^t (\alpha(s)X(s) + g(s), v)\, ds + \mathrm{i}\beta \int_0^t (X(s), v)\, dB^H(s), \tag{3.24}$$

for all $t \in [0,T]$, $v \in \mathbb{V}$, and a.e. $\omega \in \Omega$.

Let $n \in \mathbb{N}^*$. We formulate the corresponding finite-dimensional problem: A process $X_n \in L^\infty(\Omega; C(0,T;\mathbb{V}_n))$ is a solution of the *finite-dimensional stochastic nonlinear Schrödinger equation* if

$$(X_n(t), \varphi_j) = (\gamma_{0n}, \varphi_j) + \int_0^t (-\mathrm{i}AX_n(s) - P_n|X_n(s)|^2 X_n(s), \varphi_j) ds$$
$$+ \int_0^t (\alpha(s)X_n(s) + P_n g(s), \varphi_j) ds$$
$$+ \mathrm{i}\beta \int_0^t (X_n(s), \varphi_j) dB^H(s), \tag{3.25}$$

for all $t \in [0,T]$, $j \in \{1, 2, \ldots, n\}$, and a.e. $\omega \in \Omega$. Note that (3.25) can be written equivalently as an equation in $\mathbb{H}_n$:

$$X_n(t) = \gamma_{0n} + \int_0^t \left(-\mathrm{i}AX_n(s) - P_n|X_n(s)|^2 X_n(s) + P_n g(s)\right) ds$$

$$+ \mathrm{i}\beta \int_0^t X_n(s) dB^H(s), \qquad (3.26)$$

for all $t \in [0, T]$, and a.e. $\omega \in \Omega$.

3.4. Solution of the Stochastic Partial Differential Equation

First, we recall some results concerning the fBm; see Ref. [4] for more details.

The real-valued Gaussian process $\left(B^H(t)\right)_{t \geq 0}$ is a *fractional Brownian motion* with the Hurst index $H \in (0,1)$ if

$$\mathbb{P}(B^H(0) = 0) = 1,$$

$$\mathbb{E}\left(B^H(t)\right) = 0 \quad \text{for all } t \geq 0,$$

$$\mathbb{E}\left(B^H(t)B^H(s)\right) = \frac{1}{2}(t^{2H} + s^{2H} - |t-s|^{2H}) \quad \text{for all } s, t \geq 0.$$

Note that a fBm has a continuous version. The stochastic integrals below are Wick–Itô–Skorohod integrals with respect to a real-valued fBm B^H with $H \in (0,1)$ (see Ref. [4, Chapter 4]). The following Itô formula for functionals of B^H is given in Ref. [4, Theorem 4.2.6].

Theorem 3.1. *Suppose that $F : \mathbb{R} \times \mathbb{R} \to \mathbb{R}$ is a function which belongs to $C^{1,2}(\mathbb{R}_+ \times \mathbb{R})$. Assume that the random variables $F(t, B^H(t))$, $\int_0^t \frac{\partial F}{\partial s}(s, B^H(s))ds$, and $\int_0^t \frac{\partial^2 F}{\partial x^2}(s, B^H(s))s^{2H-1}ds$*

are square integrable for every $t \geq 0$. Then, the following holds:

$$F(t, B^H(t)) = F(0,0) + \int_0^t \frac{\partial F}{\partial s}(s, B^H(s))ds$$

$$+ \int_0^t \frac{\partial F}{\partial x}(s, B^H(s))dB^H(s)$$

$$+ H \int_0^t \frac{\partial^2 F}{\partial x^2}(s, B^H(s))s^{2H-1}ds,$$

for all $t \geq 0$.

We consider for $t \geq 0$ the complex-valued stochastic process

$$y(t) = \exp\left\{\int_0^t \alpha(s)ds + \frac{1}{2}\beta^2 t^{2H} + i\beta B^H(t)\right\}, \quad t \in [0,T], \text{ a.e. } \omega \in \Omega. \tag{3.27}$$

By Theorem 3.1, this process solves the linear equation

$$y(t) = 1 + \int_0^t \alpha(s)y(s)ds + i\beta \int_0^t y(s)dB^H(s), \quad t \in [0,T], \quad \text{a.e. } \omega \in \Omega.$$

Observe that

$$|y(t)| = \exp\left\{\int_0^t \operatorname{Re}\alpha(s)ds + \frac{1}{2}\beta^2 t^{2H}\right\}, \quad \text{for all } t \in [0,T],$$

and for all $t \in [0,T]$,

$$K_1 \leq |y(t)|^2 = \exp\left\{2\int_0^t \operatorname{Re}\alpha(s)ds + \beta^2 t^{2H}\right\} \leq K_2, \tag{3.28}$$

$$\frac{1}{K_2} \leq \frac{1}{|y(t)|^2} \leq \frac{1}{K_1},$$

where

$$K_1 = \exp\left\{-2\int_0^T |\alpha(s)|ds\right\}, \quad K_2 = \exp\left\{2\int_0^T |\alpha(s)|ds + \beta^2 T^{2H}\right\}.$$

For each $g \in \mathcal{G}$, we denote for a.e. $t \in (0,T)$ and a.e. $\omega \in \Omega$,

$$\hat{g}(t) = \frac{g(t)}{y(t)} = g(t)\exp\left\{-\int_0^t \alpha(s)ds - \frac{1}{2}\beta^2 t^{2H} - i\beta B^H(t)\right\}. \tag{3.29}$$

Then, by (3.2), (3.5), and (3.8), it follows that

$$\|\hat{g}(t)\|_\mathbb{V} \leq \frac{1}{\sqrt{K_1}}\|g(t)\|_\mathbb{V} \leq \frac{\rho L}{\pi\sqrt{K_1}}, \qquad (3.30)$$

and

$$\|\hat{g}(t)\| \leq \frac{L}{\pi\sqrt{K_1}}\|g(t)\|_\mathbb{V} \leq \frac{\rho L^2}{\pi^2\sqrt{K_1}}. \qquad (3.31)$$

We consider the *variational solution* $Z \in C(0,T;\mathbb{V})$ of the following nonlinear pathwise PDE:

$$(Z(t), v) = (\gamma_0, v) - \mathrm{i}\int_0^t \langle AZ(s), v\rangle ds$$

$$+ \int_0^t (-|y(s)|^2|Z(s)|^2 Z(s) + \hat{g}(s), v)ds, \qquad (3.32)$$

for all $t \in [0,T]$ and $v \in \mathbb{V}$. For fixed $n \in \mathbb{N}^*$, its corresponding pathwise finite-dimensional problem is

$$(Z_n(t), \varphi_j) = (\gamma_{0n}, \varphi_j) - \mathrm{i}\int_0^t (AZ_n(s), \varphi_j)ds$$

$$+ \int_0^t (-|y(s)|^2 P_n|Z_n(s)|^2 Z_n(s) + P_n\hat{g}(s), \varphi_j)ds, \qquad (3.33)$$

for all $t \in [0,T]$ and $j \in \{1,2,\ldots,n\}$. We will show that $Z_n \in C(0,T;\mathbb{V}_n)$. Observe that (3.33) can be written equivalently as an equation in $\mathbb{H}_n$:

$$Z_n(t) = \gamma_{0n} + \int_0^t -\mathrm{i}AZ_n(s) - |y(s)|^2 P_n|Z_n(s)|^2 Z_n(s) + P_n\hat{g}(s)ds, \qquad (3.34)$$

for all $t \in [0,T]$.

Remark 3.1.

(1) Note that X and Z depend on g, while X_n and Z_n depend also on g through $P_n g$, but we omit writing explicitly this dependence.

Only in the cases when this dependence is explicitly needed to be expressed (as in Sections 3.5 and 3.6), we write X^g, Z^g, X_n^g (instead of $X_n^{P_n g}$) and Z_n^g (instead of $Z_n^{P_n g}$) if $g \in \mathcal{G}$, and we use the same notation, X_n^g and Z_n^g, when $g \in \mathcal{G}_n$.

(2) Deterministic equations of the type as (3.32) can be found in Ref. [9] (without the term y).

Theorem 3.2.

(1) If $Z \in C(0,T;\mathbb{V})$ is a solution of (3.32), then it is unique.
(2) If $Z_n \in C(0,T;\mathbb{V}_n)$, $n \in \mathbb{N}^*$, is a solution of (3.32), then it is unique.

Proof. (1) Assume that there are two variational solutions $Z, \hat{Z} \in C(0,T;\mathbb{V})$ of the pathwise problem (3.32). By denoting $z = Z - \hat{Z}$, we get, for all $t \in [0,T]$ and all $v \in V$,

$$(z(t), v) = -\mathrm{i} \int_0^t \langle Az(s), v \rangle ds - \int_0^t |y(s)|^2 (|Z(s)|^2 Z(s)$$
$$- |\hat{Z}(s)|^2 \hat{Z}(s), v) ds.$$

Applying the energy equality, we obtain

$$\|z(t)\|^2 = 2\mathrm{Im} \int_0^t \langle Az(s), z(s) \rangle ds$$
$$- 2\mathrm{Re} \int_0^t |y(s)|^2 (|Z(s)|^2 Z(s) - |\hat{Z}(s)|^2 \hat{Z}(s), z(s)) ds,$$

for all $t \in [0,T]$. We use (3.4) and (A.1) to get $\|z(t)\|^2 \leq 0$ for all $t \in [0,T]$. Consequently,

$$Z(t,x) = \hat{Z}(t,x), \quad \text{for all } t \in [0,T] \text{ and a.e. } x \in [0,L].$$

(2) The proof is similar to (1). $\square$

Theorem 3.3. *If Z_n, $n \in \mathbb{N}^*$, is the solution of (3.33), then the following estimates hold:*

$$\sup_{t \in [0,T]} \|Z_n(t)\|^2 \leq K_3, \quad \text{where } K_3 = e^T \left(\|\gamma_{0n}\|^2 + \frac{\rho^2 T L^4}{\pi^4 K_1} \right), \quad (3.35)$$

$$\sup_{t\in[0,T]} \|Z_n(t)\|_{\mathbb{V}}^2 \le K_4, \quad \text{where } K_4 = e^T\left(\|\gamma_{0n}\|_{\mathbb{V}}^2 + \frac{\rho^2 T L^2}{\pi^2 K_1}\right), \tag{3.36}$$

$$\sup_{t\in[0,T]} \|Z_n(t)\|_{D_A}^2 \le K_5, \tag{3.37}$$

where $K_5 = e^{10K_2 K_4 K_L T}\left(\|\gamma_{0n}\|_{D_A}^2 + \frac{\rho^2 T}{K_1}\right)$, and for all $t \in (0,T)$,

$$\left\|\frac{dZ_n}{dt}(t)\right\| \le K_6, \tag{3.38}$$

where $K_6 = \sqrt{K_5} + K_2\sqrt{K_3}K_4 K_L + \frac{\rho L^2}{\pi^2 \sqrt{K_1}}$.

Proof. Recall that Z_n is the unique solution of (3.33). Then, by (3.15), we have, for all $t \in [0,T]$,

$$\|Z_n(t)\|^2 = \|\gamma_{0n}\|^2 + 2\int_0^t \mathrm{Im}\langle AZ_n(s), Z_n(s)\rangle ds$$

$$- 2\int_0^t \mathrm{Re}\,|y(s)|^2 \|Z_n(s)\|_{\mathbb{U}}^4 ds + 2\int_0^t \mathrm{Re}\,(P_n\hat{g}(s), Z_n(s)) ds,$$

which, by (3.4), becomes

$$\|Z_n(t)\|^2 + 2\int_0^t |y(s)|^2 \|Z_n(s)\|_{\mathbb{U}}^4 ds$$

$$\le \|\gamma_{0n}\|^2 + 2\int_0^t \|\hat{g}(s)\| \cdot \|Z_n(s)\| ds$$

$$\le \|\gamma_{0n}\|^2 + \int_0^t \left(\|\hat{g}(s)\|^2 + \|Z_n(s)\|^2\right) ds.$$

Then, by Gronwall's lemma and (3.31), we have

$$\sup_{t\in[0,T]} \|Z_n(t)\|^2 \le e^T\left(\|\gamma_{0n}\|^2 + \frac{\rho^2 T L^4}{\pi^4 K_1}\right).$$

From (3.33), it follows that for all $t \in [0,T]$ and $j \in \{1, ..., n\}$,

$$|(Z_n(t), \varphi_j)|^2 = |(\gamma_{0n}, \varphi_j)|^2 + 2\int_0^t \text{Im}(AZ_n(s), \varphi_j)\overline{(Z_n(s), \varphi_j)}ds$$

$$- 2\int_0^t |y(s)|^2 \text{Re}\,(|Z_n(s)|^2 Z_n(s), \varphi_j)\overline{(Z_n(s), \varphi_j)}ds$$

$$+ 2\int_0^t \text{Re}\,(P_n \hat{g}(s), \varphi_j)\overline{(Z_n(s), \varphi_j)}ds. \qquad (3.39)$$

We multiply both sides of the above equality by μ_j and add up from $j = 1$ to n to get, by (3.16),

$$(Z_n(t), AZ_n(t))$$
$$= (\gamma_{0n}, A\gamma_{0n}) - 2\int_0^t |y(s)|^2 \text{Re}\,(|Z_n(s)|^2 Z_n(s), AZ_n(s))ds$$
$$+ 2\int_0^t \text{Re}\,(P_n \hat{g}(s), AZ_n(s))ds.$$

By (3.14), (3.16), and (3.21), we get, for all $t \in [0,T]$,

$$(Z_n(t), AZ_n(t)) = \|Z_n(t)\|_V^2 \leq \|\gamma_{0n}\|_V^2 + \int_0^t \left(\|\hat{g}(s)\|_V^2 + \|Z_n(s)\|_V^2\right)ds.$$

Gronwall's lemma and (3.30) yield

$$\sup_{t \in [0,T]} \|Z_n(t)\|_V^2 \leq e^T \left(\|\gamma_{0n}\|_V^2 + \frac{\rho^2 T L^2}{\pi^2 K_1}\right).$$

We use (3.39), multiply both sides of this equality by μ_j^2, and add up from $j = 1$ to n to derive, by (3.9),

$$\|Z_n(t)\|_{D_A}^2 = \|\gamma_{0n}\|_{D_A}^2 - 2\int_0^t |y(s)|^2 \text{Re}\,(|Z_n(s)|^2 Z_n(s), Z_n(s))_{D_A} ds$$
$$+ 2\int_0^t \text{Re}\,(P_n \hat{g}(s), Z_n(s))_{D_A} ds.$$

By Lemma 3.1 and (3.22), we get for all $t \in [0,T]$

$$\|Z_n(t)\|_{D_A}^2 \le \|\gamma_{0n}\|_{D_A}^2 + 10K_L \int_0^t |y(s)|^2 \|Z_n(s)\|_{\mathbb{V}}^2 \|Z_n(s)\|_{D_A}^2 ds$$
$$+ \int_0^t (\|\hat{g}(s)\|_{D_A}^2 + \|Z_n(s)\|_{D_A}^2) ds.$$

Gronwall's lemma, (3.2), and (3.36) imply

$$\sup_{t \in [0,T]} \|Z_n(t)\|_{D_A}^2 \le e^{10K_2K_4K_LT} \left(\|\gamma_{0n}\|_{D_A}^2 + \frac{\rho^2 T}{K_1} \right).$$

Since $Z_n(t) \in D_A \subset \mathbb{V} \subset \mathbb{H}$ and $AZ_n(t) \in \mathbb{H}$, we have $\langle AZ_n(t), v \rangle = (AZ_n(t), v)$ for each $v \in \mathbb{V}$ by (3.6). Using (3.34), it follows that, for all $t \in (0, T)$,

$$\frac{dZ_n}{dt}(t) = -\mathrm{i} A Z_n(t) - |y(t)|^2 P_n |Z_n(t)|^2 Z_n(t) + P_n \hat{g}(t).$$

Then, for all $t \in (0, T)$,

$$\left\| \frac{dZ_n}{dt}(t) \right\| \le \|AZ_n(t)\| + \||y(t)|^2 |Z_n(t)|^2 Z_n(t)\| + \|P_n \hat{g}(t)\|.$$

But (3.7), (3.37), (3.13), and (3.31) imply, for all $t \in (0,T)$,

$$\left\| \frac{dZ_n}{dt}(t) \right\| \le \|Z_n(t)\|_{D_A} + K_2 \||Z_n(t)|^2 Z_n(t)\| + \frac{1}{\sqrt{K_1}} \|g(t)\|$$
$$\le \sqrt{K_5} + K_2 \sqrt{K_3} K_4 K_L + \frac{\rho L^2}{\pi^2 \sqrt{K_1}}. \qquad \Box$$

Theorem 3.4. *There exist* $Z \in C(0,T;\mathbb{V}) \cap \mathcal{L}^\infty(0,T;D_A)$ *solution of* (3.32) *and* $Z_n \in C(0,T;\mathbb{V}_n)$ *solution of* (3.33) *such that*

$$Z_n \to Z \quad \text{in } C(0,T;\mathbb{V}) \quad \text{as } n \to \infty, \qquad (3.40)$$

and

$$Z_n \overset{*}{\rightharpoonup} Z \quad \text{in } \mathcal{L}^\infty(0,T;D_A) \quad \text{as } n \to \infty.$$

Moreover, it holds that

$$\sup_{t\in[0,T]} \|Z(t)\|^2 \leq K_7,$$

where $K_7 = e^T \left(\|\gamma_0\|^2 + \frac{\rho^2 T L^4}{\pi^4 K_1}\right)$,

$$\sup_{t\in[0,T]} \|Z(t)\|_\mathbb{V}^2 \leq K_8,$$

where $K_8 = e^T \left(\|\gamma_0\|_\mathbb{V}^2 + \frac{\rho^2 T L^2}{\pi^2 K_1}\right)$,

$$\operatorname*{ess\,sup}_{t\in[0,T]} \|Z(t)\|_{D_A}^2 \leq K_9, \tag{3.41}$$

where $K_9 = e^{10 K_2 K_8 K_L T} \left(\|\gamma_0\|_{D_A}^2 + \frac{\rho^2 T}{K_1}\right)$,

$$\operatorname*{ess\,sup}_{t\in[0,T]} \left\|\frac{dZ}{dt}(t)\right\| \leq K_{10}, \tag{3.42}$$

where $K_{10} = \sqrt{K_9} + K_2 \sqrt{K_7} K_8 K_L + \frac{\rho L^2}{\pi^2 \sqrt{K_1}}$.

Proof. The existence of the solution of $Z_n \in C(0,T;\mathbb{V}_n)$ is assured by the classic theory of differential equations (with locally Lipschitz functions; see (3.12)) and the estimates from Theorem 3.3.

By (3.36) and (3.38), it follows that $(Z_n)_n$ is a bounded sequence in $\mathcal{L}^\infty(0,T;D_A)$, while $\left(\frac{dZ_n}{dt}\right)_n$ is a bounded sequence in $\mathcal{L}^\infty(0,T;\mathbb{H})$. By Theorem 3.18 ($\mathbb{B}_0 = D_A, \mathbb{B}_1 = \mathbb{V}, \mathbb{B}_2 = \mathbb{H}, q = r = \infty$) and by Theorem 3.20 ($\mathbb{B} = D_A$), it follows that there exist $Z \in C(0,T;\mathbb{V}) \cap \mathcal{L}^\infty(0,T;D_A)$ and a subsequence of $(Z_n)_n$ (for which we use the same notation) such that

$$Z_n \to Z \quad \text{in } C(0,T;\mathbb{V}), \tag{3.43}$$

and

$$Z_n \overset{*}{\rightharpoonup} Z \quad \text{in } \mathcal{L}^\infty(0,T;D_A), \quad \text{as } n \to \infty.$$

Then, by (3.5), (3.12), and (3.17), we obtain

$$\||P_n|Z_n(s)|^2 Z_n(s) - |Z(s)|^2 Z(s)\|$$
$$\leq \||P_n|Z_n(s)|^2 Z_n(s) - P_n|Z(s)|^2 Z(s)\|$$
$$+ \||Z(s)|^2 Z(s) - P_n|Z(s)|^2 Z(s)\|$$
$$\leq \||Z_n(s)|^2 Z_n(s) - |Z(s)|^2 Z(s)\| + \|(\mathbb{I} - P_n)|Z(s)|^2 Z(s)\|$$
$$\leq \frac{3K_L}{2}(\|Z_n(s)\|_{\mathbb{V}}^2 + \|Z(s)\|_{\mathbb{V}}^2)\|Z_n(s) - Z(s)\|$$
$$+ \|(\mathbb{I} - P_n)|Z(s)|^2 Z(s)\|$$
$$\leq \frac{3LK_L}{2\pi}\max\{K_4, K_8\}\|Z_n(s) - Z(s)\|_{\mathbb{V}} + \|(\mathbb{I} - P_n)|Z(s)|^2 Z(s)\|.$$

By (3.43) and (3.19) (where we use that (3.13) and $Z(s) \in \mathbb{V}$ imply $|Z(s)|^2 Z(s) \in \mathbb{H}$), it then follows that $P_n|Z_n|^2 Z_n \to |Z|^2 Z$ in $\mathcal{L}^2(0, T; \mathbb{H})$. By taking $n \to \infty$ on both sides of (3.33), using the above convergences, we deduce that Z is the solution of (3.32). The uniqueness of the solution of (3.32) (see Theorem 3.2) implies that the convergences stated in this theorem hold for the whole sequence $(Z_n)_n$ (see also Theorem 3.19).

By using (3.43), $C(0, T; \mathbb{V}) \hookrightarrow C(0, T; \mathbb{H})$, and (3.35), we obtain

$$\sup_{t \in [0,T]} \|Z(t)\|^2 = \lim_{n \to \infty} \sup_{t \in [0,T]} \|Z_n(t)\|^2 \leq K_7,$$

where $K_7 = e^T \left(\|\gamma_0\|^2 + \frac{\rho^2 T L^4}{\pi^4 K_1}\right)$, and by (3.43) and (3.36), we derive

$$\sup_{t \in [0,T]} \|Z(t)\|_{\mathbb{V}}^2 = \lim_{n \to \infty} \sup_{t \in [0,T]} \|Z_n(t)\|_{\mathbb{V}}^2 \leq K_8,$$

where $K_8 = e^T \left(\|\gamma_0\|_{\mathbb{V}}^2 + \frac{\rho^2 T L^2}{\pi^2 K_1}\right)$. By using Theorem 3.20 ($\mathbb{B} = D_A$ is a separable Hilbert space), (3.37), and the above notation K_8, we have

$$\operatorname*{ess\,sup}_{t \in [0,T]} \|Z(t)\|_{D_A}^2 = \liminf_{n \to \infty} \operatorname*{ess\,sup}_{t \in [0,T]} \|Z_n(t)\|_{D_A}^2 \leq K_9,$$

where $K_9 = e^{10K_2 K_8 K_L T}\left(\|\gamma_0\|_{D_A}^2 + \frac{\rho^2 T}{K_1}\right)$. Since $Z(t) \in D_A \subset \mathbb{V} \subset \mathbb{H}$ and $AZ(t) \in \mathbb{H}$, then $\langle AZ(t), v \rangle = (AZ(t), v)$ for each $v \in \mathbb{V}$ by (3.6).

Using (3.32), it follows that, for each $t \in (0,T)$ and all $v \in \mathbb{V}$,

$$\left(\frac{dZ}{dt}(t), v\right) = (-\mathrm{i}AZ(t) - |y(t)|^2|Z(t)|^2 Z(t) + \hat{g}(t), v).$$

Then, for a.e. $t \in (0,T)$,

$$\left\|\frac{dZ}{dt}(t)\right\| \leq \|AZ(t)\| + \||y(t)|^2|Z(t)|^2 Z(t)\| + \|\hat{g}(t)\|.$$

But (3.7), (3.13), (3.41), and (3.31) imply

$$\operatorname*{ess\,sup}_{t\in[0,T]} \left\|\frac{dZ}{dt}(t)\right\| \leq \operatorname*{ess\,sup}_{t\in[0,T]} \left(\|Z(t)\|_{D_A} + K_2\||Z(t)|^2 Z(t)\| + \frac{\|g(t)\|}{\sqrt{K_1}}\right)$$
$$\leq K_{10},$$

where $K_{10} = \sqrt{K_9} + K_2\sqrt{K_7}K_8 K_L + \frac{\rho L^2}{\pi^2 \sqrt{K_1}}$. □

Theorem 3.5. *Let y be given by (3.27) and $n \in \mathbb{N}^*$. If Z is the solution of (3.32) and Z_n is the solution of of (3.33), then*

$$X(t) = y(t)Z(t), \quad \text{for all } t \in [0,T] \text{ and a.e. } \omega \in \Omega, \tag{3.44}$$

solves the stochastic problem (3.24) and

$$X_n(t) = y(t)Z_n(t), \quad \text{for all } t \in [0,T] \text{ and a.e. } \omega \in \Omega, \tag{3.45}$$

solves the stochastic finite-dimensional problem (3.25).

Proof. We compute, for each $t \in [0,T]$,

$$d(y(t)Z(t)) = y(t)dZ(t) + Z(t)dy(t)$$
$$= (-\mathrm{i}Ay(t)Z(t) - |y(t)Z(t)|^2 y(t)Z(t) + g(t))dt$$
$$+ \mathrm{i}\beta y(t)Z(t)dB^H(t).$$

Therefore, $X = yZ$ solves (3.24). Similarly, $X_n = yZ_n$ solves (3.25). □

By (3.28) and Theorems 3.3 and 3.4, it follows that, for a.e. $\omega \in \Omega$,

$$\sup_{t\in[0,T]} \|X(t)\|_{\mathbb{V}}^2 \leq K_2 e^T \left(\|\gamma_0\|_{\mathbb{V}}^2 + \frac{\rho^2 T L^2}{\pi^2 K_1} \right), \qquad (3.46)$$

$$\operatorname*{ess\,sup}_{t\in[0,T]} \|X(t)\|_{D_A}^2 \leq K_2 e^{10 K_2 K_8 K_L T} \left(\|\gamma_0\|_{D_A}^2 + \frac{\rho^2 T}{K_1} \right),$$

$$\sup_{t\in[0,T]} \|X_n(t)\|_{\mathbb{V}}^2 \leq K_2 e^T \left(\|\gamma_{0n}\|_{\mathbb{V}}^2 + \frac{\rho^2 T L^2}{\pi^2 K_1} \right), \quad \forall\, n \in \mathbb{N}^*. \quad (3.47)$$

This implies that $X \in L^\infty(\Omega; C(0,T;\mathbb{V}) \cap \mathcal{L}^\infty(0,T;D_A))$ and X solves (3.24). $X_n \in L^\infty(\Omega; C(0,T;\mathbb{V}_n))$, for $n \in \mathbb{N}^*$, is a solution of (3.25). By (3.44), (3.45), and Theorem 3.4, it follows that

$$\sup_{t\in[0,T]} \|X(t) - X_n(t)\|_{\mathbb{V}}^2 \to 0 \quad \text{as } n \to \infty, \quad \text{for a.e. } \omega \in \Omega, \quad (3.48)$$

and

$$X_n \overset{*}{\rightharpoonup} X \quad \text{in } \mathcal{L}^\infty(0,T;D_A) \quad \text{as } n \to \infty, \quad \text{for a.e. } \omega \in \Omega.$$

This implies, for $p > 1$, it holds that

$$X_n \rightharpoonup X \quad \text{in } \mathcal{L}^p(0,T;D_A), \quad \text{as } n \to \infty, \quad \text{for a.e. } \omega \in \Omega.$$

Theorem 3.6. Let $g, g_n \in \mathcal{G}$, $n \in \mathbb{N}^*$, be such that $g_n \overset{*}{\rightharpoonup} g$ in $\mathcal{L}^\infty(0,T;\mathbb{V})$ as $n \to \infty$. Then, $Z^{g_n} \to Z^g$ in $C(0,T;\mathbb{V})$ and $Z^{g_n} \overset{*}{\rightharpoonup} Z^g$ in $\mathcal{L}^\infty(0,T;D_A)$ as $n \to \infty$.

Proof. Recall that $g_n \overset{*}{\rightharpoonup} g$ in $\mathcal{L}^\infty(0,T;\mathbb{V})$ implies

$$\int_0^T (g_n(s) - g(s), \varphi(s))_{\mathbb{V}}\, ds \to 0, \quad \text{as } n \to \infty, \quad \forall\, \varphi \in \mathcal{L}^2(0,T;\mathbb{V}), \quad (3.49)$$

since $\mathcal{L}^2(0,T;\mathbb{V}) \hookrightarrow \mathcal{L}^1(0,T;\mathbb{V})$.

By (3.41) and (3.42), it follows that $(Z^{g_n})_n$ is a bounded sequence in $\mathcal{L}^\infty(0,T;D_A)$, while $\left(\frac{dZ^{g_n}}{dt}\right)_n$ is a bounded sequence in $\mathcal{L}^\infty(0,T;\mathbb{H})$. By Theorem 3.18 ($\mathbb{B}_0 = D_A, \mathbb{B}_1 = \mathbb{V}, \mathbb{B}_2 = \mathbb{H}, q = r = \infty$)

and by Theorem 3.20, it follows that there exist $Z \in C(0,T;\mathbb{V}) \cap \mathcal{L}^\infty(0,T;D_A)$ and a subsequence of $(Z^{g_n})_n$ (for which we use the same notation) such that

$$Z^{g_n} \to Z \quad \text{in } C(0,T;\mathbb{V}) \text{ (and hence in } C(0,T;\mathbb{H})), \qquad (3.50)$$

$$Z^{g_n} \overset{*}{\rightharpoonup} Z \quad \text{in } \mathcal{L}^\infty(0,T;D_A) \quad \text{as } n \to \infty.$$

By (3.12), it holds that

$$\||Z^{g_n}(t)|^2 Z^{g_n}(t) - |Z(t)|^2 Z(t)\|$$
$$\leq \frac{3K_L}{2}\left(\|Z^{g_n}(t)\|_{\mathbb{V}}^2 + \|Z(t)\|_{\mathbb{V}}^2\right)\|Z^{g_n}(t) - Z(t)\|$$
$$\leq \frac{3K_L}{2}\left(K_8 + \|Z(t)\|_{\mathbb{V}}^2\right)\|Z^{g_n}(t) - Z(t)\|.$$

This implies, by (3.50), that

$$|Z^{g_n}|^2 Z^{g_n} \to |Z|^2 Z \quad \text{in } \mathcal{L}^2(0,T;\mathbb{H}) \quad \text{as } n \to \infty. \qquad (3.51)$$

For all $t \in [0,T]$ and $v \in \mathbb{V}$, by (3.32), the following holds:

$$(Z^{g_n}(t),v) = (\gamma_0,v) - \int_0^t i\langle AZ^{g_n}(s),v\rangle ds$$
$$- \int_0^t |y(s)|^2(|Z^{g_n}(s)|^2 Z^{g_n}(s),v)ds + \int_0^t (\hat{g}_n(s),v)ds.$$

Taking $n \to \infty$ on both sides of this equality, it follows that, by (3.50), (3.51), and (3.49),

$$(Z(t),v) = (\gamma_0,v) + \int_0^t i\langle AZ(s),v\rangle ds$$
$$+ \int_0^t (-|y(s)|^2|Z(s)|^2 Z(s) + \hat{g}(s),v)ds,$$

for all $t \in [0,T]$ and $v \in \mathbb{V}$. By the uniqueness of the solution of (3.32), it follows that $Z^g = Z$. Moreover, by the convergence principles from Theorem 3.19, the whole sequence $(Z^{g_n})_n$ converges strongly to Z^g in $C(0,T;\mathbb{V})$ and weak* to Z^g in $\mathcal{L}^\infty(0,T;D_A)$. □

Theorem 3.7. Fix $n \in \mathbb{N}^*$. Let $g, g_j \in \mathcal{G}_n$, $j \in \mathbb{N}^*$, be such that $g_j \overset{*}{\rightharpoonup} g$ in $L^\infty(0,T;\mathbb{V}_n)$ as $j \to \infty$. Then, $Z_n^{g_j} \to Z_n^g$ in $C(0,T;\mathbb{V}_n)$ and $Z_n^{g_j} \overset{*}{\rightharpoonup} Z_n^g$ in $\mathcal{L}^\infty(0,T;\mathbb{D}_n)$ as $j \to \infty$.

Proof. The compact embedding $D_A \overset{c}{\hookrightarrow} \mathbb{V} \overset{c}{\hookrightarrow} \mathbb{H}$ imply $\mathbb{D}_n \overset{c}{\hookrightarrow} \mathbb{V}_n \overset{c}{\hookrightarrow} \mathbb{H}_n$. Further, we use the same ideas as in the proof of Theorem 3.6, replacing the spaces $D_A, \mathbb{V}, \mathbb{H}$ with $\mathbb{D}_n, \mathbb{V}_n, \mathbb{H}_n$. By (3.37) and (3.38), it follows that $(Z_n^{g_j})_j$ is a bounded sequence in $\mathcal{L}^\infty(0,T;\mathbb{D}_n)$, while $\left(\frac{dZ_n^{g_j}}{dt}\right)_j$ is a bounded sequence in $\mathcal{L}^\infty(0,T;\mathbb{H}_n)$. By Theorem 3.18 ($\mathbb{B}_0 = \mathbb{D}_n, \mathbb{B}_1 = \mathbb{V}_n, \mathbb{B}_2 = \mathbb{H}_n, q = r = \infty$) and Theorem 3.20, it follows that there exist $Z_n \in C(0,T;\mathbb{V}_n) \cap \mathcal{L}^\infty(0,T;\mathbb{D}_n)$ and a subsequence of $(Z_n^{g_j})_j$ (for which we use the same notation) such that

$$Z_n^{g_j} \to Z_n \quad \text{in } C(0,T;\mathbb{V}_n),$$

and

$$Z_n^{g_j} \overset{*}{\rightharpoonup} Z_n \quad \text{in } \mathcal{L}^\infty(0,T;\mathbb{D}_n), \quad \text{as } j \to \infty.$$

Similarly, as in the proof of Theorem 3.6, it follows by the uniqueness of the solution of (3.33) that $Z_n^g = Z_n$ and by Theorem 3.19 that the whole sequence $(Z_n^{g_j})_j$ converges strongly to Z_n^g in $C(0,T;\mathbb{V}_n)$ and weak* to Z_n^g in $\mathcal{L}^\infty(0,T;\mathbb{D}_n)$. □

Theorem 3.8. Let $g, g_n \in \mathcal{G}, n \in \mathbb{N}^*$, be such that $g_n \overset{*}{\rightharpoonup} g$ in $\mathcal{L}^\infty(0,T;\mathbb{V})$ as $n \to \infty$. Then, $Z_n^{g_n} \to Z^g$ in $C(0,T;\mathbb{V})$ and $Z_n^{g_n} \overset{*}{\rightharpoonup} Z^g$ in $\mathcal{L}^\infty(0,T;D_A)$ as $n \to \infty$.

Proof. From the assumption of this theorem and the notation (3.29), we get $\hat{g}_n \overset{*}{\rightharpoonup} \hat{g}$ in $\mathcal{L}^\infty(0,T;\mathbb{V})$ as $n \to \infty$. Then,

$$\int_0^T (g_n(s) - g(s), \varphi(s))_\mathbb{V} ds \to 0, \quad \text{as } n \to \infty, \quad \forall \varphi \in \mathcal{L}^2(0,T;\mathbb{V}).$$
(3.52)

For $n \in \mathbb{N}^*$ and $\varphi \in \mathcal{L}^2(0,T;\mathbb{V})$, we use (3.18) to write, for a.e. $s \in (0,T)$,

$$(P_n g_n(s) - g(s), \varphi(s))$$
$$= (P_n g_n(s) - P_n g(s), \varphi(s)) + (P_n g(s) - g(s), \varphi(s))$$
$$= (g_n(s) - g(s), P_n \varphi(s)) + ((P_n - \mathbb{I})g(s), \varphi(s))$$
$$= (g_n(s) - g(s), \varphi(s)) + (g_n(s) - g(s), (P_n - \mathbb{I})\varphi(s))$$
$$+ ((P_n - \mathbb{I})g(s), \varphi(s)).$$

Using (3.2), (3.19), and (3.52), it follows that

$$\int_0^T (P_n g_n(s) - g(s), \varphi(s))_{\mathbb{V}} ds \to 0, \quad \text{as } n \to \infty, \quad \forall \varphi \in \mathcal{L}^2(0,T;\mathbb{V}). \tag{3.53}$$

The sequence $(Z_n^{g_n})_n$ is bounded in $\mathcal{L}^\infty(0,T;D_A)$, while $\left(\frac{dZ_n^{g_n}}{dt}\right)_n$ is a bounded sequence in $\mathcal{L}^\infty(0,T;\mathbb{H})$ (see (3.37) and (3.38)). By Theorem 3.18 ($\mathbb{B}_0 = D_A, \mathbb{B}_1 = \mathbb{V}, \mathbb{B}_2 = \mathbb{H}, q = r = \infty$) and Theorem 3.20, it follows that there exist $Z \in C(0,T;\mathbb{V}) \cap \mathcal{L}^\infty(0,T;D_A)$ and a subsequence of $(Z^{g_n})_n$ (for which we use the same notation) such that

$$Z_n^{g_n} \to Z \quad \text{in } C(0,T;\mathbb{V}) \text{ (and hence in } C(0,T;\mathbb{H})), \tag{3.54}$$

$$Z_n^{g_n} \overset{*}{\rightharpoonup} Z \quad \text{in } \mathcal{L}^\infty(0,T;D_A), \quad \text{as } n \to \infty.$$

By (3.17) and (3.12), it follows that

$$\||P_n|Z_n^{g_n}(s)|^2 Z_n^{g_n}(s) - |Z(s)|^2 Z(s)\|$$
$$\leq \frac{3K_L}{2}(\|Z_n^{g_n}(s)\|_{\mathbb{V}}^2 + \|Z(s)\|_{\mathbb{V}}^2)\|Z^{g_n}(s) - Z(s)\|$$
$$+ \|(\mathbb{I} - P_n)|Z(s)|^2 Z(s)\|$$
$$\leq \frac{3K_L}{2}(K_4 + \|Z(s)\|_{\mathbb{V}}^2)\|Z_n^{g_n}(s) - Z(s)\| + \|(\mathbb{I} - P_n)|Z(s)|^2 Z(s)\|.$$

Then, (3.54) and (3.19) yield

$$P_n|Z_n^{g_n}|^2 Z_n^{g_n} \to |Z|^2 Z \quad \text{in } \mathcal{L}^2(0,T;\mathbb{H}) \text{ as } n \to \infty. \tag{3.55}$$

By (3.32), we have

$$(Z_n^{g_n}(t), v) = (\gamma_{0n}, v) - \int_0^t i\langle AZ_n^{g_n}(s), v\rangle ds$$

$$- \int_0^t |y(s)|^2 (P_n|Z_n^{g_n}(s)|^2 Z_n^{g_n}(s), v) ds$$

$$+ \int_0^t (P_n \hat{g}_n(s), v) ds,$$

for all $t \in [0, T]$ and $v \in \mathbb{V}$ (where we used (3.18)). Taking $n \to \infty$ on both sides of the above equality, we get, by (3.54), (3.55), and (3.53),

$$(Z(t), v) = (\gamma_0, v) + \int_0^t i\langle AZ(s), v\rangle ds$$

$$+ \int_0^t (-|y(s)|^2 |Z(s)|^2 Z(s) + \hat{g}(s), v) ds,$$

for all $t \in [0, T]$ and $v \in \mathbb{V}$. By the uniqueness of the solution of (3.32), it follows that $Z^g = Z$. Finally, by Theorem 3.19, the whole sequence $(Z_n^{g_n})_n$ converges strongly to Z^g in $C(0, T; \mathbb{V})$ and weak* to Z^g in $\mathcal{L}^\infty(0, T; D_A)$. □

3.5. Linearized Finite-Dimensional Equations

In the first part of this section, $n \in \mathbb{N}^*$ and $g \in \mathcal{G}$ are fixed. In the finite-dimensional space $\mathbb{H}_n$, we write successively the following linearized stochastic equation and the corresponding pathwise equation.

Let $X_{n,0} = \gamma_{0n}$, and for each $k \geq 1$, if $X_{n,k-1} \in L^\infty(\Omega; C(0, T; \mathbb{V}_n))$, we consider $X_{n,k} \in L^\infty(\Omega; C(0, T; \mathbb{V}_n))$ to be the solution of the following stochastic linearized n-dimensional Schrödinger equation (written in $\mathbb{H}_n$):

$$X_{n,k}(t) = \gamma_{0n} + \int_0^t -iAX_{n,k}(s) - P_n|X_{n,k-1}(s)|^2 X_{n,k}(s) + P_n g(s) ds$$

$$+ i\beta \int_0^t X_{n,k}(s) dB^H(s), \quad (3.56)$$

for all $t \in [0,T]$, and a.e. $\omega \in \Omega$. We will derive in Theorem 3.12-(1), later in this section, that $X_{n,k}$, for $k \in \mathbb{N}^*$, "successively approximates" X_n, the solution of (3.26).

Let $Z_{n,0} = \gamma_{0n}$, and for each $k \geq 1$, if $Z_{n,k-1} \in C(0,T;\mathbb{V}_n)$, we consider successively $Z_{n,k} \in C(0,T;\mathbb{V}_n)$ to be the solution of the following linearized pathwise n-dimensional equation (written in $\mathbb{H}_n$):

$$Z_{n,k}(t) = \gamma_{0n} + \int_0^t \big(-iAZ_{n,k}(s) - |y(s)|^2 P_n |Z_{n,k-1}(s)|^2 Z_{n,k}(s) + P_n \hat{g}(s) \big) ds, \qquad (3.57)$$

for all $t \in [0,T]$. We will prove in Theorem 3.9 in the following that $Z_{n,k}$, for $k \in \mathbb{N}^*$, "successively approximates" Z_n, the solution of (3.34).

Later in this section and in Section 3.6, we will use the notations $X_{n,k}^g$ and $Z_{n,k}^g$ to point out the dependence on g.

We have, for each $t \in [0,T]$,

$$d\big(y(t) \cdot Z_{n,k}(t)\big) = y(t) dZ_{n,k}(t) + Z_{n,k}(t) dy(t) = dX_{n,k}(t).$$

Therefore,

$$X_{n,k}(t) = y(t) Z_{n,k}(t), \quad \text{for all } t \in [0,T] \text{ and a.e. } \omega \in \Omega, \qquad (3.58)$$

solves (3.56).

Theorem 3.9. *Fix $n \in \mathbb{N}^*$. For each $k \in \mathbb{N}^*$, there exists a unique solution $Z_{n,k} \in C(0,T;\mathbb{V}_n)$ of (3.57). The following estimates are valid:*

$$\sup_{t \in [0,T]} \|Z_{n,k}(t)\|_{\mathbb{V}}^2 \leq \mu_n e^T \left(\|\gamma_{0n}\|^2 + \frac{\rho^2 T L^4}{\pi^4 K_1} \right), \qquad (3.59)$$

$$\sup_{t \in [0,T]} \|Z_{n,k}(t)\|_{D_A}^2 \leq \mu_n^2 e^T \left(\|\gamma_{0n}\|^2 + \frac{\rho^2 T L^4}{\pi^4 K_1} \right), \qquad (3.60)$$

$$\sup_{t \in [0,T]} \|Z_{n,k}(t) - Z_n(t)\|_{\mathbb{V}}^2 \leq \mu_n \cdot \frac{3 K_2 K_4 K_L T}{2^{k-2}} \big(\|\gamma_0\|^2 + K_3 \big) e^{10 K_2 K_4 K_L T}, \qquad (3.61)$$

and

$$\sup_{t\in[0,T]} \|Z_{n,k}(t) - Z_n(t)\|_{D_A}^2$$
$$\leq \mu_n^2 \cdot \frac{3K_2K_4K_LT}{2^{k-2}}(\|\gamma_0\|^2 + K_3)e^{10K_2K_4K_LT}. \tag{3.62}$$

The following convergences hold:

$$\sup_{t\in[0,T]} \|Z_{n,k}(t) - Z_n(t)\|_{\mathbb{V}}^2 \to 0$$

and $\quad \sup_{t\in[0,T]} \|Z_{n,k}(t) - Z_n(t)\|_{D_A}^2 \to 0 \quad$ as $k \to \infty$. (3.63)

Proof. The result is obtained successively: If $Z_{n,k-1} \in C(0,T;\mathbb{V}_n)$, then there exists $Z_{n,k} \in C(0,T;\mathbb{V}_n)$ unique solution of (3.57), which is proved using standard methods for finite-dimensional equations involving mappings that depend linearly on the solution; (3.57) is an equation which depends linearly on unknown $Z_{n,k}$. By the structure of (3.57) and by (3.4), we obtain (similar to (3.35)) the estimate

$$\|Z_{n,k}(t)\|^2 \leq e^T\left(\|\gamma_{0n}\|^2 + \frac{\rho^2 TL^4}{\pi^4 K_1}\right).$$

By the relations between $\|\cdot\|$, $\|\cdot\|_{\mathbb{V}}$, and $\|\cdot\|_{D_A}$ in the spaces $\mathbb{V}_n$ and $\mathbb{D}_n$ and by $\mu_n = \max\{\mu_1, \ldots, \mu_n\}$, it holds that

$$\|Z_{n,k}(t)\|_{\mathbb{V}}^2 \leq \mu_n \|Z_{n,k}(t)\|^2 \quad \text{and} \quad \|Z_{n,k}(t)\|_{D_A}^2 \leq \mu_n^2 \|Z_{n,k}(t)\|^2. \tag{3.64}$$

Therefore, (3.59) and (3.60) hold.

The first convergence in (3.63) is proved as in Ref. [7]: We use (3.57), (3.33), and (3.4) to write

$$\|Z_{n,k}(t) - Z_n(t)\|^2 = -2\text{Re}\int_0^t |y(s)|^2(|Z_{n,k-1}(s)|^2 Z_{n,k}(s)$$
$$- |Z_n(s)|^2 Z_n(s), Z_{n,k}(s) - Z_n(s))\,ds,$$

for all $t \in [0,T]$. We define

$$\theta(t) = \exp\left(-10K_L \int_0^t |y(s)|^2 \|Z_n(s)\|_{\mathbb{V}}^2 ds\right), \quad \text{for all } t \in [0,T].$$

Then, we obtain for all $t \in [0, T]$,

$$\theta(t)\|Z_{n,k}(t) - Z_n(t)\|^2$$
$$= -2\operatorname{Re} \int_0^t \theta(s)|y(s)|^2 \bigl(|Z_{n,k-1}(s)|^2 Z_{n,k}(s)$$
$$\quad - |Z_n(s)|^2 Z_n(s), Z_{n,k}(s) - Z_n(s)\bigr) ds$$
$$\quad - 10K_L \int_0^t \theta(s)|y(s)|^2 \|Z_n(s)\|_V^2 \|Z_{n,k}(s) - Z_n(s)\|^2 ds. \quad (3.65)$$

We use inequality (A.3) for $c = Z_n(s), c_1 = Z_{n,k}(s), c_2 = Z_{n,k-1}(s)$, and (3.10) to derive

$$-2\operatorname{Re}\bigl(|Z_{n,k-1}(s)|^2 Z_{n,k}(s) - |Z_n(s)|^2 Z_n(s), Z_{n,k}(s) - Z_n(s)\bigr)$$
$$\leq \int_0^L \bigl(4|Z_n(s)|^2 |Z_{n,k}(s) - Z_n(s)|^2$$
$$\quad + 3|Z_n(s)|^2 |Z_{n,k-1}(s) - Z_n(s)|^2\bigr) dx$$
$$\leq 4K_L \|Z_n(s)\|_V^2 \|Z_{n,k}(s) - Z_n(s)\|^2$$
$$\quad + 3K_L \|Z_n(s)\|_V^2 \|Z_{n,k-1}(s) - Z_n(s)\|^2.$$

Therefore, by (3.10) and (3.65),

$$\theta(t)\|Z_{n,k}(t) - Z_n(t)\|^2$$
$$\leq +3K_L \int_0^t \theta(s)|y(s)|^2 \|Z_n(s)\|_V^2 \|Z_{n,k-1}(s) - Z_n(s)\|^2 ds$$
$$\quad - 6K_L \int_0^t \theta(s)|y(s)|^2 \|Z_n(s)\|_V^2 \|Z_{n,k}(s) - Z_n(s)\|^2 ds. \quad (3.66)$$

This implies

$$\int_0^T \theta(s)|y(s)|^2 \|Z_n(s)\|_V^2 \|Z_{n,k}(s) - Z_n(s)\|^2 ds$$
$$\leq \frac{1}{2} \int_0^T \theta(s)|y(s)|^2 \|Z_n(s)\|_V^2 \|Z_{n,k-1}(s) - Z_n(s)\|^2 ds.$$

Successively for each $k \in \mathbb{N}^*$, we get

$$\int_0^T \theta(s)|y(s)|^2 \|Z_n(s)\|_V^2 \|Z_{n,k}(s) - Z_n(s)\|^2 ds$$

$$\leq \frac{1}{2^k} \int_0^T \theta(s)|y(s)|^2 \|Z_n(s)\|_V^2 \|\gamma_{0n} - Z_n(s)\|^2 ds$$

$$\leq \frac{K_2}{2^{k-1}} \left(\|\gamma_{0n}\|^2 + \sup_{s\in[0,T]} \|Z_n(s)\|^2 \right) \int_0^T \|Z_n(s)\|_V^2 ds$$

$$\leq \frac{1}{2^{k-1}} K_2 K_4 T (\|\gamma_{0n}\|^2 + K_3) \leq \frac{1}{2^{k-1}} K_2 K_4 T (\|\gamma_0\|^2 + K_3).$$

Note that, above, we took into consideration (3.28), (3.35), and (3.36). Using the above result in (3.66), we obtain for each $k \in \mathbb{N}^*$,

$$\sup_{t\in[0,T]} \|Z_{n,k}(t) - Z_n(t)\|^2 \leq \frac{3K_2 K_4 K_L T}{2^{k-2}\theta(T)} (\|\gamma_0\|^2 + K_3)$$

$$\leq \frac{3K_2 K_4 K_L T}{2^{k-2}} (\|\gamma_0\|^2 + K_3) e^{10 K_2 K_4 K_L T}.$$

Then, by (3.64), it follows that (3.61) and (3.62) hold and (3.63) is valid. □

Theorem 3.10. *Arbitrarily set $k, n \in \mathbb{N}^*$. Let $g, g_j \in \mathcal{G}_n$, $j \in \mathbb{N}^*$ be such that $g_j \overset{*}{\rightharpoonup} g$ in $\mathcal{L}^\infty(0,T;\mathbb{V}_n)$ as $j \to \infty$. Then, the following convergences hold:*

(1) $Z_{n,k}^{g_j} \to Z_{n,k}^{g}$ in $C(0,T;\mathbb{V}_n)$ and $Z_{n,k}^{g_j} \overset{*}{\rightharpoonup} Z_{n,k}^{g}$ in $\mathcal{L}^\infty(0,T;\mathbb{D}_n)$ as $j \to \infty$;

(2) $Z_{n,n}^{g_j} \to Z_{n,n}^{g}$ in $C(0,T;\mathbb{V}_n)$ and $Z_{n,n}^{g_j} \overset{*}{\rightharpoonup} Z_{n,n}^{g}$ in $\mathcal{L}^\infty(0,T;\mathbb{D}_n)$ as $j \to \infty$.

Proof. (1) The result follows inductively with respect to $k \in \mathbb{N}^*$:

- First, $g_j \overset{*}{\rightharpoonup} g$ in $\mathcal{L}^\infty(0,T;\mathbb{V}_n)$ implies $Z_{n,1}^{g_j} \to Z_{n,1}^{g}$ in $C(0,T;\mathbb{V}_n)$ and $Z_{n,1}^{g_j} \overset{*}{\rightharpoonup} Z_{n,1}^{g}$ in $\mathcal{L}^\infty(0,T;\mathbb{D}_n)$ as $j \to \infty$.
- Second, $g_j \overset{*}{\rightharpoonup} g$ in $\mathcal{L}^\infty(0,T;\mathbb{V}_n)$ and $Z_{n,k-1}^{g_j} \to Z_{n,k-1}^{g}$ in $C(0,T;\mathbb{V}_n)$ imply $Z_{n,k}^{g_j} \to Z_{n,k}^{g}$ in $C(0,T;\mathbb{V}_n)$ and $Z_{n,k}^{g_j} \overset{*}{\rightharpoonup} Z_{n,k}^{g}$ in $\mathcal{L}^\infty(0,T;\mathbb{D}_n)$ as $j \to \infty$.

Adapting the proof of Theorem 3.7 and using that $Z^{g_j}_{n,k-1} \to Z^g_{n,k-1}$ in $C(0,T;\mathbb{V}_n)$ implies $|Z^{g_j}_{n,k-1}|^2 \to |Z^g_{n,k-1}|^2$ in $C(0,T;\mathbb{V}_n)$ as $j \to \infty$, we get the stated convergences.

(2) We consider $k = n$ and apply the results stated in (1). □

Theorem 3.11. *Let $n \in \mathbb{N}^*$ and $g, g_k \in \mathcal{G}_n, k \in \mathbb{N}^*$, be such that $g_k \overset{*}{\rightharpoonup} g$ in $\mathcal{L}^\infty(0,T;\mathbb{V}_n)$ as $k \to \infty$. Then, $Z^{g_k}_{n,k} \to Z^g_n$ in $C(0,T;\mathbb{V}_n)$ and $Z^{g_k}_{n,k} \overset{*}{\rightharpoonup} Z^g_n$ in $\mathcal{L}^\infty(0,T;\mathbb{D}_n)$ as $k \to \infty$.*

Proof. The right-hand sides of (3.61) and (3.62) do not depend on the control (only on ρ); therefore,

$$\sup_{t \in [0,T]} \|Z^{g_k}_{n,k}(t) - Z^{g_k}_n(t)\|^2_{\mathbb{V}} \to 0, \quad \|Z^{g_k}_{n,k}(t) - Z^{g_k}_n(t)\|^2_{D_A} \to 0 \quad \text{as } k \to \infty.$$

By Theorem 3.7, we have

$$\sup_{t \in [0,T]} \|Z^{g_k}_n(t) - Z^g_n(t)\|^2_{\mathbb{V}} \to 0 \quad \text{as } k \to \infty,$$

and $Z^{g_k}_n \overset{*}{\rightharpoonup} Z^g_n$ in $\mathcal{L}^\infty(0,T;\mathbb{D}_n)$ as $k \to \infty$. Then, it follows that $Z^{g_k}_{n,k} \to Z^g_n$ in $C(0,T;\mathbb{V}_n)$ and $Z^{g_k}_{n,k} \overset{*}{\rightharpoonup} Z^g_n$ in $\mathcal{L}^\infty(0,T;\mathbb{D}_n)$ as $k \to \infty$. □

Theorem 3.12.

(1) *Let $n \in \mathbb{N}^*$ and $g, g_k \in \mathcal{G}_n$, $k \in \mathbb{N}^*$, be such that $g_k \overset{*}{\rightharpoonup} g$ in $\mathcal{L}^\infty(0,T;\mathbb{V}_n)$ as $k \to \infty$. Then, for a.e. $\omega \in \Omega$, it holds that*

$$\sup_{t \in [0,T]} \|X^g_{n,k}(t) - X^g_n(t)\|^2_{\mathbb{V}} \to 0 \quad \text{as } k \to \infty, \tag{3.67}$$

and

$$\sup_{t \in [0,T]} \|X^{g_k}_{n,k}(t) - X^g_n(t)\|^2_{\mathbb{V}} \to 0 \quad \text{as } k \to \infty. \tag{3.68}$$

(2) *Let $g \in \mathcal{G}, g_n \in \mathcal{G}_n$, $n \in \mathbb{N}^*$, be such that $g_n \overset{*}{\rightharpoonup} g$ in $\mathcal{L}^\infty(0,T;\mathbb{V})$ as $n \to \infty$. Then, for a.e. $\omega \in \Omega$, it holds that*

$$\sup_{t \in [0,T]} \|X^g_{n,n}(t) - X^g(t)\|^2_{\mathbb{V}} \to 0 \quad \text{as } n \to \infty, \tag{3.69}$$

and

$$\sup_{t \in [0,T]} \|X^{g_n}_{n,n}(t) - X^g(t)\|^2_{\mathbb{V}} \to 0, \quad \text{as } n \to \infty. \tag{3.70}$$

Proof. (1) The convergence (3.67) follows by using (3.61), (3.45), and (3.58). Theorem 3.11 implies (3.68).

(2) We consider $k = n$ in the estimates (3.61) and (3.62). Since

$$\lim_{n\to\infty} \mu_n \cdot \frac{3K_2 K_4 K_L T}{2^{n-2}} (\|\gamma_0\|^2 + K_3) e^{10 K_2 K_4 K_L T} = \lim_{n\to\infty} \frac{n^2}{2^{n-1}} = 0,$$

and

$$\lim_{n\to\infty} \mu_n^2 \cdot \frac{3K_2 K_4 K_L T}{2^{n-2}} (\|\gamma_0\|^2 + K_3) e^{10 K_2 K_4 K_L T} = \lim_{n\to\infty} \frac{n^4}{2^{n-1}} = 0,$$

it follows by (3.61) and (3.62) (the right-hand sides of (3.61) and (3.62) do not depend explicitly on g or g_n) that

$$\sup_{t\in[0,T]} \|Z_{n,n}^g(t) - Z_n^g(t)\|_{\mathbb{V}}^2 \to 0, \quad \sup_{t\in[0,T]} \|Z_{n,n}^g(t) - Z_n^g(t)\|_{D_A}^2 \to 0,$$

and

$$\sup_{t\in[0,T]} \|Z_{n,n}^{g_n}(t) - Z_n^{g_n}(t)\|_{\mathbb{V}}^2 \to 0, \quad \sup_{t\in[0,T]} \|Z_{n,n}^{g_n}(t) - Z_n^{g_n}(t)\|_{D_A}^2 \to 0,$$

as $n \to \infty$. We use Theorem 3.4 to conclude that

$$\lim_{n\to\infty} \sup_{t\in[0,T]} \|Z_{n,n}^g(t) - Z^g(t)\|_{\mathbb{V}}^2$$

$$\leq 2 \lim_{n\to\infty} \sup_{t\in[0,T]} \|Z^g(t) - Z_n^g(t)\|_{\mathbb{V}}^2$$

$$+ 2 \lim_{n\to\infty} \sup_{t\in[0,T]} \|Z_{n,n}^g(t) - Z_n^g(t)\|_{\mathbb{V}}^2 = 0.$$

By Theorem 3.8, we have

$$\lim_{n\to\infty} \sup_{t\in[0,T]} \|Z_{n,n}^{g_n}(t) - Z^g(t)\|_{\mathbb{V}}^2$$

$$\leq 2 \lim_{n\to\infty} \sup_{t\in[0,T]} \|Z_{n,n}^{g_n}(t) - Z_n^{g_n}(t)\|_{\mathbb{V}}^2$$

$$+ 2 \lim_{n\to\infty} \sup_{t\in[0,T]} \|Z_n^{g_n}(t) - Z^g(t)\|_{\mathbb{V}}^2 = 0.$$

Taking into consideration (3.44) and (3.58), it follows that (3.69) and (3.70) hold. □

3.6. Optimal Control Problems

For $g \in \mathcal{G}$ (see (3.2)), we consider the cost function

$$J(g, X) = \mathbb{E} \int_0^T J_1(t, X(t))dt + \mathbb{E}J_2(X(T)) + \int_0^T J_3(t, g(t))dt,$$

such that J_1, J_2, and J_3 satisfy the following hypotheses:

($\mathcal{H}_1$) $J_1 : (0,T) \times \mathbb{V} \to \mathbb{R}_+$ is such that, for a.e. $t \in (0,T)$, the function $v \in \mathbb{V} \mapsto J_1(t,v)$ is continuous; for every $v \in \mathbb{V}$, the function $t \in (0,T) \mapsto J_1(t,v)$ is measurable (i.e., J_1 is a Carathéodory function); for each $r > 0$, there exists $a_r \in \mathcal{L}^1(0,T)$ such that $J_1(t,v) \leq a_r(t)$ for all $\|v\|_\mathbb{V} \leq r$ and a.e. $t \in (0,T)$.

($\mathcal{H}_2$) $J_2 : \mathbb{V} \to \mathbb{R}_+$ is a continuous function, and for each $r > 0$, there exists $b_r > 0$ such that $J_2(v) \leq b_r$ for all $\|v\|_\mathbb{V} \leq r$.

($\mathcal{H}_3$) $J_3 : (0,T) \times \mathbb{V} \to \mathbb{R}_+$ is such that: for a.e. $t \in (0,T)$, the function $z \in \mathbb{V} \mapsto J_3(t,z)$ is convex and continuous; for every $z \in \mathbb{V}$, the function $t \in (0,T) \mapsto J_3(t,z)$ is measurable; for each $r > 0$, there exists $c_r \in \mathcal{L}^1(0,T)$ such that $J_3(t,z) \leq c_r(t)$ for all $\|z\| \leq r$ and a.e. $t \in (0,T)$.

Consider the following optimal control problem:

$$(\mathcal{P}) \quad \begin{cases} J(g, X^g) \to \min, \\ g \in \mathcal{G}, \end{cases}$$

where X^g is the process given by (3.44) and solves (3.24).

Let $k, n \in \mathbb{N}^*$. We introduce the following finite-dimensional optimal control problems:

$$(\mathcal{P}_n) \quad \begin{cases} J(g, X_n^g) \to \min, \\ g \in \mathcal{G}_n, \end{cases}$$

where X_n^g is the process given by (3.45) and solves (3.25), where the control function is $g \in \mathcal{G}_n$;

$$(\hat{\mathcal{P}}_{n,k}) \quad \begin{cases} J(g, X_{n,k}^g) \to \min, \\ g \in \mathcal{G}_n, \end{cases}$$

where $X_{n,k}^g = yZ_{n,k}^g$ (see (3.58)) solves the finite-dimensional linearized equation (3.56) and $Z_{n,k}^g$ is the solution of (3.57);

$$(\hat{\mathcal{P}}_n) \quad \begin{cases} J(g, X_{n,n}^g) \to \min, \\ g \in \mathcal{G}_n, \end{cases}$$

where $X_{n,n}^g = yZ_{n,n}^g$, with $Z_{n,n}^g = Z_{n,k}^g\big|_{k=n}$ and $Z_{n,k}^g$ being the solution of (3.57).

Theorem 3.13. *Let $q \geq 1$. Each sequence $(g_m)_m$ in $\mathcal{G}$ admits a subsequence $(g_{m_j})_j$ having the property that there exists $g \in \mathcal{G}$ such that $g_{m_j} \stackrel{*}{\rightharpoonup} g$ in $\mathcal{L}^\infty(0,T;\mathbb{V})$,*

$$\int_0^T (g_{m_j}(s) - g(s), \varphi(s))_{\mathbb{V}} ds \to 0, \quad \text{as } j \to \infty, \quad \forall \, \varphi \in \mathcal{L}^q(0,T;\mathbb{V}), \tag{3.71}$$

and

$$\sup_{t \in [0,T]} \|X^{g_{m_j}}(t) - X^g(t)\|_{\mathbb{V}} \to 0, \quad \text{as } j \to \infty, \quad \text{for a.e. } \omega \in \Omega. \tag{3.72}$$

For each fixed $n, k \in \mathbb{N}^*$, the Galerkin approximations satisfy

$$\sup_{t \in [0,T]} \|X_n^{g_{m_j}}(t) - X_n^g(t)\|_{\mathbb{V}} \to 0, \quad \text{as } j \to \infty, \quad \text{for a.e. } \omega \in \Omega, \tag{3.73}$$

and for the linearized approximations, it holds that

$$\sup_{t \in [0,T]} \|X_{n,k}^{g_{m_j}}(t) - X_{n,k}^g(t)\|_{\mathbb{V}} \to 0, \quad \text{as } j \to \infty, \quad \text{for a.e. } \omega \in \Omega, \tag{3.74}$$

$$\sup_{t \in [0,T]} \|X_{n,n}^{g_{m_j}}(t) - X_{n,n}^g(t)\|_{\mathbb{V}} = 0, \quad \text{as } j \to \infty, \quad \text{for a.e. } \omega \in \Omega. \tag{3.75}$$

Proof. The sequence $(g_m)_m \subset \mathcal{G} \subset \mathcal{L}^\infty(0,T;\mathbb{V})$ is bounded. Then, by using Theorem 3.20 (recall that $\mathbb{V}$ is a separable Hilbert space), there exist a subsequence $(g_{m_j})_j$ and $g \in \mathcal{L}^\infty(0,T;\mathbb{V})$ such that

$$g_{m_j} \overset{*}{\rightharpoonup} g. \tag{3.76}$$

The limit in (3.76) is equivalent to

$$\int_0^T (g_{m_j}(s) - g(s), \varphi(s))_\mathbb{V} ds \to 0 \quad \text{as } j \to \infty, \quad \forall \varphi \in \mathcal{L}^1(0,T;\mathbb{V}).$$

By the properties of the weak* convergence (see Theorem 3.20), it follows that

$$\|g\|_{\mathcal{L}^\infty(0,T;\mathbb{V})} \leq \liminf_{j \to \infty} \|g_{m_j}\|_{\mathcal{L}^\infty(0,T;\mathbb{V})} \leq \rho.$$

Therefore, $g \in \mathcal{G}$. Since $\varphi \in \mathcal{L}^q(0,T;\mathbb{V}) \hookrightarrow \mathcal{L}^1(0,T;\mathbb{V})$, it follows by (3.76) that (3.71) holds.

For a.e. $\omega \in \Omega$, we have:

- by Theorem 3.6,

$$\sup_{t \in [0,T]} \|X^{g_{m_j}}(t) - X^g(t)\|_\mathbb{V} = \sup_{t \in [0,T]} |y(t)| \|Z^{g_{m_j}}(t) - Z^g(t)\|_\mathbb{V}$$

$$\leq \sqrt{K_2} \sup_{t \in [0,T]} \|Z^{g_{m_j}}(t) - Z^g(t)\|_\mathbb{V} \to 0;$$

- by Theorem 3.7,

$$\sup_{t \in [0,T]} \|X_n^{g_{m_j}}(t) - X_n^g(t)\|_\mathbb{V} \leq \sqrt{K_2} \sup_{t \in [0,T]} \|Z_n^{g_{m_j}}(t) - Z_n^g(t)\|_\mathbb{V} \to 0;$$

- by Theorem 3.10,

$$\sup_{t \in [0,T]} \|X_{n,k}^{g_{m_j}}(t) - X_{n,k}^g(t)\|_\mathbb{V}$$

$$\leq \sqrt{K_2} \sup_{t \in [0,T]} \|Z_{n,k}^{g_{m_j}}(t) - Z_{n,k}^g(t)\|_\mathbb{V} \to 0,$$

and

$$\sup_{t\in[0,T]} \|X_{n,n}^{g_{m_j}}(t) - X_{n,n}^{g}(t)\|_{\mathbb{V}}$$
$$\leq \sqrt{K_2} \sup_{t\in[0,T]} \|Z_{n,n}^{g_{m_j}}(t) - Z_{n,n}^{g}(t)\|_{\mathbb{V}} \to 0.$$

Hence, (3.72), (3.73), (3.74), and (3.75) hold. □

Theorem 3.14. *Let $n, k \in \mathbb{N}^*$. If $(\mathcal{H}_1)$, $(\mathcal{H}_2)$, and $(\mathcal{H}_3)$ are satisfied, then each of the control problems $(\mathcal{P})$, $(\mathcal{P}_n)$, $(\hat{\mathcal{P}}_{n,k})$, and $(\hat{\mathcal{P}}_n)$ admits at least one solution.*

Proof. The proof follows the standard steps for these types of problems: Let $(g_m)_m$ be a minimizing sequence in $\mathcal{G}$ for problem $(\mathcal{P})$, i.e.,

$$\lim_{m\to\infty} J(g_m, X^{g_m}) = \inf_{\tilde{g}\in\mathcal{G}} J(\tilde{g}, X^{\tilde{g}}).$$

Theorem 3.13 implies the existence of a subsequence $(g_{m_j})_j$ and $g \in \mathcal{G}$ such that (3.71) and (3.72) hold.

By the weak* convergence of $(g_{m_j})_j$ to g (see (3.71)), by $(\mathcal{H}_3)$, and by Theorem 3.21 (for $\mathbb{B} = \mathbb{V}$), we deduce the sequentially weak, lower semicontinuity property

$$\int_0^T J_3(t, g(t))dt \leq \liminf_{j\to\infty} \int_0^T J_3(t, g_{m_j}(t))dt. \qquad (3.77)$$

By (3.72), we have for a.e. $\omega \in \Omega$ that

$$\sup_{t\in[0,T]} \|X^{g_{m_j}}(t) - X^g(t)\|_{\mathbb{V}} + \|X^{g_{m_j}}(T) - X^g(T)\|_{\mathbb{V}} \to 0.$$

The continuity assumptions on J_1 and J_2 imply, for a.e. $\omega \in \Omega$,

$$J_1(t, X^{g_{m_j}}(t)) \to J_1(t, X^g(t)), \quad \text{a.e. } t \in (0, T)$$
$$J_2(X^{g_{m_j}}(T)) \to J_2(X^g(T)).$$

By Fatou's lemma, we get

$$\mathbb{E}\int_0^T J_1(t, X^g(t))dt + \mathbb{E}J_2(X^g(T))$$
$$\leq \liminf_{j\to\infty}\left(\mathbb{E}\int_0^T J_1(t, X^{g_{m_j}}(t))dt + \mathbb{E}J_2(X^{g_{m_j}}(T))\right). \quad (3.78)$$

Then, by (3.77) and (3.78), it follows that

$$J(g, X^g) \leq \liminf_{j\to\infty} J(g_{m_j}, X^{g_{m_j}}) = \inf_{\tilde{g}\in\mathcal{G}} J(\tilde{g}, X^{\tilde{g}}) \leq J(g, X^g),$$

and hence $g \in \mathcal{G}$ is a solution of $(\mathcal{P})$.

Fix $n, k \in \mathbb{N}^*$. Using (3.73), (3.74), and (3.75) and taking similar steps as above, it follows that each of the problems $(\mathcal{P}_n)$, $(\hat{\mathcal{P}}_{n,k})$, and $(\hat{\mathcal{P}}_n)$ admits at least one solution (where we use the strong convergence results from (3.73), (3.74), and (3.75)). $\square$

Further, we formulate the ε-optimality results.

Theorem 3.15. *Assume that* $(\mathcal{H}_1)$, $(\mathcal{H}_2)$, *and* $(\mathcal{H}_3)$ *hold. Let* $g \in \mathcal{G}$ *be a solution of* $(\mathcal{P})$, *and let* $\varepsilon > 0$. *Then, there exists a sequence* $(g_n)_n$ *such that* $g_n \in \mathcal{G}_n$, *for each* $n \in \mathbb{N}^*$, *and* $n_\varepsilon \in \mathbb{N}^*$ *such that, for all* $n \geq n_\varepsilon$, g_n *is a solution of* $(\mathcal{P}_n)$ *and*

$$|J(g_n, X_n^{g_n}) - J(g, X^g)| < \varepsilon, \quad 0 \leq J(g_n, X^{g_n}) - J(g, X^g) < \varepsilon,$$

i.e., g_n *is an* ε-*optimal solution for* $(\mathcal{P})$ *for* $n \geq n_\varepsilon$.

Proof. By (3.48), we have, for a.e. $\omega \in \Omega$,

$$\sup_{t\in[0,T]} \|X_n^g(t) - X^g(t)\|_\mathbb{V} + \|X_n^g(T) - X^g(T)\|_\mathbb{V} \to 0 \quad \text{as } n \to \infty.$$

The continuity assumptions on J_1, J_2, and J_3 yield

$$J_1(t, X_n^g(t)) \to J_1(t, X^g(t)) \quad \text{as } n \to \infty,$$
$$\text{for a.e. } t \in (0, T), \quad \text{a.e. } \omega \in \Omega,$$
$$J_2(X_n^g(T)) \to J_2(X^g(T)) \quad \text{as } n \to \infty, \quad \text{for a.e. } \omega \in \Omega,$$
$$J_3(t, P_n g(t)) \to J_3(t, g(t)) \quad \text{as } n \to \infty,$$
$$\text{a.e. } t \in (0, T) \text{ (where we use (3.20))}.$$

By $(\mathcal{H}_1)$, $(\mathcal{H}_2)$, (3.46), and (3.47) and by the dominated convergence theorem, it follows that

$$\mathbb{E}\int_0^T J_1(t, X_n^g(t))dt + \mathbb{E}J_2(X_n^g(T)) + \int_0^T J_3(t, P_n g(t))dt$$

$$\to \mathbb{E}\int_0^T J_1(t, X^g(t))dt + \mathbb{E}J_2(X^g(T)) + \int_0^T J_3(t, g(t))dt \quad (3.79)$$

as $n \to \infty$. Then, there exists an $m_\varepsilon \in \mathbb{N}^*$ such that

$$|J(P_n g, X_n^g) - J(g, X^g)| < \frac{\varepsilon}{2}, \quad \forall\, n \geq m_\varepsilon. \quad (3.80)$$

For each $n \in \mathbb{N}^*$, let $g_n \in \mathcal{G}_n$ be a solution of $(\mathcal{P}_n)$ (see Theorem 3.14). By Theorem 3.13, $(g_n)_n$ has a subsequence $(g_{n_j})_j$ which converges weak* to $\tilde{g} \in \mathcal{G}$ such that, for a.e. $\omega \in \Omega$,

$$\sup_{t \in [0,T]} \|X^{g_{n_j}}(t) - X^{\tilde{g}}(t)\|_V + \|X^{g_{n_j}}(T) - X^{\tilde{g}}(T)\|_V \to 0 \quad \text{as } j \to \infty.$$

Similar to (3.79), we get

$$\mathbb{E}\int_0^T J_1(t, X^{g_{n_j}}(t))dt + \mathbb{E}J_2(X^{g_{n_j}}(T))$$

$$\to \mathbb{E}\int_0^T J_1(t, X^{\tilde{g}}(t))dt + \mathbb{E}J_2(X^{\tilde{g}}(T)) \quad \text{as } j \to \infty. \quad (3.81)$$

By (3.44) and (3.45), we have, for each $t \in [0, T]$,

$$\|X_{n_j}^{g_{n_j}}(t) - X^{\tilde{g}}(t)\|_V \leq |y(t)| \cdot \|Z_{n_j}^{g_{n_j}}(t) - Z^{\tilde{g}}(t)\|.$$

Using Theorem 3.8, we get

$$\sup_{t \in [0,T]} \|X_{n_j}^{g_{n_j}}(t) - X^{\tilde{g}}(t)\|_V \to 0 \quad \text{as } j \to \infty.$$

Then, analogously to (3.79), we get

$$\mathbb{E}\int_0^T J_1(t, X_{n_j}^{g_{n_j}}(t))dt + \mathbb{E}J_2(X_{n_j}^{g_{n_j}}(T))$$

$$\to \mathbb{E}\int_0^T J_1(t, X^{\tilde{g}}(t))dt + \mathbb{E}J_2(X^{\tilde{g}}(T)).$$

By (3.81) and the above limit, there exist $j_\varepsilon \in \mathbb{N}^*$ and $j_\varepsilon \geq m_\varepsilon$ such that

$$|J(g_{n_j}, X^{g_{n_j}}) - J(g_{n_j}, X_{n_j}^{g_{n_j}})| < \frac{\varepsilon}{2}, \quad \forall\, j \geq j_\varepsilon. \tag{3.82}$$

Fix $j \geq j_\varepsilon$. Since $g_{n_j} \in \mathcal{G}_{n_j}$ is a solution of $(\mathcal{P}_{n_j})$, we have, by (3.80),

$$J(g_{n_j}, X_{n_j}^{g_{n_j}}) - J(g, X^g) \leq J(P_{n_j}g, X_{n_j}^g) - J(g, X^g) < \frac{\varepsilon}{2}.$$

$g \in \mathcal{G}$ is a solution of $(\mathcal{P})$, then by (3.82), we get

$$J(g, X^g) - J(g_{n_j}, X_{n_j}^{g_{n_j}}) \leq J(g_{n_j}, X^{g_{n_j}}) - J(g_{n_j}, X_{n_j}^{g_{n_j}}) < \frac{\varepsilon}{2}.$$

Hence,

$$|J(g_{n_j}, X_{n_j}^{g_{n_j}}) - J(g, X^g)| < \frac{\varepsilon}{2} < \varepsilon, \quad \forall\, j \geq j_\varepsilon.$$

Then, by (3.82) and the above inequality, we conclude that, for all $j \geq j_\varepsilon$,

$$0 \leq J(g_{n_j}, X^{g_{n_j}}) - J(g, X^g)$$
$$\leq |J(g_{n_j}, X^{g_{n_j}}) - J(g_{n_j}, X_{n_j}^{g_{n_j}})| + |J(g_{n_j}, X_{n_j}^{g_{n_j}}) - J(g, X^g)| < \varepsilon.$$

Therefore, the sequence $(g_{n_j})_j$ and $j_\varepsilon \in \mathbb{N}^*$ satisfy the statement of this theorem. □

Theorem 3.16. *Let $n \in \mathbb{N}^*$ and $\varepsilon > 0$. Consider $g_n \in \mathcal{G}_n$ to be a solution of $(\mathcal{P}_n)$. Then, there exist a sequence $(g_{n,k})_k$ in $\mathcal{G}_n$ and $k_\varepsilon \in \mathbb{N}^*$ such that, for all $k \geq k_\varepsilon$, $g_{n,k}$ is a solution of $(\hat{\mathcal{P}}_{n,k})$ and*

$$|J(g_{n,k}, X_{n,k}^{g_{n,k}}) - J(g_n, X_n^{g_n})| < \varepsilon, \quad 0 \leq J(g_{n,k}, X_n^{g_{n,k}}) - J(g_n, X_n^{g_n}) < \varepsilon,$$

i.e., $g_{n,k}$ is an ε-optimal solution for $(\mathcal{P}_n)$ for $k \geq k_\varepsilon$.

Proof. This result is proved analogously to Theorem 3.15 but in n-dimensional Hilbert spaces. One also uses Theorems 3.12, 3.13, and 3.14. □

Theorem 3.17. *Let $\varepsilon > 0$. Consider $g \in \mathcal{G}$ to be a solution of $(\mathcal{P})$. Then, there exists a sequence $(g_n)_n$ in $\mathcal{G}$ and $n_\varepsilon \in \mathbb{N}^*$ such that, for all $n \geq n_\varepsilon$, $g_n \in \mathcal{G}_n$ is a solution of $(\hat{\mathcal{P}}_n)$ and*

$$|J(g_n, X_{n,n}^{g_n}) - J(g, X^g)| < \varepsilon, \quad 0 \leq J(g_n, X^{g_n}) - J(g, X^g) < \varepsilon,$$

i.e., g_n is an ε-optimal solution for $(\mathcal{P})$ for $n \geq n_\varepsilon$.

Proof. This theorem is proved analogously to Theorem 3.15.

By Theorem 3.12(2), it holds that, for a.e. $\omega \in \Omega$,

$$\sup_{t \in [0,T]} \|X_{n,n}^g(t) - X^g(t)\|_{\mathbb{V}}^2 + \|X_{n,n}^g(T) - X^g(T)\|_{\mathbb{V}}^2 \to 0 \quad \text{as } n \to \infty.$$

The continuity assumptions on J_1, J_2, and J_3, as well as $(\mathcal{H}_1)$, $(\mathcal{H}_2)$, (3.59), and the dominated convergence theorem imply

$$\mathbb{E} \int_0^T J_1(t, X_{n,n}^g(t)) dt + \mathbb{E} J_2(X_{n,n}^g(T)) + \int_0^T J_3(t, P_n g(t)) dt$$

$$\to \mathbb{E} \int_0^T J_1(t, X^g(t)) dt + \mathbb{E} J_2(X^g(T)) + \int_0^T J_3(t, g(t)) dt \quad (3.83)$$

as $n \to \infty$. Then, there exists $m_\varepsilon \in \mathbb{N}^*$ such that

$$|J(P_n g, X_{n,n}^g) - J(g, X^g)| < \frac{\varepsilon}{2}, \quad \forall n \geq m_\varepsilon. \quad (3.84)$$

For each $n \in \mathbb{N}^*$, let $g_n \in \mathcal{G}_n$ be a solution of $(\hat{\mathcal{P}}_n)$ (by Theorem 3.14). Theorem 3.13 implies that there exists a subsequence $(g_{k_j})_j$ of $(g_n)_n$ converging weak* to $g \in \mathcal{G}$ and

$$\sup_{t \in [0,T]} \|X^{g_{k_j}}(t) - X^g(t)\| + \|X^{g_{k_j}}(T) - X^g(T)\| \to 0$$

as $j \to \infty$, for a.e. $\omega \in \Omega$.

Hence, as in (3.83), we obtain

$$E \int_0^T J_1(t, X^{g_{k_j}}(t)) dt + \mathbb{E} J_2(X^{g_{k_j}}(T))$$

$$\to \mathbb{E} \int_0^T J_1(t, X^g(t)) dt + \mathbb{E} J_2(X^g(T)) \quad (3.85)$$

as $j \to \infty$. Using Theorem 3.12, we get

$$\sup_{t \in [0,T]} \|X_{k_j,k_j}^{g_{k_j}}(t) - X^g(t)\| \to 0, \quad \text{as } j \to \infty, \quad \text{for a.e. } \omega \in \Omega.$$

Then, as in (3.83), we derive

$$\mathbb{E}\int_0^T J_1(t, X_{k_j,k_j}^{g_{k_j}}(t))dt + \mathbb{E}J_2(X_{k_j,k_j}^{g_{k_j}}(T))$$

$$\to \mathbb{E}\int_0^T J_1(t, X^g(t))dt + \mathbb{E}J_2(X^g(T)) \quad \text{as } j \to \infty.$$

By (3.85) and by the above convergence, it follows that there exist $j_\varepsilon \in \mathbb{N}^*$ and $j_\varepsilon \geq m_\varepsilon$ such that

$$|J(g_{k_j}, X^{g_{k_j}}) - J(g_{k_j}, X_{k_j,k_j}^{g_{k_j}})| < \frac{\varepsilon}{2}, \quad \forall j \geq j_\varepsilon. \tag{3.86}$$

Fix $j \geq j_\varepsilon$. Recall that $g_{k_j} \in \mathcal{G}_{k_j}$ is a solution of $(\hat{\mathcal{P}}_{k_j})$. Then, by (3.84), we have

$$J(g_{k_j}, X_{k_j,k_j}^{g_{k_j}}) - J(g, X^g) \leq J(P_{k_j}g, X_{k_j,k_j}^g) - J(g, X^g) < \frac{\varepsilon}{2}.$$

Using that $g \in \mathcal{G}$ is a solution of $(\mathcal{P})$, by (3.86), we write

$$J(g, X^g) - J(g_{k_j}, X_{k_j,k_j}^{g_{k_j}}) \leq J(g_{k_j}, X^{g_{k_j}}) - J(g_{k_j}, X_{k_j,k_j}^{g_{k_j}}) < \frac{\varepsilon}{2}.$$

These inequalities yield

$$|J(g_{k_j}, X_{k_j,k_j}^{g_{k_j}}) - J(g, X^g)| < \frac{\varepsilon}{2} < \varepsilon, \quad \forall j \geq j_\varepsilon.$$

Finally, by (3.86) and the above inequality, we obtain, for all $j \geq j_\varepsilon$,

$$0 \leq J(g_{k_j}, X^{g_{k_j}}) - J(g, X^g)$$

$$\leq |J(g_{k_j}, X^{g_{k_j}}) - J(g_{k_j}, X_{k_j,k_j}^{g_{k_j}})| + |J(g_{k_j}, X_{k_j,k_j}^{g_{k_j}}) - J(g, X^g)| < \varepsilon.$$

The statement of this theorem is satisfied by the sequence $(g_{k_j})_j$ and $j_\varepsilon \in \mathbb{N}^*$. □

Example 3.1.

(1) Let $x_1 \in \mathcal{L}^2(\Omega; \mathcal{L}^2(0,T;\mathbb{V}))$, $x_2 \in \mathcal{L}^2(\Omega;\mathbb{V})$, $x_3 \in \mathcal{L}^2(0,T;\mathbb{V})$ be given:

$$J(g, X^g) = \mathbb{E}\int_0^T \|X^g(t) - x_1(t)\|_\mathbb{V}^2 dt + \mathbb{E}\|X^g(T) - x_2\|_\mathbb{V}^2$$
$$+ \int_0^T \|g(t) - x_3(t)\|_\mathbb{V}^2 dt.$$

(2) Let $p, q > 1$, $x_1 \in \mathcal{L}^p(\Omega; \mathcal{L}^p(0,T;\mathbb{V}))$, and $x_3 \in \mathcal{L}^q(0,T;\mathbb{V})$ be given:

$$J(g, X^g) = \mathbb{E}\int_0^T \|X^g(t) - x_1(t)\|_\mathbb{V}^p dt + \int_0^T \|g(t) - x_3(t)\|_\mathbb{V}^q dt.$$

(3) We mention an example including the norm in D_A: Consider

$$J(g, X^g) = \mathbb{E}\int_0^T \|X^g(t) - x_1(t)\|_{D_A}^2 dt,$$

where $x_1 \in \mathcal{L}^2(\Omega; \mathcal{L}^2(0,T;D_A))$ is given. Using the weak* convergences proved in the theorems from Sections 3.4 and 3.5, we can derive the existence of at least one solution for each of the control problems $(\mathcal{P})$, $(\mathcal{P}_n)$, $(\hat{\mathcal{P}}_{n,k})$, and $(\hat{\mathcal{P}}_n)$. But in order to obtain the ε-optimality properties, we need further assumptions.

Appendix A.

A.1. Results from Functional Analysis

Theorem 3.18 ([6, Theorem II.5.16]). *Let $\mathbb{B}_0 \subset \mathbb{B}_1 \subset \mathbb{B}_2$ be Banach spaces such that the embedding $\mathbb{B}_1 \hookrightarrow \mathbb{B}_2$ is continuous and the embedding $\mathbb{B}_0 \overset{c}{\hookrightarrow} \mathbb{B}_1$ is compact embedding. Let $q, r \in [1,\infty]$, and define*

$$W_{q,r} = \left\{ v \in \mathcal{L}^q(0,T;\mathbb{B}_0), \frac{dv}{dt} \in \mathcal{L}^r(0,T;\mathbb{B}_2) \right\}.$$

(1) If $q < \infty$, the embedding $W_{q,r} \hookrightarrow \mathcal{L}^q(0,T;\mathbb{B}_1)$ is compact.
(2) If $q = \infty$ and $r > 1$, the embedding $W_{q,r} \hookrightarrow C(0,T;\mathbb{B}_1)$ is compact.

Theorem 3.19. *Let $(x_n)_n$ be a sequence in a Banach space $\mathbb{B}$, and let $(y_n)_n$ be a sequence in $\mathbb{B}^*$. Then, the following assertions hold:*

(1) *If every subsequence of $(x_n)_n$ has a subsequence which converges strongly to the same limit $x \in \mathbb{B}$, then the original sequence $(x_n)_n$ converges strongly to x.*
(2) *Let $x \in \mathbb{B}$. If every subsequence of $(x_n)_n$ has, in turn, a subsequence which converges weakly to x, then the original sequence $(x_n)_n$ converges weakly to x.*
(3) *Let $y \in \mathbb{B}^*$. If every subsequence of $(y_n)_n$ has, in turn, a subsequence which converges weak* to y, then the original sequence $(y_n)_n$ converges weak* to y.*

Proof. For (1) and (2), we use the convergence principles from [23, Proposition 10.13]. We prove statement (3) by contradiction, similar to the proof of Proposition 10.13-(2) in Ref. [23]. □

Theorem 3.20. *We assume that $(\mathbb{B}, (\cdot,\cdot)_\mathbb{B})$ is a separable Hilbert space. Let $(y_n)_n$ be a bounded sequence in $\mathcal{L}^\infty(0,T;\mathbb{B})$. Then, there exist $y \in \mathcal{L}^\infty(0,T;\mathbb{B})$ and a subsequence $(y_{n_j})_j$ that converges weak* to y, i.e.,*

$$\int_0^T (y_{n_j}(t) - y(t), \phi(t))_\mathbb{B} dt \to 0, \quad \text{as } j \to \infty, \quad \forall \phi \in \mathcal{L}^1(0,T;\mathbb{B}).$$

Moreover,

$$\|y\|_{\mathcal{L}^\infty(0,T;\mathbb{B})} \leq \liminf_{j\to\infty} \|y_{n_j}\|_{\mathcal{L}^\infty(0,T;\mathbb{B})}.$$

Proof. Recall $\mathbb{B} \cong \mathbb{B}^*$. We use Problems 23.12(d) and 23.12(e) in Ref. [24] and the property

$$\left(\mathcal{L}^1(0,T;\mathbb{B})\right)^* \cong \mathcal{L}^\infty(0,T;\mathbb{B}).$$

For the inequality from the statement of the theorem, we use Ref. [24, Proposition 21.26(b)]. For duality results in Bochner spaces, we also refer to Ref. [12, Theorem 1 and Corollary 4 in Chapter IV]. □

One can easily adapt the proof of Theorem 10.16 from Ref. [22, see pp. 349–351] (Tonelli-type theorem) for the case of Hilbert space valued functions to get the following result.

Theorem 3.21. *Let $q > 1$, and let $\mathbb{B}$ be a Hilbert space. For functions $u : (0,T) \to \mathbb{B}$, define*

$$F(u) = \int_0^T f(t, u(t))dt,$$

where $f : (0,T) \times \mathbb{B} \to \mathbb{R}_+$ is such that:

- *for a.e. $t \in (0,T)$, the function $z \in \mathbb{B} \mapsto f(t,z)$ is convex and continuous;*
- *for all $z \in \mathbb{B}$, the function $t \in (0,T) \mapsto f(t,z)$ is measurable;*
- *$\int_0^T f(t, u(t))dt < \infty$ for each $u \in \mathcal{L}^q(0,T;\mathbb{B})$.*

Then, the function F is sequentially weakly lower semicontinuous on $\mathcal{L}^q(0,T;\mathbb{B})$, i.e., if $u_n \rightharpoonup u$ in $\mathcal{L}^q(0,T;\mathbb{B})$, then it follows that

$$F(u) \leq \liminf_{n \to \infty} F(u_n).$$

A.2. Useful Inequalities

For all $c_1, c_2 \in \mathbb{C}$, it is easy to prove that

$$\operatorname{Re}\left(|c_1|^2 c_1 - |c_2|^2 c_2\right)(\bar{c}_1 - \bar{c}_2) \geq 0 \tag{A.1}$$

and

$$\left||c_1|^2 c_1 - |c_2|^2 c_2\right| \leq \frac{3}{2}\left(|c_1|^2 + |c_2|^2\right)|c_1 - c_2|. \tag{A.2}$$

Lemma 3.2. *For all $c, c_1, c_2 \in \mathbb{C}$, it holds that*

$$-2\operatorname{Re}\left((|c_2|^2 c_1 - |c|^2 c)(\bar{c}_1 - \bar{c})\right) \leq 4|c|^2 |c_1 - c|^2 + 3|c|^2 |c_2 - c|^2. \tag{A.3}$$

Proof. Observe that

$$-2\operatorname{Re}\left((|c_2|^2 c_1 - |c|^2 c)(\bar{c}_1 - \bar{c})\right)$$
$$= -2|c_2|^2 |c_1 - c|^2 - 2\operatorname{Re}\left((|c_2|^2 - |c|^2)c(\bar{c}_1 - \bar{c})\right).$$

First, we write

$$-2|c_2|^2|c_1-c|^2 = -2\Big(|c_2-c|^2+|c|^2+2\mathrm{Re}\left((c_2-c)\bar{c}\right)\Big)|c_1-c|^2$$
$$\leq -2|c_2-c|^2|c_1-c|^2 - 2|c|^2|c_1-c|^2 + 4|c_2-c||c||c_1-c|^2$$
$$\leq -|c_2-c|^2|c_1-c|^2 + 2|c|^2|c_1-c|^2.$$

Second, we compute

$$-2\mathrm{Re}\left((|c_2|^2-|c|^2)c(\bar{c}_1-\bar{c})\right) \leq 2\big||c_2|^2-|c|^2\big||c||c_1-c|$$
$$\leq 2|c_2-c|\big(|c_2-c|+2|c|\big)|c||c_1-c|$$
$$= 2|c_2-c|^2|c||c_1-c| + 4|c_2-c||c|^2|c_1-c|$$
$$\leq |c_2-c|^2|c_1-c|^2 + 2|c|^2|c_1-c|^2 + 3|c|^2|c_2-c|^2.$$

Then, we easily get (A.3). □

References

[1] N. W. Ashcroft and N. D. Mermin, *Solid State Physics*. Holt, Rinehart and Winstor, New York (1976).

[2] V. Barbu, M. Röckner, and D. Zhang, Stochastic nonlinear Schrödinger equations with linear multiplicative noise: Rescaling approach. *Journal of Nonlinear Science*. **24**, 383–409 (2014).

[3] V. Barbu, M. Röckner, and D. Zhang, Optimal bilinear control of nonlinear stochastic Schrödinger equations driven by linear multiplicative noise. *Annals of Probability*. **46**, 1957–1999 (2018).

[4] F. Biagini, Y. Hu, B. Øksendal, and T. Zhang, *Stochastic Calculus for Fractional Brownian Motion and Applications*. Springer-Verlag, London (2008).

[5] M. Blencowe, Quantum electromechanical systems. *Physics Reports*. **39**(5), 159–222 (2004).

[6] F. Boyer and P. Fabrie, *Mathematical Tools for the Study of the Incompressible Navier-Stokes Equations and Related Models*. Springer-Verlag, New York (2013).

[7] B. E. Breckner and H. Lisei, Approximation of the solution of a stochastic Ginzburg-Landau equation. *Studia Universitatis Babeş-Bolyai Mathematica*. **66**, 307–319 (2021).

[8] B. E. Breckner, H. Lisei, and G. I. Şimon, Optimal control results for a class of stochastic Schrödinger equations. *Applied Mathematics and Computation.* **407**(126310), 1–17 (2021).

[9] R. Dautray and J.-L. Lions, *Mathematical Analysis and Numerical Methods for Science and Technology. Volume 5: Evolution Problems I.* Springer Verlag, Berlin (2000).

[10] R. Dautray and J.-L. Lions, *Mathematical Analysis and Numerical Methods for Science and Technology. Volume 2: Evolution Problems I.* Springer Verlag, Berlin (1988).

[11] A. De Bouard and A. Debussche, A stochastic nonlinear Schrödinger equation. *Communications in Mathematical Physics.* **205**, 161–181 (1999).

[12] J. Diestel and J. J. Uhl Jr., *Vector Measures.* American Mathematical Society, Providence (1977).

[13] D. E. Edmunds and W. D. Evans, *Spectral Theory and Differential Operators*, 2nd edn. Oxford University Press, Oxford (2018).

[14] H. Gajewski, Über Näherungsverfahren zur Lösung der nichtlinearen Schrödinger-Gleichung. *Mathematische Nachrichten.* **85**, 283–302 (1978).

[15] W. Grecksch and H. Lisei, Stochastic nonlinear equations of Schrödinger type. *Stochastic Analysis and Applications.* **29**, 631–651 (2011).

[16] W. Grecksch and H. Lisei, Linear approximation of nonlinear Schrödinger equations driven by cylindrical wiener processes. *Discrete and Continuous Dynamical Systems - Series B.* **21**, 3095–3114 (2016).

[17] W. Grecksch and H. Lisei, Stochastic Schrödinger equations. In W. Grecksch and H. Lisei (eds.), *Infinite Dimensional and Finite Dimensional Stochastic Equations and Applications in Physics.* World Scientific Publishing, Singapore (2020), pp. 115–160.

[18] W. Grecksch, H. Lisei, and B. E. Breckner, Optimal control for a nonlinear Schrödinger problem perturbed by multiplicative fractional noise. *Optimization.* **73**(11), 3411–3435 (2024).

[19] D. Keller and H. Lisei, Variational solution of stochastic Schrödinger equations with power-type nonlinearity. *Stochastic Analysis and Applications.* **33**, 653–672 (2015).

[20] P. I. Kelley, Self-focusing of Optical beams. *Physical Review Letters.* **15**, 1005–1008 (1965).

[21] H. Lisei and D. Keller, A stochastic nonlinear Schrödinger problem in variational formulation. *Nonlinear Differential Equations and Applications (NoDEA).* **23**, 1–27 (2016). Article 22.

[22] M. Renardy and R. C. Rogers, *An Introduction to Partial Differential Equations*. Texts in Applied Mathematics, Vol. 13. Springer-Verlag, New York (2004).
[23] E. Zeidler, *Nonlinear Functional Analysis and Its Applications. I: Fixed Point Theorems*. Springer-Verlag, New York (1986).
[24] E. Zeidler, *Nonlinear Functional Analysis and Its Applications. II/A: Linear Monotone Operators*. Springer-Verlag, New York (1990).
[25] E. Zeidler, *Nonlinear Functional Analysis and Its Applications. II/B: Nonlinear Monotone Operators*. Springer-Verlag, New York (1990).

© 2025 World Scientific Publishing Company
https://doi.org/10.1142/9789819802104_0004

Chapter 4

Calibration of Non-Semimartingale Models: An Adjoint Approach

Christian Bender[*] and Matthias Thiel[†]

*Department of Mathematics, Saarland University,
Postfach 151150, D-66041 Saarbrücken, Germany*
[*]*bender@math.uni-saarland.de*
[†]*thiel@math.uni-saarland.de*

We design and analyze a Monte Carlo algorithm for calibrating a financial model, in which some quantities (e.g., volatility) are represented in terms of a stochastic differential equation driven by a continuous p-variation process for $p \in (1, 2)$ (e.g., a fractional Brownian motion with the Hurst parameter larger than one half). The p-variation process can be correlated to the Brownian motion, which drives the stock prices, in order to capture the so-called leverage effect. The key tool is an adjoint gradient representation via a new type of anticipating backward stochastic differential equation, which is formulated in terms of the Russo–Vallois forward integral. We provide rates of convergence for an Euler approximation of this adjoint equation. Finally, the results are illustrated by a case study calibrating a fractional Heston model to market data.

4.1. Introduction

Stochastic volatility models enjoy great popularity in financial engineering since they are able to capture several features observed in market data, such as the implied volatility smile and the leverage effect (i.e., negative correlation between asset prices and volatility);

see Ref. [1] for a discussion of stylized facts in stock returns. In classical volatility models of the 20th century [2–5], asset price and volatility are governed by stochastic differential equations (SDEs) driven by correlated Brownian motions. Motivated by the phenomenon of volatility persistence (where large absolute changes in stock returns tend to be followed by large absolute changes), continuous-time models with volatility processes driven by fractional Brownian motion with the Hurst parameter larger than one half have been suggested [6–10] — exploiting the long memory effect of fractional Brownian motion for this range of the Hurst parameter. More recently, the smile expansions in Ref. [11] led the authors of Ref. [12] to introduce rough volatility models, which correspond to fractional Brownian motion with the Hurst parameter smaller than one half; see also Refs. [13–15].

In this chapter, we extend the Monte Carlo algorithm of [16] for calibrating stochastic volatility models driven by Brownian motions to a wide class of models, including fractional volatility models with the Hurst parameter $H > 1/2$ (and hence in the long memory regime). More generally, we consider models consisting of two systems of SDEs. The first one is driven by a continuous p-variation process for some $p \in (1, 2)$, e.g., fractional Brownian motion with the Hurst parameter larger than one half. This SDE is interpreted in the sense of the Young integration [17] and may be used to model non-tradable quantities, such as the volatilities of stocks or a short rate process governing the term structure of interest rates. The second SDE system is driven by a Brownian motion and is interpreted in the classical Itô sense. Its coefficients depend on the solution of the first system of SDEs, and we may think of the solution of the second system as the prices of tradable assets in the market. We assume that the model is not fully specified in the sense that the two systems of SDEs depend on a parameter vector. The objective is to minimize the quadratic error between the model prices and observed market prices for some liquidly traded options over the parameter vector. Borrowing ideas from Ref. [16], we design a gradient-based adjoint Monte Carlo algorithm for tackling this problem. However, in contrast to Ref. [16], who first discretize the optimization problem through sample average approximation [18], we follow the optimize-then-discretize approach and study the optimization problem in continuous time.

The chapter is organized as follows. In Section 4.2, we discuss the main results. After setting the model dynamics and the optimization problem in continuous time in Section 4.2.1, the gradient of the cost functional with respect to the parameter vector is studied in Section 4.2.2. The main result (Theorem 4.3) is a new adjoint representation of the gradient in terms of an anticipating backward stochastic differential equation (BSDE), which is jointly driven by the p-variation process and the Brownian motion. In order to formulate this equation in a proper way, we make use of the forward integral of Russo and Vallois [19], which encompasses the Itô integral and the Young integral as special cases and, at the same time, extends the Itô integral to non-adapted integrands. The key advantage of the adjoint equation is that its dimension is independent of the number of parameters, while the Fréchet derivative of the original SDE system with respect to the parameter vector solves a linear SDE (sometimes called the sensitivity equation; see Ref. [16]) whose dimension increases linearly in the number of parameters. Hence, algorithms based on the adjoint equation are expected to be more efficient than those based on the sensitivity equation if the dimension of the parameter vector is large. This general advantage of adjoint techniques has been found to be useful in various applications; see, e.g., Refs. [20–22].

Section 4.2.3 is devoted to the Euler discretization of the original SDE system, its sensitivity equation, and the adjoint anticipating BSDE. We provide rates of convergence for the error measured in the supremum norm in time and the L^p-norm in the sample paths (Theorem 4.4). The key technical difficulty is to control the growth of the Euler scheme for the pathwise Young differential equations. Compared to the literature on Young differential equations (e.g., Ref. [23]), we have to keep track of the dependence of the constants on the sample paths. To this end, we adapt the greedy sequence technique [24,25] to Euler partitions in a suitable way to come up with a variant of Gronwall's lemma (Lemma 4.9), which is tailor-made for our purposes. The results on the Euler schemes are then applied to estimate the error of approximations to the cost functional and to the adjoint gradient representation.

Section 4.3 provides some background information on Young integration and Russo–Vallois forward integration, which is required for the proofs of the main results in Section 4.4. We sketch the proofs of all main results, emphasizing the key ideas rather than providing all the technical details. At times, we focus on simplified versions of the equations which already contain the key difficulties. Detailed proofs of all the results in full generality can be found in the second author's PhD thesis [26].

Finally, in Section 4.5, we present a Monte Carlo algorithm for model calibration based on the discretized adjoint gradient representation, replacing all expectations with empirical means over simulated sample paths. The algorithm is then applied to calibrate a fractional version of Heston's model to call option price data on the EUROSTOXX 50. Our case study suggests that a Hurst parameter of about $H = 0.65$ yields the best fit to the data. Additional numerical experiments illustrate the rates of convergence established in Section 4.2.3.

4.2. Discussion of the Main Results

4.2.1. *The model dynamics and the cost functional*

In this section, we introduce the general setting consisting of two parameter-dependent systems of SDEs. The first SDE is driven by a multivariate stochastic process $(w_t)_{t \in [0,T]}$ which has continuous paths of finite p-variation for some $p \in (1, 2)$, which includes, e.g., a fractional Brownian motion with Hurst parameter $H \in (1/2, 1)$. We here recall that a *fractional Brownian motion* $(B_t^H)_{t \in [0,T]}$ with Hurst parameter $H \in (0, 1)$ is a centered Gaussian process with covariance structure

$$\mathrm{E}[B_t^H B_s^H] = \frac{1}{2}\left(|t|^{2H} + |s|^{2H} - |t-s|^{2H}\right).$$

In financial applications, this first SDE may model several factors which are not directly tradable and storable, such as the volatility of primary assets or the short rate of a money market account. The second SDE is driven by a multi-dimensional Brownian motion and may be considered to represent, e.g., the price processes of the primary assets traded in the market.

We fix a time horizon $T > 0$ and positive integers $n_1, m_1, n_2, m_2, d \in \mathbb{N} = \{1, 2, \dots\}$. Let $(\Omega, \mathcal{F}, \mathbb{F}, P)$ be a filtered probability space (satisfying the usual conditions) carrying an m_1-dimensional stochastic process $(w_t)_{t\in[0,T]}$, whose paths are almost surely continuous and have finite p-variation for $p \in (1,2)$, and an m_2-dimensional standard Brownian motion $(B_t)_{t\in[0,T]}$, both adapted to the filtration $\mathbb{F} = (\mathcal{F}_t)_{t\in[0,T]}$ but possibly dependent. The dependence between the two driving processes is crucial to capture, e.g., the leverage effect between stock prices and volatility [27].

Furthermore, let $\mathcal{U}$ be an open, convex, and bounded subset of $\mathbb{R}^d$, which represents the parameter set. We consider the parameter-dependent systems of SDEs

$$\xi_t^u = \xi_0(u) + \int_0^t b(r, \xi_r^u, u)\, dr + \sum_{j=1}^{m_1} \int_0^t \sigma^j(r, \xi_r^u, u)\, dw_r^j, \qquad (4.1)$$

$$x_t^u = x_0(u) + \int_0^t \hat{b}(r, \xi_r^u, x_r^u, u)\, dr + \sum_{j=1}^{m_2} \int_0^t \hat{\sigma}^j(r, \xi_r^u, x_r^u, u)\, dB_r^j,$$
$$(4.2)$$

where $\xi_0 : \mathcal{U} \to \mathbb{R}^{n_1}$, $b : [0,T] \times \mathbb{R}^{n_1} \times \mathcal{U} \to \mathbb{R}^{n_1}$, $\sigma = (\sigma^1, \ldots, \sigma^{m_1}) :$ $[0,T] \times \mathbb{R}^{n_1} \times \mathcal{U} \to \mathbb{R}^{n_1 \times m_1}$ and $x_0 : \mathcal{U} \to \mathbb{R}^{n_2}$, $\hat{b} : [0,T] \times \mathbb{R}^{n_1} \times \mathbb{R}^{n_2} \times \mathcal{U} \to \mathbb{R}^{n_2}$, $\hat{\sigma} = (\hat{\sigma}^1, \ldots, \hat{\sigma}^{m_2}) : [0,T] \times \mathbb{R}^{n_1} \times \mathbb{R}^{n_2} \times \mathcal{U} \to \mathbb{R}^{n_2 \times m_2}$. Conditions on the coefficient functions which guarantee the well-posedness of (4.1)–(4.2) for each parameter choice, $u \in \mathcal{U}$, will be imposed at the end of this section.

The stochastic integral with respect to w in (4.1) can be understood in the sense of Young integration [17, 28, 29], while the stochastic integral in (4.2) can be interpreted as a classical Itô integral (e.g., Ref. [30]). It is, however, more convenient to work with a unifying notion of stochastic integration, which generalizes the Young integral and the Itô integral, namely with the *forward integral* of Russo and Vallois [19, 31], which is defined as follows: Let the integrator $(X_t)_{t\in[0,T]}$ be a continuous process and the integrand $(Y_t)_{t\in[0,T]}$ be integrable in the time variable, i.e., $\int_0^T |Y_s|\, ds < \infty$, P-almost surely. For every $\varepsilon > 0$, the ε-*forward integral*

$$I^-(\varepsilon, Y, dX)(t) = \int_0^t Y_s \frac{X_{(s+\varepsilon)\wedge T} - X_s}{\varepsilon}\, ds$$

is then well defined and may be considered a regularized version of a forward Riemann sum (i.e., a Riemann sum with tag point at the left interval boundary point of the subintervals of the partition) of Y with respect to X. The *forward integral* of Y with respect to X is said to exist and is denoted by $(\int_0^t Y_s\, d^- X_s)_{t \in [0,T]}$ if

$$\lim_{\varepsilon \to 0} \sup_{t \in [0,T]} \left| \int_0^t Y_s\, d^- X_s - I^-(\varepsilon, Y, dX)(t) \right| = 0 \quad \text{in probability,} \tag{4.3}$$

i.e., it is the uniform limit in probability of the ε-forward integral. More background information on the Russo–Vallois forward integral and the Young integral will be provided in Section 4.3.

With this notation at hand, we may rewrite the system (4.1)–(4.2) in more compact form as

$$\mathcal{X}_t^u = \begin{pmatrix} \xi_0(u) \\ x_0(u) \end{pmatrix} + \int_0^t \begin{pmatrix} b\left(r, \mathcal{X}_r^{u,1:n_1}, u\right) \\ \hat{b}\left(r, \mathcal{X}_r^u, u\right) \end{pmatrix} dr$$

$$+ \sum_{j=1}^{m_1} \int_0^t \begin{pmatrix} \sigma^j\left(r, \mathcal{X}_r^{u,1:n_1}, u\right) \\ 0 \end{pmatrix} d^- w_r^j + \sum_{j=1}^{m_2} \begin{pmatrix} 0 \\ \hat{\sigma}^j\left(r, \mathcal{X}_r^u, u\right) \end{pmatrix} d^- B_r^j, \tag{4.4}$$

where $\mathcal{X}_t^u = (\mathcal{X}_t^{u,1}, \ldots, \mathcal{X}_t^{u,n_1+n_2})^\top$, $\mathcal{X}_t^{u,1:n_1} := (\mathcal{X}_t^{u,1}, \ldots, \mathcal{X}_t^{u,n_1})^\top = \xi_t^u$, and $\mathcal{X}_t^{u,n_1+1:n_2} := (\mathcal{X}_t^{u,n_1+1}, \ldots, \mathcal{X}_t^{u,n_2})^\top = x_t^u$.

We now introduce the cost functional

$$J(u) = \frac{1}{2} \sum_{\mu=1}^M \mathrm{E}\left[g_\mu(\mathcal{X}_{T_\mu}^u)\right]^2, \quad u \in \mathcal{U}, \tag{4.5}$$

where $0 < T_1 \leq \cdots \leq T_M = T$ is a finite sequence of time points. Each of the functions $g_\mu : \mathbb{R}^{n_1+n_2} \to \mathbb{R}$ may represent the difference between the payoff function of an option with maturity T_μ and its observed market price when calibrating a financial model.

We aim at minimizing the cost functional J by a gradient descent, for which we derive two gradient representations in the next section. Before doing so, we collect the assumptions required for the results of this chapter, as follows.

For the *p*-variation process w, we assume the following exponential moment bound:

(W) There exists $K > 0$ such that

$$E\left[e^{K\|w\|^2_{p,0,T}}\right] < \infty. \tag{4.6}$$

Here, $\|x\|_{p,s,t}$ denotes the p-variation norm of a path x over the interval $[s,t]$.

Remark 4.1. Let us recall some standard notation concerning p-variation functions for $p \geq 1$. We define $\mathcal{P}([s,t])$ as the set of all finite partitions of the interval $[s,t]$. For a partition $\Pi_k = (t_i)_{i=0,\ldots,k}$ of $[s,t]$ into k subintervals (i.e., $s = t_0 < t_1 < \cdots < t_k = t$), we call $|\Pi_k| = \max_{i=0,\ldots,k-1}\{|t_{i+1} - t_i|\}$ the mesh of the partition, and for $i = 0,\ldots,k-1$, we call $[t_i, t_{i+1}]$ a subinterval of the partition. If the number of subintervals of a partition does not need to be specified, we will omit the index k. For $1 \leq p < \infty$, the p-variation seminorm of a function $x : [s,t] \to \mathbb{R}^{n \times m}$ is then given by

$$|x|_{p,s,t} := \sup_{k \in \mathbb{N}, \Pi_k \in \mathcal{P}([s,t])} \left(\sum_{i=0}^{k-1} |x_{t_{i+1}} - x_{t_i}|^p\right)^{\frac{1}{p}},$$

where $|\cdot|$ denotes the Frobenius norm of a matrix. x is said to be *of finite p-variation* over the interval $[s,t]$ if $|x|_{p,s,t} < \infty$. We write $W^p([s,t], \mathbb{R}^{n \times m})$ for the space of finite p-variation functions over $[s,t]$, which, endowed with the p-variation norm $\|x\|_{p,s,t} := |x_s| + |x|_{p,s,t}$, becomes a Banach space. The subspace of continuous functions in $W^p([s,t], \mathbb{R}^{n \times m})$ will be denoted by $C^p([s,t], \mathbb{R}^{n \times m})$.

Remark 4.2. If $(w_t)_{t \in [0,T]}$ is a Gaussian process with continuous paths, which are of bounded p-variation for some $p \in (1,2)$, then the exponential moment bound in condition (W) is satisfied by Theorem 2.3 in Ref. [32].

Concerning the coefficients of the SDE (4.1), we suppose the following:

(H_1) Let $\xi_0 : \mathcal{U} \to \mathbb{R}^{n_1}$ be continuously differentiable such that ξ_0 and its Jacobian $D\xi_0$ are bounded.
(H_2) Let $b : [0,T] \times \mathbb{R}^{n_1} \times \mathcal{U} \to \mathbb{R}^{n_1}$ be a continuous function which satisfies that:

○ $b(t,\xi,u)$ is bounded and twice continuously differentiable with respect to ξ and u with bounded partial derivatives.

(H_3) Let $\sigma := (\sigma^1,\ldots,\sigma^{m_1}) : [0,T] \times \mathbb{R}^{n_1} \times \mathcal{U} \to \mathbb{R}^{n_1 \times m_1}$ be a continuous function which satisfies the following:

○ $\sigma(t,\xi,u)$ is bounded and three times continuously differentiable with respect to ξ and u with bounded partial derivatives;

○ $\sigma(t,\xi,u)$ and all its partial derivatives with respect to ξ and u up to order two are Hölder continuous in t with Hölder exponent $\beta \in [\frac{1}{2}, 1]$.

For the coefficients of SDE (4.2), we assume the following:

(B_1) Let $x_0 : \mathcal{U} \to \mathbb{R}^{n_2}$ be a continuously differentiable deterministic function such that x_0 and its Jacobian Dx_0 are bounded.

(B_2) Let $\hat{b} : [0,T] \times \mathbb{R}^{n_1} \times \mathbb{R}^{n_2} \times \mathcal{U} \to \mathbb{R}^{n_2}$ be a continuous function which satisfies that:

○ $\hat{b}(t,\xi,x,u)$ is continuously differentiable with respect to x, ξ, and u with bounded partial derivatives.

(B_3) Let $\hat{\sigma} = (\hat{\sigma}^1,\ldots,\hat{\sigma}^{m_2}) : [0,T] \times \mathbb{R}^{n_1} \times \mathbb{R}^{n_2} \times \mathcal{U} \to \mathbb{R}^{n_2 \times m_2}$ be a continuous function which satisfies that:

○ $\hat{\sigma}(t,\xi,x,u)$ is continuously differentiable with respect to x, ξ, and u with bounded partial derivatives.

Finally, on the cost functional, we impose the following condition:

(G) For every $\mu = 1,\ldots,M$, the function $g_\mu : \mathbb{R}^{n_1+n_2} \to \mathbb{R}$ is bounded and continuously differentiable with a bounded and Lipschitz continuous first derivative.

We refer to (W), (H_1)–(H_3), (B_1)–(B_3), and (G) as the *standing assumptions*. They are supposed to be in force for the rest of this chapter.

4.2.2. Gradient representations

In this section, we derive two representations for the gradient of the cost functional J introduced in (4.5). The first one is a simple consequence of the Fréchet differentiability of the state dynamics (4.4) with respect to the parameter vector u and the chain rule. The second one is an adjoint gradient representation in terms of an anticipating BSDE in the sense of forward integration and constitutes one of the main results of this chapter.

Formally differentiating the state dynamics $\mathcal{X}^u$ in (4.4) with respect to u (cp. Section 4.4.1) suggests that its Fréchet derivative $D\mathcal{X}^u$ is an $\mathbb{R}^{(n_1+n_2)\times d}$-valued stochastic process, which solves the linear matrix SDE

$$\mathcal{Y}^u_t = \begin{pmatrix} D\xi_0(u) \\ Dx_0(u) \end{pmatrix}$$

$$+ \int_0^t \begin{pmatrix} b_\xi\left(r, \mathcal{X}^{u,1:n_1}_r, u\right) & 0 \\ \hat{b}_\xi\left(r, \mathcal{X}^u_r, u\right) & \hat{b}_x\left(r, \mathcal{X}^u_r, u\right) \end{pmatrix} \mathcal{Y}^u_r + \begin{pmatrix} b_u\left(r, \mathcal{X}^{u,1:n_1}_r, u\right) \\ \hat{b}_u\left(r, \mathcal{X}^u_r, u\right) \end{pmatrix} dr$$

$$+ \sum_{j=1}^{m_1} \int_0^t \begin{pmatrix} \sigma^j_\xi\left(r, \mathcal{X}^{u,1:n_1}_r, u\right) & 0 \\ 0 & 0 \end{pmatrix} \mathcal{Y}^u_r + \begin{pmatrix} \sigma^j_u\left(r, \mathcal{X}^{u,1:n_1}_r, u\right) \\ 0 \end{pmatrix} d^- w^j_r$$

$$+ \sum_{j=1}^{m_2} \int_0^t \begin{pmatrix} 0 & 0 \\ \hat{\sigma}^j_\xi\left(r, \mathcal{X}^u_r, u\right) & \hat{\sigma}^j_x\left(r, \mathcal{X}^u_r, u\right) \end{pmatrix} \mathcal{Y}^u_r + \begin{pmatrix} 0 \\ \hat{\sigma}^j_u\left(r, \mathcal{X}^u_r, u\right) \end{pmatrix} d^- B^j_r.$$

(4.7)

We refer to (4.7) as the *sensitivity equation*; cp., e.g., Ref. [16].

The following theorem proves the existence and uniqueness of the SDEs (4.4) and (4.7) under the standing assumptions. We make use of the space $L^l_\mathbb{F}(\Omega, C^{p,0}[0,T], \mathbb{R}^{(n_1+k)\times m})$ of $\mathbb{F}$-adapted processes $(x_t)_{t\in[0,T]}$ satisfying $\mathrm{E}[\|x\|^l_{\infty,0,T}] < \infty$ and taking values in $\mathbb{R}^{(n_1+k)\times m}$ such that P-almost every path is continuous and the component paths in the first n_1 lines of x are of finite p-variation. Here, $\|x\|_{\infty,s,t} := \sup_{r\in[s,t]} |x_r|$ denotes the supremum norm over the interval $[s,t]$.

Theorem 4.1. *For every $u \in \mathcal{U}$, the SDEs (4.4) and (4.7) have unique solutions $\mathcal{X}^u \in L^l_\mathbb{F}(\Omega, C^{p,0}[0,T], \mathbb{R}^{n_1+n_2})$ and $\mathcal{Y}^u \in L^l_\mathbb{F}(\Omega, C^{p,0}[0,T], \mathbb{R}^{(n_1+n_2)\times d})$, respectively, for every $l \geq 1$. Moreover, there is a constant $C_{\mathcal{X},\mathcal{Y},l}$ independent of $u \in \mathcal{U}$ such that*

$$\mathrm{E}\left[\|\mathcal{X}^u\|^l_{\infty,0,T}\right] + \mathrm{E}\left[\|\mathcal{Y}^u\|^l_{\infty,0,T}\right] \leq C_{\mathcal{X},\mathcal{Y},l}, \quad (l \geq 1).$$

Finally, for every $l \geq 1$, the map

$$\mathcal{U} \to L^l_{\mathbb{F}}(\Omega, C[0,T], \mathbb{R}^{n_1+n_2}), \quad u \mapsto \mathcal{X}^u,$$

is Fréchet differentiable with the Fréchet derivative $D\mathcal{X}^u = \mathcal{Y}^u$.

The details of the technical proof can be found in Ref. [26]. We will comment in Section 4.4.1 on the key steps of the proof and on related results in the literature.

By applying the previous theorem in conjunction with assumption (G) and the chain rule for Fréchet derivatives (see Proposition 1.1.4 in Ref. [33]), we obtain the following representation for the gradient of the cost functional under the standing assumptions:

$$\nabla J(u) = \sum_{\mu=1}^{M} \mathrm{E}[g_\mu(\mathcal{X}^u_{T_\mu})] \mathrm{E}[g'_\mu(\mathcal{X}^u_{T_\mu}) \mathcal{Y}^u_{T_\mu}]. \tag{4.8}$$

In order to obtain the adjoint gradient representation, we first derive a variation-of-constants formula for the linear matrix SDE (4.7) in Theorem 4.2.

Notation 4.1. To simplify the notation for the rest of this chapter, we define for $r \in [0,T]$ and $u \in \mathcal{U}$: $a^u(r) := a(r, \mathcal{X}^{u,1:n_1}_r, \mathcal{X}^{u,n_1+1:n_2}_r, u)$, for some generic function a mapping from $[0,T] \times \mathbb{R}^{n_1} \times \mathbb{R}^{n_2} \times \mathcal{U}$ to some Euclidean space.

With this notation at hand, we consider the following homogeneous matrix-valued SDEs with an initial condition equal to the unit matrix $I_{n_1+n_2}$ in $\mathbb{R}^{(n_1+n_2) \times (n_1+n_2)}$ at time $s_0 \in [0,T]$:

$$\Phi^{s_0}_t = I_{n_1+n_2} + \int_{s_0}^{t} \begin{pmatrix} b^u_\xi(r) & 0 \\ \hat{b}^u_\xi(r) & \hat{b}^u_x(r) \end{pmatrix} \Phi^{s_0}_r \, dr$$

$$+ \sum_{j=1}^{m_1} \int_{s_0}^{t} \begin{pmatrix} \sigma^{u,j}_\xi(r) & 0 \\ 0 & 0 \end{pmatrix} \Phi^{s_0}_r \, d^- w^j_r$$

$$+ \sum_{j=1}^{m_2} \int_{s_0}^{t} \begin{pmatrix} 0 & 0 \\ \hat{\sigma}^{u,j}_\xi(r) & \hat{\sigma}^{u,j}_x(r) \end{pmatrix} \Phi^{s_0}_r \, d^- B^j_r, \tag{4.9}$$

and

$$\Psi_t^{s_0} = I_{n_1+n_2} - \int_{s_0}^t \Psi_r^{s_0} \left[\begin{pmatrix} b_\xi^u(r) & 0 \\ \hat{b}_\xi^u(r) & \hat{b}_x^u(r) \end{pmatrix} - \sum_{j=1}^{m_2} \begin{pmatrix} 0 & 0 \\ \hat{\sigma}_\xi^{u,j}(r) & \hat{\sigma}_x^{u,j}(r) \end{pmatrix}^2 \right] dr$$
$$- \sum_{j=1}^{m_1} \int_{s_0}^t \Psi_r^{s_0} \begin{pmatrix} \sigma_\xi^{u,j}(r) & 0 \\ 0 & 0 \end{pmatrix} d^- w_r^j$$
$$- \sum_{j=1}^{m_2} \int_{s_0}^t \Psi_r^{s_0} \begin{pmatrix} 0 & 0 \\ \hat{\sigma}_\xi^{u,j}(r) & \hat{\sigma}_x^{u,j}(r) \end{pmatrix} d^- B_r^j, \qquad (4.10)$$

for $t \in [s_0, T]$, suppressing the dependence on u by abbreviating $\Phi_t^{s_0} = \Phi_t^{s_0,u}$ and $\Psi_t^{s_0} = \Psi_t^{s_0,u}$.

Theorem 4.2. *For every $u \in \mathcal{U}$ and $s_0 \in [0,T]$, the matrix-valued SDEs (4.9) and (4.10) have a unique solution $\Phi^{s_0,u}$ and $\Psi^{s_0,u}$, respectively, in $L_\mathbb{F}^l(\Omega, C^{p,0}[s_0,T], \mathbb{R}^{(n_1+n_2)\times(n_1+n_2)})$ for every $l \geq 1$, such that*

$$\mathrm{E}\left[\|\Phi^{s_0,u}\|_{\infty,s_0,T}^l\right] + \mathrm{E}\left[\|\Psi^{s_0,u}\|_{\infty,s_0,T}^l\right] \leq C_{\Phi,\Psi,l},$$

where the positive constant $C_{\Phi,\Psi,l}$ is independent of u and s_0. Moreover, $\Psi_t^{s_0,u} = (\Phi_t^{s_0,u})^{-1}$ for $t \in [s_0,T]$, P-almost surely. Furthermore, the solution $\mathcal{Y}_t^u$ to (4.7) is, for every $t \in [0,T]$, given by the following variation-of-constants formula (where we set the initial time of the homogeneous equations to $s_0 = 0$ and skip the superscripts s_0 and u):

$$\mathcal{Y}_t^u = \Phi_t \begin{pmatrix} D\xi_0(u) \\ Dx_0(u) \end{pmatrix} + \Phi_t \int_0^t \Phi_r^{-1} \left[\begin{pmatrix} b_u^u(r) \\ \hat{b}_u^u(r) \end{pmatrix} - \sum_{j=1}^{m_2} \begin{pmatrix} 0 \\ \hat{\sigma}_x^{u,j}(r)\hat{\sigma}_u^{u,j}(r) \end{pmatrix} \right] dr$$
$$+ \sum_{j=1}^{m_1} \Phi_t \int_0^t \Phi_r^{-1} \begin{pmatrix} \sigma_u^{u,j}(r) \\ 0 \end{pmatrix} d^- w_r^j$$
$$+ \sum_{j=1}^{m_2} \Phi_t \int_0^t \Phi_r^{-1} \begin{pmatrix} 0 \\ \hat{\sigma}_u^{u,j}(r) \end{pmatrix} d^- B_r^j. \qquad (4.11)$$

A sketch of the proof will be provided in Section 4.4.2. Inserting the variation-of-constants formula (4.11) into the gradient representation (4.8), we obtain, by manipulating the forward integrals as detailed in Lemma 4.8 in Section 4.4.3,

$$\nabla J(u) = E\left[\Lambda_0 \begin{pmatrix} D\xi_0(u) \\ Dx_0(u) \end{pmatrix} + \int_0^T \Lambda_r \left[\begin{pmatrix} b_u^u(r) \\ \hat{b}_u^u(r) \end{pmatrix} - \sum_{j=1}^{m_2} \begin{pmatrix} 0 \\ \hat{\sigma}_x^{u,j}(r)\hat{\sigma}_u^{u,j}(r) \end{pmatrix}\right] dr \right. \\ \left. + \sum_{j=1}^{m_1} \int_0^T \Lambda_r \begin{pmatrix} \sigma_u^{u,j}(r) \\ 0 \end{pmatrix} d^-w_r^j + \sum_{j=1}^{m_2} \int_0^T \Lambda_r \begin{pmatrix} 0 \\ \hat{\sigma}_u^{u,j}(r) \end{pmatrix} d^-B_r^j\right]. \tag{4.12}$$

Here, $(\Lambda_t)_{t\in[0,T]} = (\Lambda_t^u)_{t\in[0,T]}$ is the $\mathbb{R}^{1\times(n_1+n_2)}$-valued (i.e., row-vector-valued) process defined by

$$\Lambda_t = \sum_{\mu;\, T_\mu \geq t} E[g_\mu(\mathcal{X}_{T_\mu}^u)] g_\mu'(\mathcal{X}_{T_\mu}^u) \Phi_{T_\mu} \Phi_t^{-1} \tag{4.13}$$

(where again $\Phi = \Phi^{0,u}$).

We emphasize that the process Λ is not $\mathbb{F}$-adapted but anticipates future information through the factors $g_\mu'(\mathcal{X}_{T_\mu}^u)\Phi_{T_\mu}$. In particular, the integrals with respect to the Brownian motions in (4.12) cannot be interpreted as Itô integrals, but are "true" forward integrals. These anticipating forward integrals, in general, do not have zero expectation and, hence, contribute to the *adjoint gradient representation* (4.12).

Theorem 4.3. *The process* $\Lambda = (\Lambda^u)$ *satisfies* $E[\|\Lambda\|_{\infty,0,T}^l] < \infty$ *for every* $l \geq 1$ *and solves the anticipating BSDE* ($t \in [0,T]$)

$$\Lambda_t = \sum_{T_\mu \geq t} E[g_\mu(\mathcal{X}_{T_\mu}^u)] g_\mu'(\mathcal{X}_{T_\mu}^u)$$

$$+ \int_t^T \Lambda_r \left[\begin{pmatrix} b_\xi^u(r) & 0 \\ \hat{b}_\xi^u(r) & \hat{b}_x^u(r) \end{pmatrix} - \sum_{j=1}^{m_2} \begin{pmatrix} 0 & 0 \\ \hat{\sigma}_\xi^{u,j}(r) & \hat{\sigma}_x^{u,j}(r) \end{pmatrix}^2\right] dr$$

$$+ \sum_{j=1}^{m_1} \int_t^T \Lambda_r \begin{pmatrix} \sigma_\xi^{u,j}(r) & 0 \\ 0 & 0 \end{pmatrix} d^-w_r^j$$

$$+ \sum_{j=1}^{m_2} \int_t^T \Lambda_r \begin{pmatrix} 0 & 0 \\ \hat{\sigma}_\xi^{u,j}(r) & \hat{\sigma}_x^{u,j}(r) \end{pmatrix} d^-B_r^j. \tag{4.14}$$

The proof will be given in Section 4.4.3.

We call (4.14) the *adjoint equation*, as it resembles the adjoint equation in the Pontryagin maximum principle for optimal control problems:

(1) *Deterministic case*: $\sigma \equiv 0$, $b \equiv 0$, $\xi_0 \equiv 0$, $\hat{\sigma} \equiv 0$, and $T_\mu = T$ for every $\mu = 1, \ldots, M$:
Then, (4.14) reduces to the terminal value problem for the ordinary differential equation (ignoring the first n_1 components of Λ)

$$\dot{\Lambda}_t = -\Lambda_t \hat{b}_x^u(t), \quad \Lambda_T = \sum_{\mu=1}^{M} E[g_\mu(\mathcal{X}_{T_\mu}^u)] g_\mu'(\mathcal{X}_{T_\mu}^u),$$

which corresponds to the adjoint equation in the Pontryagin maximum principle for the control of ordinary differential equations; cp. e.g., Chapter 3.2 in Ref. [34]. Note that the local optimality condition in terms of the Hamiltonian in the optimal control situation, e.g., Eq. (3.2.7) in Ref. [34], turns into the global condition

$$0 = \nabla J(u) = Dx_0(u) + \int_0^T \Lambda_r \hat{b}_u^u(r) dr$$

in our case (see (4.12)) because we are minimizing over constant parameters (in contrast to optimizing dynamically over functions).

(2) *Brownian motion case*: $\sigma \equiv 0$, $b \equiv 0$, $\xi_0 \equiv 0$, and $T_\mu = T$ for every $\mu = 1, \ldots, M$: In order to avoid working with the anticipating process Λ, one can project Λ (ignoring the first n_1 components of Λ, again) on the available information, leading to

$$p_t := E[\Lambda_t | \mathcal{F}_t].$$

By Theorem 7.2.2 in Ref. [34] and its proof, there is a matrix-valued process $q = (q^1, \ldots, q^{m_2})$ such that the pair (p, q) solves the following nonanticipating BSDE in terms of Itô integration:

$$dp_t = -\left(p_t \hat{b}_x^u(t) + \sum_{j=1}^{m_2} q_t^j \hat{\sigma}_x^{u,j}(t)\right) + \sum_{j=1}^{m_2} q_t^j dB_t,$$

$$p_T = \sum_{\mu=1}^{M} E[g_\mu(\mathcal{X}_{T_\mu}^u)] g_\mu'(\mathcal{X}_{T_\mu}^u).$$

The process q can be constructed via the martingale representation theorem. This BSDE corresponds to the first adjoint equation in the maximum principle for controlled SDEs; see Eq. (3.3.8) in Ref. [34].

Note that the techniques related to moving from Λ to p, by conditioning in the Brownian motion case, heavily rely on the martingale property of a Brownian motion. As these martingale techniques are not at our disposal in the presence of the second driving process $(w_t)_{t \in [0,T]}$, we decided to work directly with Λ and derive the adjoint equation in terms of a new type of an anticipating BSDE in Theorem 4.3.

4.2.3. *Euler discretization of the gradient representations*

Any numerical resolution of the gradient representations (4.8) or (4.12) requires a time-discretization scheme either for the pair $(\mathcal{X}^u, \mathcal{Y}^u)$ or for the pair $(\mathcal{X}^u, \Lambda^u)$. From a computational point of view, it is beneficial to approximate the pair $(\mathcal{X}^u, \Lambda^u)$ because both processes are $\mathbb{R}^{n_1+n_2}$-dimensional, while $\mathcal{Y}^u$ is a matrix-valued, precisely $\mathbb{R}^{(n_1+n_2) \times d}$-dimensional process, where d equals the number of parameters. However, Λ^u solves, by Theorem 4.3, a new type of anticipating BSDE in terms of the forward integral. Devising an Euler scheme for Λ^u and analyzing its rate of convergence constitute the main results of this section.

For the discretization of the state dynamics (4.4) and the sensitivity equation (4.7), we apply a continuously interpolated Euler scheme along a partition $\Pi^E = (t_i)_{i=0,\ldots,n}$ of $[0,T]$, which is not necessarily equidistant. We always assume that the time points T_μ, connected to the cost functional (4.5), are elements of Π^E.

Precisely, we consider, for $t \in (t_i, t_{i+1}]$,

$$\mathcal{X}_t^{n,u} = \begin{pmatrix} \xi_t^{n,u} \\ x_t^{n,u} \end{pmatrix} = \begin{pmatrix} \xi_{t_i}^{n,u} \\ x_{t_i}^{n,u} \end{pmatrix} + \begin{pmatrix} b(t_i, \xi_{t_i}^{n,u}, u) \\ \hat{b}(t_i, \xi_{t_i}^{n,u}, x_{t_i}^{n,u}, u) \end{pmatrix} (t - t_i)$$
$$+ \sum_{j=1}^{m_1} \begin{pmatrix} \sigma^j(t_i, \xi_{t_i}^{n,u}, u) \\ 0 \end{pmatrix} \left(w_t^j - w_{t_i}^j \right)$$
$$+ \sum_{j=1}^{m_2} \begin{pmatrix} 0 \\ \hat{\sigma}^j(t_i, \xi_{t_i}^{n,u}, x_{t_i}^{n,u}, u) \end{pmatrix} \left(B_t^j - B_{t_i}^j \right) \qquad (4.15)$$

initialized at $\mathcal{X}_0^{n,u}$ via $\xi_0^{n,u} = \xi_0(u)$ and $x_0^{n,u} = x_0(u)$, as well as

$$\mathcal{Y}_t^{n,u} = \mathcal{Y}_{t_i}^{n,u} + \eta_{t_i,t}^{n,u} \mathcal{Y}_{t_i}^{n,u} + \begin{pmatrix} b_u\left(t_i, \xi_{t_i}^{n,u}, u\right) \\ \hat{b}_u\left(t_i, \xi_{t_i}^{n,u}, x_{t_i}^{n,u}, u\right) \end{pmatrix} (t - t_i)$$

$$+ \sum_{j=1}^{m_1} \begin{pmatrix} \sigma_u^j(t_i, \xi_{t_i}^{n,u}, u) \\ 0 \end{pmatrix} \left(w_t^j - w_{t_i}^j\right)$$

$$+ \sum_{j=1}^{m_2} \begin{pmatrix} 0 \\ \hat{\sigma}_u^j\left(t_i, \xi_{t_i}^{n,u}, x_{t_i}^{n,u}, u\right) \end{pmatrix} \left(B_t^j - B_{t_i}^j\right), \qquad (4.16)$$

$$\mathcal{Y}_0^{n,u} = \begin{pmatrix} D\xi_0(u) \\ Dx_0(u) \end{pmatrix}, \qquad (4.17)$$

where

$$\eta_{t_i,t}^{n,u} = \begin{pmatrix} b_\xi(t_i, \xi_{t_i}^{n,u}, u) & 0 \\ \hat{b}_\xi(t_i, \xi_{t_i}^{n,u}, x_{t_i}^{n,u}, u) & \hat{b}_x(t_i, \xi_{t_i}^{n,u}, x_{t_i}^{n,u}, u) \end{pmatrix} (t - t_i)$$

$$+ \sum_{j=1}^{m_1} \begin{pmatrix} \sigma_\xi^j(t_i, \xi_{t_i}^{n,u}, u) & 0 \\ 0 & 0 \end{pmatrix} (w_t^j - w_{t_i}^j)$$

$$+ \sum_{j=1}^{m_2} \begin{pmatrix} 0 & 0 \\ \hat{\sigma}_\xi^j(t_i, \xi_{t_i}^{n,u}, x_{t_i}^{n,u}, u) & \hat{\sigma}_x^j(t_i, \xi_{t_i}^{n,u}, x_{t_i}^{n,u}, u) \end{pmatrix} (B_t^j - B_{t_i}^j).$$

$$(4.18)$$

The discretization of the adjoint equation (4.12) is initialized at terminal time $t_n = T$ via

$$\Lambda_{t_n}^{n,u} = \sum_{\mu;\, T_\mu = T} E[g_\mu(\mathcal{X}_T^{n,u})] g_\mu'(\mathcal{X}_T^{n,u}) \qquad (4.19)$$

and then follows a backward recursion along the grid points

$$\Lambda_{t_i}^{n,u} = \Lambda_{t_{i+1}}^{n,u} + \Lambda_{t_{i+1}}^{n,u} \eta_{t_i,t_{i+1}}^{n,u} + \sum_{\mu;\, T_\mu = t_i} E[g_\mu(\mathcal{X}_{t_i}^{n,u})] g_\mu'(\mathcal{X}_{t_i}^{n,u}). \qquad (4.20)$$

We apply a piecewise constant interpolation on the interval $[0, T]$, i.e.,

$$\Lambda_t^{n,u} = \Lambda_{t_{i+1}}^{n,u}$$

for $t \in (t_i, t_{i+1})$.

Remark 4.3. In order to motivate the discretization for the adjoint equation, let us look at a one-dimensional equation of the form

$$\lambda_t = \lambda_T + \int_t^T \lambda_r (b_r - \hat{\sigma}_r^2)\, dt + \int_t^T \lambda_r \sigma_r\, d^-w_r + \int_t^T \lambda_r \hat{\sigma}_r\, d^- B_r,$$

where b, σ, and $\hat{\sigma}$ are $\mathbb{F}$-adapted processes. As forward integrals are based on Riemann sums with the tag point at the left interval boundary point, we get

$$\lambda_{t_i} \approx \lambda_{t_{i+1}} + \lambda_{t_i}\left[(b_{t_i} - \hat{\sigma}_{t_i}^2)\Delta_i + \sigma_{t_i}\Delta w_i + \hat{\sigma}_{t_i}\Delta B_i\right] =: \lambda_{t_{i+1}} + \lambda_{t_i}\widetilde{\eta}_{t_i},$$

for $\Delta_i := t_{i+1} - t_i$, $\Delta w_i := w_{t_{i+1}} - w_{t_i}$, and $\Delta B_i := B_{t_{i+1}} - B_{t_i}$. Rearranging terms and performing a second-order Taylor expansion (dropping higher-order terms), we arrive at

$$\lambda_{t_i} \approx \lambda_{t_{i+1}}(1 - \widetilde{\eta}_{t_i})^{-1} \approx \lambda_{t_{i+1}}(1 + \widetilde{\eta}_{t_i} + \widetilde{\eta}_{t_i}^2)$$
$$\approx \lambda_{t_{i+1}}(1 + (b_{t_i} - \hat{\sigma}_{t_i}^2)\Delta_i + \sigma_{t_i}\Delta w_i + \hat{\sigma}_{t_i}\Delta B_i + \hat{\sigma}_{t_i}^2 \Delta B_i^2)$$
$$\approx \lambda_{t_{i+1}}(1 + b_{t_i}\Delta_i + \sigma_{t_i}\Delta w_i + \hat{\sigma}_{t_i}\Delta B_i).$$

In order to measure the error of the Euler schemes under the assumed p-variation regularity, we consider

$$\delta(\omega; \Pi^E) := \max_{i=0,\ldots,n-1} |t_{i+1} - t_i| + |w(\omega)|_{p,t_i,t_{i+1}},$$

which depends on the p-variation seminorm of the realized path of w over the subintervals of the Euler partition (cp. Remark 4.1), as well as the "averaged error measure" in the lth mean,

$$\delta_l(\Pi^E) := \mathrm{E}\left[\delta(\Pi^E)^l\right]^{\frac{1}{l}}, \quad l \geq 1, \tag{4.21}$$

which is finite because w satisfies the exponential moment condition (4.6).

For the following theorem, we require two additional conditions:

(E_1): The Hölder exponent β from condition (H_3) is an element of the interval $[\frac{1}{p}, 1]$. Moreover, the function b from condition (H_2) and its partial derivatives b_ξ and b_u are Hölder continuous in t with Hölder exponent β.

(E_2): Let $\hat{b}$ and $\hat{\sigma}$ be the coefficient functions from conditions (B_2) and (B_3), respectively. There exists a constant $L > 0$ such that for all $x \in \mathbb{R}^{n_2}$, $\xi \in \mathbb{R}^{n_1}$, $u \in \mathcal{U}$, and $s \leq t \in [0, T]$,

$$|\hat{b}(t, \xi, x, u) - \hat{b}(s, \xi, x, u)| + |\hat{\sigma}(t, \xi, x, u) - \hat{\sigma}(s, \xi, x, u)|$$
$$\leq L(1 + |x| + |\xi|)(t - s)^{\frac{1}{2}}.$$

Theorem 4.4. *Suppose that, next to the standing assumptions, (E_1)–(E_2) are in force. Then, there is a constant $C_{E,l}$ depending on $l \geq 2$ (but independent of u and n) such that, for every $u \in \mathcal{U}$ and $l \geq 2$,*

$$\mathrm{E}\left[\|\mathcal{X}^u - \mathcal{X}^{n,u}\|_{\infty,0,T}^l\right]^{\frac{1}{l}} + \mathrm{E}\left[\|\mathcal{Y}^u - \mathcal{Y}^{n,u}\|_{\infty,0,T}^l\right]^{\frac{1}{l}}$$
$$+ \sup_{t \in [0,T]} \mathrm{E}\left[|\Lambda_t^u - \Lambda_t^{n,u}|^l\right]^{\frac{1}{l}} \leq C_{E,l}(\delta_{4l}(\Pi^E))^{(2-p) \wedge \frac{1}{2}}.$$

The key steps of the proof will be discussed in Section 4.4.4.

Remark 4.4. If we assume (instead of the p-variation regularity) that w has Hölder continuous paths with Hölder index $H \in (1/2, 1)$ and replace the p-variation norm with the Hölder norm in the exponential moment bound (4.6), then the error of the Euler approximations can be bounded in terms of the mesh size $|\Pi^E|$ of the partition. Precisely, the upper bound in Theorem 4.4 can then be replaced by $C'_{E,l}|\Pi^E|^{(2H-1) \wedge \frac{1}{2}}$ for some (possibly different) constant $C'_{E,l}$. Note that the p-variation norm $|w|_{p,t_i,t_{i+1}}$ is $\mathcal{O}(|t_{i+1} - t_i|^H)$ if w is H-Hölder continuous and $p = 1/H$, so that the rates in the p-variation case and in the Hölder case fit each other.

Remark 4.5. The convergence analysis of Euler schemes is a classical topic in the literature on numerical SDEs. We mention some results which are closely related to Theorem 4.4. Lejay [23] considers differential equations driven by a (deterministic) p-variation function w and derives a convergence rate of the order $(\max_{i=0,\ldots,n-1}|w|_{p,t_i,t_{i+1}})^{2-p}$ under the Lipschitz conditions. In order to achieve estimates in the lth mean, as for the convergence to $\mathcal{X}^{u,1:n_1}$

in Theorem 4.4, we additionally need to control the dependence of the constants on the realization of w, for which we apply the greedy sequence technique [25]. As in Ref. [35] for the fractional Brownian motion case, the boundedness of the coefficients significantly helps to carry out the analysis. In the case of Hölder paths, Euler schemes for SDEs driven by a fractional Brownian motion with the Hurst parameter $H > 1/2$ are well studied. Under the Lipschitz conditions, a rate of convergence of the order $2H - 1$ in the sense of almost-sure convergence is known to hold and to be sharp [36–39]. The equation solved using $\mathcal{X}^{u,n_1+1:n_2}$ is driven by a Brownian motion. Although the coefficients depend on $\mathcal{X}^{u,1:n_1}$, the error estimate follows along classical lines. Recall that strong convergence of the Euler scheme for SDEs driven by a Brownian of the order $1/2$ in the mesh size is well known under the Lipschitz conditions; see, e.g., Ref. [40]. Hence, the results mentioned above indicate that the convergence rate of the Euler scheme for $\mathcal{X}^u$ derived in Theorem 4.4 and Remark 4.4 are the best ones that one could expect. The key difficulty in analyzing the Euler scheme for the linear matrix SDE $\mathcal{Y}^u$ is to control the growth of the Euler scheme driven by the p-variation process $(w_t)_{t \in [0,T]}$ in the presence of time-dependent coefficients. We are not aware of a related convergence result in the lth mean in the p-variation context, but we refer to Ref. [41] for a study of linear equations driven by fractional Brownian motion in the Hölder space setting. For the analysis of the Euler scheme for the adjoint equation (4.14), the proof will be based on explicit variations-of-constants formulas for the continuous-time solution Λ^u and its Euler discretization $\Lambda^{n,u}$.

Having the convergence results in Theorem 4.4 at hand, we can apply them for the approximation of the cost functional (4.5) and the two representations (4.8) and (4.12) of its gradient.

Let $\Pi^{\mathrm{E}} = (t_i)_{i=0,\ldots,n}$ be a partition of the interval $[0,T]$ such that all the time points T_μ, $\mu = 1, \ldots, M$ are included in the partition. Then, the discretized cost function and the discretization of the gradient representation (4.8) are defined via

$$J^n(u) := \frac{1}{2} \sum_{\mu=1}^{M} \mathrm{E}\left[g_\mu(\mathcal{X}_{T_\mu}^{n,u})\right]^2,$$

$$(\nabla J)^n(u) := \sum_{\mu=1}^{M} \mathrm{E}\left[g_\mu(\mathcal{X}_{T_\mu}^{n,u})\right] \mathrm{E}\left[g'_\mu(\mathcal{X}_{T_\mu}^{n,u})\mathcal{Y}_{T_\mu}^{n,u}\right]. \qquad (4.22)$$

Given assumption (G) on the functions g_μ, it is easy to check that, for every $u \in \mathcal{U}$,

$$|J(u) - J^n(u)| + |(\nabla J)(u) - (\nabla J)^n(u)|$$
$$\leq C_J \left(\mathrm{E}\left[\|\mathcal{X}^u - \mathcal{X}^{n,u}\|^2_{\infty,0,T}\right]^{\frac{1}{2}} + \mathrm{E}\left[\|\mathcal{Y}^u - \mathcal{Y}^{n,u}\|^2_{\infty,0,T}\right]^{\frac{1}{2}} \right)$$

for some constant C_J (independent of u), and thus, Theorem 4.4 and Remark 4.4 provide rates of convergence under the different regularity assumptions on $(w_t)_{t \in [0,T]}$.

The following theorem establishes an alternative representation of the discretized gradient $(\nabla J)^n(u)$ in terms of the Euler scheme of the adjoint equation. It can be considered the natural discretization of the adjoint gradient representation (4.12).

Theorem 4.5. *For every $u \in \mathcal{U}$, the discretized gradient $(\nabla J)^n(u)$ (see (4.22)) can be represented by*

$$(\nabla J)^n(u) = \mathrm{E}\left[\Lambda_0^{n,u} D\mathcal{X}_0^u + \sum_{i=0}^{n-1} \Lambda_{t_{i+1}}^{n,u} \hat{\eta}_{t_i,t_{i+1}}^{n,u}\right],$$

where (suppressing the dependence of ξ^n and x^n on u)

$$\hat{\eta}_{t_i,t_{i+1}}^{n,u} := \begin{pmatrix} b_u(t_i, \xi_{t_i}^n, u) \\ \hat{b}_u(t_i, \xi_{t_i}^n, x_{t_i}^n, u) \end{pmatrix} (t_{i+1} - t_i)$$
$$+ \sum_{j=1}^{m_1} \begin{pmatrix} \sigma_u^j(t_i, \xi_{t_i}^n, u) \\ 0 \end{pmatrix} (w_{t_{i+1}}^j - w_{t_i}^j)$$
$$+ \sum_{j=1}^{m_2} \begin{pmatrix} 0 \\ \hat{\sigma}_u^j(t_i, \xi_{t_i}^n, x_{t_i}^n, u) \end{pmatrix} (B_{t_{i+1}}^j - B_{t_i}^j)$$

for all $i = 0, \ldots, n-1$.

Proof. Recall that, by (4.16),

$$\mathcal{Y}_{t_{i+1}}^{n,u} = \mathcal{Y}_{t_i}^{n,u} + \eta_{t_i,t_{i+1}}^{n,u} \mathcal{Y}_{t_i}^n + \hat{\eta}_{t_i,t_{i+1}}^{n,u}.$$

and, by (4.20),

$$\Lambda_{t_i}^{n,u} = \Lambda_{t_{i+1}}^{n,u} + \Lambda_{t_{i+1}}^{n,u} \eta_{t_i,t_{i+1}}^{n,u} + \sum_{\mu;\, T_\mu = t_i} E[g_\mu(\mathcal{X}_{T_\mu}^{n,u})] g_\mu'(\mathcal{X}_{T_\mu}^{n,u}).$$

Hence,

$$\Lambda^{n,u}_{t_{i+1}}\mathcal{Y}^{n,u}_{t_{i+1}} = \Lambda^{n,u}_{t_{i+1}}\mathcal{Y}^{n,u}_{t_i} + \Lambda^{n,u}_{t_{i+1}}\eta^{n,u}_{t_i,t_{i+1}}\mathcal{Y}^n_{t_i} + \Lambda^{n,u}_{t_{i+1}}\hat{\eta}^{n,u}_{t_i,t_{i+1}}$$

$$= \Lambda^{n,u}_{t_i}\mathcal{Y}^{n,u}_{t_i} + \Lambda^{n,u}_{t_{i+1}}\hat{\eta}^{n,u}_{t_i,t_{i+1}}$$

$$- \sum_{\mu;\, T_\mu = t_i} E[g_\mu(\mathcal{X}^{n,u}_{T_\mu})] g'_\mu(\mathcal{X}^{n,u}_{T_\mu})\mathcal{Y}^{n,u}_{t_i}.$$

Therefore,

$$\Lambda^{n,u}_{t_n}\mathcal{Y}^{n,u}_{t_n} - \Lambda^{n,u}_{t_0}\mathcal{Y}^{n,u}_{t_0} = \sum_{i=0}^{n-1} \Lambda^{n,u}_{t_{i+1}}\hat{\eta}^{n,u}_{t_i,t_{i+1}}$$

$$- \sum_{\mu;\, T_\mu < t_n} E[g_\mu(\mathcal{X}^{n,u}_{T_\mu})] g'_\mu(\mathcal{X}^{n,u}_{T_\mu})\mathcal{Y}^{n,u}_{T_\mu}.$$

Inserting the terminal condition (4.19) for $\Lambda^{n,u}$ and the initial condition (4.17) for $\mathcal{Y}^{n,u}$, we obtain

$$\sum_{\mu=1}^{M} E[g_\mu(\mathcal{X}^{n,u}_{T_\mu})] g'_\mu(\mathcal{X}^{n,u}_{T_\mu})\mathcal{Y}^{n,u}_{T_\mu} = \Lambda^{n,u}_0 D\mathcal{X}^u_0 + \sum_{i=0}^{n-1} \Lambda^{n,u}_{t_{i+1}}\hat{\eta}^{n,u}_{t_i,t_{i+1}}.$$

Recalling the definition (4.22) of $(\nabla J)^n(u)$, the proof is completed by taking expectation. □

4.3. On the Young Integral and the Russo–Vallois Forward Integral

4.3.1. *Background on Young integration*

The Young integral [17] can be considered a Riemann–Stieltjes integral in the context of p-variation functions. Suppose $[s,t]$ is a compact interval, $x : [s,t] \to \mathbb{R}^{n\times m}$ and $w : [s,t] \to \mathbb{Y}$, where either $\mathbb{Y} = \mathbb{R}$ or $\mathbb{Y} = \mathbb{R}^{m\times d}$. Given a partition $\Pi_k = (t_i)_{i=0,\ldots,k}$ of $[s,t]$ and a finite sequence of tag points $\Theta_k = (\theta_i)_{i=0,\ldots,k-1}$, where $t_i \leq \theta_i \leq t_{i+1}$, the pair (Π_k, Θ_k) is said to be a *tagged partition*. The *Riemann–Stieltjes sum* of x with respect to w on the tagged partition (Π_k, Θ_k) is defined to be

$$RS(x, dw, (\Pi_k, \Theta_k)) = \sum_{i=0}^{k-1} x_{\theta_i}(w_{t_{i+1}} - w_{t_i}).$$

The *Riemann–Stieltjes integral* is said to exist and is then denoted by $\int_s^t x_r dw_r$ if, for every $\epsilon > 0$, there is a $\delta > 0$ such that

$$\left| \int_s^t x_r dw_r - RS(x, dw, (\Pi, \Theta)) \right| < \epsilon$$

for every tagged partition (Π, Θ) with mesh size $|\Pi| < \delta$. In the context of p-variation functions, the Riemann–Stieltjes integral exists, e.g., under the following conditions.

Theorem 4.6 (Young Integral). *For $1 \leq p$, $1 \leq q$ such that $\alpha = \frac{1}{p} + \frac{1}{q} > 1$, let $x \in W^q([s,t], \mathbb{R}^{n \times m})$ and $w \in C^p([s,t], \mathbb{Y})$. Then, the Riemann–Stieltjes integral $\int_s^t x_r dw_r$ exists, and the inequality*

$$\left| \int_s^t x_r dw_r - x_\theta(w_t - w_s) \right| \leq C_{p,q} |x|_{q,s,t} |w|_{p,s,t} \tag{4.23}$$

holds for every $\theta \in [s,t]$, where $C_{p,q} = \zeta(\alpha)$ for $\zeta(y) = \sum_{i=1}^{\infty} \left(\frac{1}{i}\right)^y$ ($y > 1$). Moreover, we have

$$\left| \int_s^t x_r dw_r \right| \leq C_{p,q} \|x\|_{q,s,t} |w|_{p,s,t}. \tag{4.24}$$

In this situation, we use the term Young integral *and call inequality (4.24) the* Love–Young *estimate.*

Proof. As the integrator w is continuous and taking Theorem 2.42 in Ref. [29] into account, this result is a special case of Corollary 3.91 in Ref. [29]. □

As a limit of the Riemann–Stieltjes sums, the Young integral inherits, e.g., bilinearity as an operator in integrand and integrator.

Control functions are well known to be useful tools for estimating p-variation (semi-)norms; see, e.g., Ref. [42].

Definition 4.1. A continuous map φ taking values in the nonnegative real numbers, defined on the simplex $\Delta([s,t]) = \{(u,v) \in \mathbb{R}^2 \mid 0 \leq u \leq v \leq t\}$, is called a *control function* on $[s,t]$ if it satisfies the following conditions:

(1) For all $r \in [s,t]$, $\varphi(r,r) = 0$.
(2) For all $u \leq r \leq v$ in $[s,t]$, $\varphi(u,r) + \varphi(r,v) \leq \varphi(u,v)$.

The following lemma is a variant of Proposition 5.10 in Ref. [42].

Lemma 4.1. *Let $\varphi_1, \ldots, \varphi_m$ be superadditive functions on $[s,t]$ (i.e., they satisfy property (2) in Definition 4.1), $p \geq 1$, $C_1, \ldots, C_k$ be positive constants, and $x : [s,t] \to \mathbb{R}^{n \times m}$ be a function on $[s,t]$. The pointwise estimate*

$$|x_v - x_u| \leq \sum_{j=1}^{m} C_j \varphi_j(u,v)^{\frac{1}{p}} \quad \text{for all } u \leq v \text{ in } [s,t]$$

implies the p-variation estimate

$$|x|_{p,u,v} \leq \sum_{j=1}^{m} C_j \varphi_j(u,v)^{\frac{1}{p}} \quad \text{for all } u \leq v \text{ in } [s,t].$$

If φ_j is a control function on $[s,t]$ for all $j = 1, \ldots, m$, then x is continuous on $[s,t]$.

Note that p-variation estimates lead to estimates in the sup-norm via the relation

$$\|x\|_{\infty,s,t} \leq |x|_s + |x|_{p,s,t} = \|x\|_{p,s,t}. \qquad (4.25)$$

The following proposition states that the pth power of the p-variation seminorm constitutes a control. For a proof, we refer to Proposition 5.8 in Ref. [42].

Proposition 4.1. *Let $p \geq 1$ and $x : [s,t] \to \mathbb{R}^{n \times m}$ be a continuous function of finite p-variation. Then,*

$$\varphi(u,v) = |x|^p_{p,u,v}$$

defines a control function on $[s,t]$.

Calibration of Non-Semimartingale Models   173

We state two elementary lemmas (without proof), which are useful for estimating p-variation (semi-)norms.

Lemma 4.2. *Let $p \geq 1$, $B \in W^p([s,t], \mathbb{R}^{n \times n})$, $x \in W^p([s,t], \mathbb{R}^{n \times m})$, and assume that $f : \mathbb{R}^{n \times m} \to \mathbb{R}^k$ is Lipschitz continuous with constant L. Then, we have*

$$\|Bx\|_{p,s,t} \leq |B_s x_s| + \|B\|_{\infty,s,t} |x|_{p,s,t} + \|x\|_{\infty,s,t} |B|_{p,s,t}$$
$$\leq 2\|B\|_{p,s,t} \|x\|_{p,s,t}$$

and

$$|f(x)|_{p,s,t} \leq L|x|_{p,s,t}.$$

Lemma 4.3. *Let $x \in W^p([s,t], \mathbb{R}^{n \times m})$, $p \geq 1$. If $s = t_0 < t_1 < \cdots < t_k = t$, then*

$$\sum_{i=0}^{k-1} |x|_{p,t_i,t_{i+1}}^p \leq |x|_{p,s,t}^p \leq k^{p-1} \sum_{i=0}^{k-1} |x|_{p,t_i,t_{i+1}}^p.$$

The following lemma (see, e.g., Theorem 3.92 in Ref. [29]) is devoted to the indefinite integral

$$I_Y(x,w)(u) = \int_s^u x_r \, dw_r \quad \forall u \in [s,t].$$

Lemma 4.4. *Let $1 \leq p$, $1 \leq q$ such that $\alpha = \frac{1}{p} + \frac{1}{q} > 1$, $x \in W^q([s,t], \mathbb{R}^{n \times m})$ and $w \in C^p([s,t], \mathbb{Y})$. The indefinite integral $I_Y(x,w)$ exists and is an element of $C^p([s,t], \mathbb{X})$, where $\mathbb{X} = \mathbb{R}^{n \times d}$ or $\mathbb{X} = \mathbb{R}^{n \times m}$ depending on the choice of $\mathbb{Y}$. Furthermore, we have*

$$\|I_Y(x,w)\|_{p,s,t} = |I_Y(x,w)|_{p,s,t} = \left| \int_s^\cdot x_r \, dw_r \right|_{p,s,t} \leq C_{p,q} \|x\|_{q,s,t} |w|_{p,s,t}.$$

We provide a proof in order to illustrate the control function technique.

Proof. For every $r \in [s,t]$, the indefinite integral $I_Y(x,w)(r)$ exists by Theorem 4.6. Let $u < v \in [s,t]$. Then, we have by additivity of the Young integral and the Love–Young estimate

$$|I_Y(x,w)(v) - I_Y(x,w)(u)| = \left|\int_u^v x_r \, dw_r\right| \leq C_{p,q}\|x\|_{q,u,v}|w|_{p,u,v}$$

$$\leq C_{p,q}\|x\|_{q,s,t}|w|_{p,u,v}.$$

Since $\varphi(u,v) = |w|_{p,u,v}^p$ is a control function on $[s,t]$, we conclude the proof by applying Lemma 4.1 and by noting that $I_Y(x,w)(s) = 0$. □

A crucial tool for the study of Young differential equations is the following variant of Gronwall's lemma.

Lemma 4.5. *Let $1 \leq p \leq q$ satisfy $\frac{1}{p} + \frac{1}{q} > 1$ and fix $T > 0$. Assume that $y \in W^q([0,T], \mathbb{R}^{n \times m})$ and $w \in C^p([0,T], \mathbb{Y})$ satisfy the following condition: There exist constants $K_1, K_2 > 0$ such that for all $[s,t] \subset [0,T]$, which satisfy $|t-s| + |w|_{p,s,t} \leq K_2$, we have*

$$|y|_{q,s,t} \leq K_1 + |y_s|. \tag{4.26}$$

Then,

$$|y|_{q,0,T} \leq (K_1 + |y_0|)e^{2^p K_2^{-p}(T^p + |w|_{p,0,T}^p)} \tag{4.27}$$

and

$$\|y\|_{\infty,0,T} \leq \|y\|_{q,0,T} \leq (K_1 + 2|y_0|)e^{2^p K_2^{-p}(T^p + |w|_{p,0,T}^p)}.$$

If the right-hand side of (4.26) only consists of the constant K_1, then the estimates simplify to

$$|y|_{q,0,T} \leq K_1 2^{p-1} K_2^{-p}(T^p + |w|_{p,0,T}^p) \tag{4.28}$$

and

$$\|y\|_{\infty,0,T} \leq \|y\|_{q,0,T} \leq |y_0| + K_1 2^{p-1} K_2^{-p}(T^p + |w|_{p,0,T}^p). \tag{4.29}$$

This lemma is a matrix-valued variant of results in Ref. [25] (see their Lemma 3.3, Remark 3.4, and Corollary 3.5). We include the proof in order to illustrate the greedy sequence technique in Refs. [24]

and [25]. A *greedy sequence* is an increasing sequence of time points $(\tau_i)_{i=0,\ldots,N}$ of the interval $[0,T]$ with $\tau_N = T$ satisfying

$$|\tau_{i+1} - \tau_i| + |w|_{p,\tau_i,\tau_{i+1}} = \mu \quad \text{for } i = 0, \ldots, N-2,$$
$$|\tau_N - \tau_{N-1}| + |w|_{p,\tau_{N-1},\tau_N} \leq \mu, \tag{4.30}$$

for given $\mu > 0$, $p \geq 1$. For the construction of such a sequence, one can first define $\tau_0 = 0$. Note that $\kappa(t) = t + |w|_{p,0,t}$ is continuous and strictly increasing with respect to t, with $\kappa(0) = 0$ and $\kappa(T) = T + |w|_{p,0,T}$. The intermediate value theorem ensures that there exists a unique $t > 0$ such that $t + |w|_{p,0,t} = \mu$, if $\mu < T + |w|_{p,0,T}$. In this case, we let $\tau_1 = \sup\{0 \leq t \leq T \mid t + |w|_{p,0,t} \leq \mu\}$. Otherwise, let $\tau_1 = T$. This construction can be continued inductively. For $T > 0$ and $0 \leq s < t \leq T$, denote

$$\overline{N}(t) = \sup_{k \in \mathbb{N}_0} \{\tau_k \leq t\}, \quad \underline{N}(t) = \inf_{k \in \mathbb{N}_0} \{\tau_k \geq t\} \quad \text{and}$$
$$N(s,t) = \overline{N}(t) - \underline{N}(s).$$

It has been shown in Ref. [25, Lemma 2.6] that the number $N(s,t)$ of subintervals defined by the greedy sequence in an interval $[s,t] \subset [0,T]$ is bounded by

$$N(s,t) \leq \frac{2^{p-1}}{\mu^p}\left((t-s)^p + |w|_{p,s,t}^p\right). \tag{4.31}$$

In particular, one obtains a finite partition of the interval $[0,T]$ using this construction.

Proof of Lemma 4.5. We denote by $0 = \tau_0 < \cdots < \tau_N = T$ the greedy sequence of times with $\mu = K_2$. Hence,

$$(\tau_{i+1} - \tau_i) + |w|_{p,\tau_i,\tau_{i+1}} \leq K_2$$

for $i = 0, \ldots, N-1$, where $N = N(0,T)$ satisfies (4.31). Then, by (4.26), we have

$$|y|_{q,s,t} \leq K_1 + |y_s| \tag{4.32}$$

for all $s, t \in [\tau_i, \tau_{i+1}]$, $s \leq t$. In view of (4.25), this yields

$$|y_{\tau_{i+1}}| \leq \|y\|_{\infty,\tau_i,\tau_{i+1}} \leq K_1 + 2|y_{\tau_i}|$$

for all $i = 0, \ldots, N-1$. If $N = 1$, then (4.27) trivially holds. Now, let $N \geq 2$, and fix $i \in \{0, \ldots, N-1\}$ such that $\tau_i < t \leq \tau_{i+1}$. Inductively, we get

$$K_1 + |y_{\tau_i}| \leq K_1 + K_1 + 2|y_{\tau_{i-1}}|$$
$$\leq 2(K_1 + |y_{\tau_{i-1}}|) \leq \cdots \leq 2^i(K_1 + |y_0|).$$

Hence,

$$|y|_{q,\tau_i,\tau_{i+1}} \leq K_1 + |y_{\tau_i}| \leq 2^i(K_1 + |y_0|).$$

By Lemma 4.3, we obtain

$$|y|_{q,0,T} \leq N^{\frac{q-1}{q}} \left(\sum_{i=0}^{N-1} |y|_{q,\tau_i,\tau_{i+1}}^q \right)^{\frac{1}{q}} \leq N^{\frac{q-1}{q}} (K_1 + |y_0|) \left(\sum_{i=0}^{N-1} 2^{iq} \right)^{\frac{1}{q}}$$
$$\leq (K_1 + |y_0|)e^{2N}. \qquad (4.33)$$

Taking (4.31) into account, we observe that

$$|y|_{q,0,T} \leq (K_1 + |y_0|)e^{2^p K_2^{-p}(|T|^p + |w|_{p,0,T}^p)}.$$

In view of the inequality (4.25), we conclude that

$$\|y\|_{\infty,0,T} \leq \|y\|_{q,0,T} \leq (K_1 + 2|y_0|)e^{2^p K_2^{-p}(T^p + |w|_{p,0,T}^p)}.$$

Now, suppose (4.32) simplifies to

$$|y|_{q,s,t} \leq K_1.$$

Then, we can directly apply the first inequality in (4.33) to get

$$|y|_{q,0,T} \leq N K_1.$$

By (4.31), the assertions in (4.28) and (4.29) follow. $\square$

4.3.2. Background on Russo–Vallois forward integration

We now turn to the Russo–Vallois forward integral [19, 31], which is defined as in (4.3) above. The key tool from the theory of forward integration in our context is the integration-by-parts formula. It involves the following notion of a generalized covariation. Suppose that $(X_t)_{t\in[0,T]}$ and $(Y_t)_{t\in[0,T]}$ are continuous stochastic processes (extended by constant extrapolation to $t > T$ if necessary). For every $\varepsilon > 0$, the ε-*covariation* is defined as

$$C(\varepsilon, Y, X)(t) = \int_0^t \frac{(X_{s+\varepsilon} - X_s)(Y_{s+\varepsilon} - Y_s)}{\varepsilon} ds.$$

The *generalized covariation* is then defined to be the limit in the sense of uniform convergence in probability, as ε goes to zero, of $C(\varepsilon, Y, X)$. In the case of existence, it is denoted by $[X, Y]_t$. We write $[X]_t = [X, X]_t$ for the *generalized quadratic variation* and note that (provided all terms exist)

$$|[X, Y]_t| \leq ([X]_t [Y]_t)^{1/2}. \tag{4.34}$$

The *integration-by-parts formula* (Proposition 1 in Ref. [31]) now states that

$$X_t Y_t = X_0 Y_0 + \int_0^t X_s d^- Y_s + \int_0^t Y_s d^- X_s + [X, Y]_s \tag{4.35}$$

(provided all terms on the right-hand side exist).

The following theorem relates forward integration to the Itô integration and to the Young integration.

Theorem 4.7.

(1) *Suppose* $(H_t)_{t\in[0,T]}$ *is an* $\mathbb{F}$-*adapted process satisfying* $\int_0^T |H_s|^2 ds < \infty$, P-*a.s., and* $(B_t)_{t\in[0,T]}$ *is an* $\mathbb{F}$-*adapted Brownian motion. Then, the forward integral* $(\int_0^t H_s d^- B_s)_{t\in[0,T]}$ *exists and coincides with the Itô integral* $(\int_0^t H_s dB_s)_{t\in[0,T]}$.

(2) *Suppose* $(X_t)_{t\in[0,T]}$ *is a stochastic process with paths in* $C^p([0,T], \mathbb{R})$ *and* $(H_t)_{t\in[0,T]}$ *is a stochastic process with paths*

in $W^q([0,T], \mathbb{R})$ for $p, q \geq 1$ such that $\frac{1}{p} + \frac{1}{q} > 1$. Then, the forward integral $(\int_0^t H_s d^- X_s)_{t \in [0,T]}$ exists and coincides with the (pathwise) Young integral $(\int_0^t H_s dX_s)_{t \in [0,T]}$. Moreover, $[\int_0^\cdot H_s d^- X_s] = 0$ on $[0, T]$.

Proof. Part (1) is Theorem 2 in Ref. [31]. Part (2) is proved in Proposition 3 in Ref. [31] under additional Hölder assumptions. We next provide a proof for the p-variation case.

Step 1: We show that the forward integral exists and coincides with the Young integral.

For $\varepsilon > 0$, let

$$X_t^{\varepsilon-} = \frac{1}{\varepsilon} \int_0^t X_{r+\varepsilon} - X_r \, dr.$$

Then, $X^{\varepsilon-}$ is continuously differentiable with derivative $\dot{X}_t^{\varepsilon-} = \varepsilon^{-1}(X_{t+\varepsilon} - X_t)$ and

$$I^-(\varepsilon, H, dX)(t) = \int_0^t H_s \dot{X}_s^{\varepsilon-} ds = \int_0^t H_s dX_s^{\varepsilon-},$$

where the integral on the right-hand side is the Riemann–Stieltjes integral with respect to the smooth integrator $X^{\varepsilon-}$ and, thus, equals the Young integral of H with respect to $X^{\varepsilon-}$. Fix some $p' > p$ such that $1/p' + 1/q > 1$. Then, by (4.25) and by the Love–Young inequality in the form of Lemma 4.4,

$$\left\| I^-(\varepsilon, H, dX) - \int_0^\cdot H_s dX_s \right\|_{\infty, 0, T} \leq \left\| \int_0^\cdot H_s d(X^{\varepsilon-} - X)_s \right\|_{p', 0, T}$$

$$\leq C_{p', q} \|H\|_{q, 0, T} |X^{\varepsilon-} - X|_{p', 0, T}.$$

Hence, the forward integral $(\int_0^t H_s d^- X_s)_{t \in [0,T]}$ exists and coincides with the Young integral if

$$\lim_{\varepsilon \to 0} |X^{\varepsilon-} - X|_{p', 0, T} = 0, \quad P\text{-a.s.} \tag{4.36}$$

In order to establish (4.36), we define $Z_t^\varepsilon = X_t^{\varepsilon-} - X_t$ for $t \in [0, T]$. Since

$$X_t^{\varepsilon-} = \frac{1}{\varepsilon} \int_0^t X_{r+\varepsilon} - X_r \, dr = \frac{1}{\varepsilon} \int_t^{t+\varepsilon} X_r \, dr - \frac{1}{\varepsilon} \int_0^\varepsilon X_r \, dr,$$

we obtain, for $0 \leq s \leq t \leq T$,

$$Z_t^\varepsilon - Z_s^\varepsilon = \frac{1}{\varepsilon} \int_t^{t+\varepsilon} X_r - X_t \, dr - \frac{1}{\varepsilon} \int_s^{s+\varepsilon} X_r - X_s \, dr$$

$$= \frac{1}{\varepsilon} \int_0^\varepsilon (X_{t+r} - X_t) - (X_{s+r} - X_s) \, dr.$$

Thus, by Jensen's inequality,

$$|Z_t^\varepsilon - Z_s^\varepsilon|$$

$$\leq \frac{1}{\varepsilon} \int_0^\varepsilon |X_{t+r} - X_t - (X_{s+r} - X_s)| \, dr$$

$$\leq \left(\frac{1}{\varepsilon} \int_0^\varepsilon |X_{t+r} - X_t - (X_{s+r} - X_s)|^{p'-p} \right.$$

$$\left. \cdot |X_{t+r} - X_t - (X_{s+r} - X_s)|^p \, dr \right)^{\frac{1}{p'}}$$

$$\leq 2^{1-\frac{p}{p'}} \sup_{\substack{r,u \in [0,T] \\ |u-r| \leq \varepsilon}} |X_r - X_u|^{1-\frac{p}{p'}}$$

$$\cdot \left(2^{p-1} \frac{1}{\varepsilon} \int_0^\varepsilon |X_{t+r} - X_{s+r}|^p + |X_t - X_s|^p \, dr \right)^{\frac{1}{p'}}$$

$$\leq 2^{1-\frac{1}{p'}} \sup_{\substack{r,u \in [0,T] \\ |u-r| \leq \varepsilon}} |X_r - X_u|^{1-\frac{p}{p'}} \left(\frac{1}{\varepsilon} \int_0^\varepsilon |X_{(\cdot+r)}|_{p,s,t}^p \, dr + |X|_{p,s,t}^p \right)^{\frac{1}{p'}}.$$

In view of Proposition 4.1, it is easy to check that

$$\varphi(s,t) = \frac{1}{\varepsilon} \int_0^\varepsilon |X_{(\cdot+r)}|_{p,s,t}^p \, dr + |X|_{p,s,t}^p$$

is superadditive on $\Delta([0,T])$. Hence, Lemma 4.1 implies

$$|Z^\varepsilon|_{p',0,T}$$

$$\leq 2^{1-\frac{1}{p'}} \sup_{\substack{r,u\in[0,T]\\|u-r|\leq\varepsilon}} |X_r - X_u|^{1-\frac{p}{p'}} \left(\frac{1}{\varepsilon}\int_0^\varepsilon |X_{(\cdot+r)}|^p_{p,0,T}\, dr + |X|^p_{p,0,T}\right)^{\frac{1}{p'}}$$

$$\leq 2^{1-\frac{1}{p'}} \left(2|X|^p_{p,0,T}\right)^{\frac{1}{p'}} \sup_{\substack{r,u\in[0,T]\\|u-r|\leq\varepsilon}} |X_r - X_u|^{1-\frac{p}{p'}}.$$

As the paths of X are uniformly continuous on $[0,T]$, we obtain (4.36).

Step 2: We show that $[X]_t = 0$ for every process X with paths in $C^p([0,T])$.

Choosing $p' \in (p,2)$, the same Love–Young inequality argument as above with $X_{s+\varepsilon} - X_s$ in place of H_s and $X^{\varepsilon-}$ in place of $X^{\varepsilon-} - X$ shows

$$\|C(\varepsilon, X, X)\|_{\infty,0,T} \leq C_{p',p'} \|X_{\cdot+\varepsilon} - X\|_{p',0,T} |X^{\varepsilon-}|_{p',0,T}.$$

By (4.36), the term $|X^{\varepsilon-}|_{p',0,T}$ remains P-a.s. bounded as $\varepsilon \to 0$. Moreover, the proof of (4.36) can be modified in a straightforward way to show that $\|X_{\cdot+\varepsilon} - X\|_{p',0,T} \to 0$ P-a.s. as $\varepsilon \to 0$. Hence, $[X] = 0$ on $[0,T]$.

Step 3: We show that $[\int_0^\cdot H_s d^- X_s] = 0$ on $[0,T]$.

By Step 1 and Lemma 4.4, the process $Z_t = \int_0^t H_s d^- X_s$ has paths in $C^p([0,T])$. Thus, Step 2 applies to Z. □

In the proof of Theorem 4.2, we make use of the integration-by-parts formula in the following form.

Theorem 4.8. *Suppose $(B_t)_{t\in[0,T]}$ is an m_2-dimensional Brownian motion and $(w_t)_{t\in[0,T]}$ is an $\mathbb{F}$-adapted process with paths in $C^p([0,T], \mathbb{R}^{m_1})$ as in the general setting of Section 4.2.1. Suppose A, $\hat{A}$, C^j, $\hat{C}^j$, D^i, and $\hat{D}^i$ ($j = 1, \ldots, m_1$, $i = 1, \ldots, m_2$) are $\mathbb{F}$-adapted processes, taking values in $\mathbb{R}^{m\times n}$ (without hat), and resp.,* in $\mathbb{R}^{n\times k}$ (with hat). Assume that

$$\int_0^T \left(|A_s| + |\hat{A}_s| + \sum_{i=1}^{m_2}\left(|D_s^i|^2 + |\hat{D}_s^i|^2\right)\right) ds < \infty, \quad P\text{-a.s.},$$

and that the processes C^j, $\hat{C}^j$ have paths of bounded q-variation for some $q \geq 1$ satisfying $\frac{1}{p} + \frac{1}{q} > 1$. Let

$$X_t = X_0 + \int_0^t A_s ds + \sum_{j=1}^{m_1} \int_0^t C_s^j d^- w_s^j + \sum_{j=1}^{m_2} \int_0^t D_s^j d^- B_s^j,$$

$$Y_t = Y_0 + \int_0^t \hat{A}_s ds + \sum_{j=1}^{m_1} \int_0^t \hat{C}_s^j d^- w_s^j + \sum_{j=1}^{m_2} \int_0^t \hat{D}_s^j d^- B_s^j.$$

Then, for every $t \in [0,T]$,

$$X_t Y_t = X_0 Y_0 + \int_0^t \left(X_s \hat{A}_s + A_s Y_s + \sum_{j=1}^{m_2} D_s^j \hat{D}_s^j\right) ds$$

$$+ \sum_{j=1}^{m_1}(X_s \hat{C}_s^j + C_s^j Y_s) d^- w_s^j + \sum_{j=1}^{m_2}(X_s \hat{D}_s^j + D_s^j Y_s) d^- B_s^j$$

Sketch of the proof. We consider the scalar-valued case $m = n = k = 1$ only, but note that the extension to the matrix-valued case is straightforward. By Theorem 4.7, all forward integrals exist. By Corollary 2 in Ref. [31] and by polarization, the generalized covariation of two continuous local martingales coincides with the usual cross-variation of local martingales (see, e.g., Chapter 1.1.5 in Ref. [30]). Then, by bilinearity of the generalized covariation in conjunction with the zero quadratic variation property of the Young integrals (Theorem 4.7, (2)) and (4.34),

$$[X,Y]_t = \sum_{j=1}^{m_2}\sum_{i=1}^{m_2}\left[\int_0^\cdot D_s^i dB_s^i, \int_0^\cdot \hat{D}_s^j dB_s^j\right]_t = \sum_{j=1}^{m_2} \int_0^t D_s^j \hat{D}_s^j ds.$$

Thus, (4.35) applies. $\square$

Remark 4.6. Suppose that the forward integral $\int_0^t H_s d^- X_s$ exists and that Z is a random variable. Then, by the definition of the forward integral, it easily follows that

$$\int_0^t Z H_s d^- X_s = Z \int_0^t H_s d^- X_s.$$

In particular, under the assumptions of Theorem 4.7, (1),

$$Z \int_0^t H_s dB_s = \int_0^t H_s d^- B_s = \int_0^t Z H_s d^- B_s.$$

The integral on the right-hand side cannot be interpreted as an Itô integral because the integrand $(ZH_s)_{s\in[0,T]}$ is not $\mathbb{F}$-adapted, unless Z is $\mathcal{F}_0$-measurable.

4.4. Proofs

4.4.1. *On the proof of Theorem 4.1*

In this section, we briefly explain some of the key arguments leading to Theorem 4.1. We first consider the SDE system for $\mathcal{X}^u$ and recall that it can be decomposed into one subsystem driven by the p-variation process w and another one driven by the Brownian motion B. We mainly concentrate on the first subsystem, which reads

$$\xi_t^u = \xi_0(u) + \int_0^t b(r, \xi_r^u, u)\, dr + \sum_{j=1}^{m_1} \int_0^t \sigma^j(r, \xi_r^u, u)\, d^- w_r^j.$$

By Theorem 4.7, the integral with respect to w is a Young integral. Then, existence and uniqueness under (H_1)–(H_3) are direct consequences of Theorem 3.6 in Ref. [25]. Note that Ref. [25] is concerned with Young differential equations driven by a deterministic p-variation function. This is the typical framework for Young differential equations (cp. also Ref. [23]). We, thus, apply their results pathwise, i.e., for a fixed realization $w(\omega)$ of the p-variation process w. As the cost functional J in (4.5) averages over the realizations by taking an expectation, we need to control the growth of the solutions depending on w. This is the reason to impose the boundedness assumptions on b and σ, which are, in fact, not required for existence and uniqueness.

Lemma 4.6. *Under (H1)–(H3), there is a constant C_1 independent of u and (the realization of) w such that for every $0 \leq s \leq t \leq T$,*

$$|\xi^u|_{p,s,t} \leq \frac{1}{2C_1}(1 + |\xi^u|_{p,s,t})((t-s) + |w|_{p,s,t}).$$

A routine proof, which relies on the Love–Young inequality (4.24) and standard Lipschitz estimates, can be found in Ref. [26, Lemma 2.27]. As a consequence of the previous lemma, we observe that

$$(t-s) + |w|_{p,s,t} \leq C_1 \quad \Rightarrow \quad |\xi^u|_{p,s,t} \leq 1. \tag{4.37}$$

Combining (4.37) with (4.29) yields

$$\|\xi^u\|_{p,0,T} \leq L + 2^{p-1}C_1^{-p}\left(T^p + |w|_{p,0,T}^p\right), \tag{4.38}$$

where L is any upper bound for $u \mapsto |\xi_0(u)|$. Then, by (4.6),

$$\sup_{u \in \mathcal{U}} \mathrm{E}[\|\xi^u\|_{\infty,0,T}^l] < \infty$$

for every $l \geq 1$. Summarizing, the boundedness assumption on b and σ ensures that the p-variation norm of ξ^u grows linearly in $|w|_{p,0,T}^p$, while without the boundedness assumption the solutions can grow exponentially in $|w|_{p,0,T}^p$; cp. Proposition 1 in Ref. [23]. The p-variation estimate for ξ^u obtained in (4.38) will turn out to be crucial for controlling the growth of the Fréchet derivative of ξ to which we turn now.

We first provide a heuristic derivation of the SDE for the Fréchet derivative of ξ in the parameter. To this end, fix $u \in \mathcal{U}$, a vector $\bar{u} \in \mathbb{R}^d$ of length 1, and choose ϵ sufficiently small such that $u_\epsilon := u + \epsilon\bar{u} \in \mathcal{U}$. Then, the difference quotient for the directional derivative reads

$$\frac{\xi_t^{u_\epsilon} - \xi_t^u}{\epsilon} := \frac{\xi_0(u_\epsilon) - \xi_0(u)}{\epsilon} + \int_0^t \frac{b(r, \xi_r^{u_\epsilon}, u_\epsilon) - b(r, \xi_r^u, u_\epsilon)}{\epsilon}$$
$$+ \frac{b(r, \xi_r^u, u_\epsilon) - b(r, \xi_r^u, u)}{\epsilon} dr$$
$$+ \sum_{j=1}^{m_1} \int_0^t \frac{\sigma^j(r, \xi_r^{u_\epsilon}, u_\epsilon) - \sigma^j(r, \xi_r^u, u_\epsilon)}{\epsilon}$$
$$+ \frac{\sigma^j(r, \xi_r^u, u_\epsilon) - \sigma^j(r, \xi_r^u, u)}{\epsilon} d^- w_r^j.$$

Passing formally to the limit $\epsilon \to 0$ suggests that the directional derivative of ξ in direction $\bar{u}$ at u is given by the solution $y^{u,\bar{u}}$ of the linear SDE

$$y_t^{u,\bar{u}} = D\xi_0(u)\bar{u} + \int_0^t \left(b_\xi(r, \xi_r^u, u) y_r^{u,\bar{u}} + b_u(r, \xi_r^u, u) \bar{u} \right) dr$$
$$+ \sum_{j=1}^{m_1} \int_0^t \left(\sigma_\xi^j(r, \xi_r^u, u) y_r^{u,\bar{u}} + \sigma_u^j(r, \xi_r^u, u) \bar{u} \right) d^- w_r^j.$$

Assuming, for the moment, that the Fréchet derivative y^u of ξ at u exists, we obtain $y^{u,\bar{u}} = y^u \cdot \bar{u}$. Thus, y^u solves

$$y_t^u = D\xi_0(u) + \int_0^t \left(b_\xi(r, \xi_r^u, u) y_r^u + b_u(r, \xi_r^u, u) \right) dr$$
$$+ \sum_{j=1}^{m_1} \int_0^t \left(\sigma_\xi^j(r, \xi_r^u, u) y_r^u + \sigma_u^j(r, \xi_r^u, u) \right) d^- w_r^j, \quad (4.39)$$

corresponding to the first n_1 lines of the matrix-valued SDE (4.7) for $\mathcal{Y}^u$. This heuristic argument can be made rigorous in a similar way as, e.g., in Ref. [43] (in the framework of controlled SDEs driven by a fractional Brownian motion, where Hölder norms are applied) or in Proposition 8 in Ref. [23] (where parameter dependence in the initial condition in a p-variation setting is considered); see Section 2.1.3 in Ref. [26] for the details. As before, all arguments leading to the differentiability of the Young SDE ξ^u are applied pathwise. Hence, in order to interchange differentiation and expectation when deriving the gradient representation (4.8), uniform integrability of the difference quotients is required. In view of the mean-value theorem and the de la Vallée–Poussin criterion for uniform integrability, this problem can be reduced to bounding the $L^l(\Omega, P)$-norm of $\|y^u\|_{\infty,0,T}$ uniformly in $u \in \mathcal{U}$. The following lemma explains how to derive suitable bounds for a simplified equation (to avoid unnecessary technicalities).

Lemma 4.7. *Suppose that $m_1 = 1$, $z_0 : \mathcal{U} \to \mathbb{R}^{n_1}$ is bounded, and $f : \mathbb{R}^{n_1} \to \mathbb{R}^{n_1 \times n_1}$ is bounded and Lipschitz continuous and that the $\mathbb{R}^{n_1}$-valued process z^u solves*

$$z_t^u = z_0(u) + \int_0^t f(\xi_s^u) z_s^u \, dw_s, \quad 0 \le t \le T$$

(in the sense of Young integration). Then, there is a constant C independent of u and (the realization of) w such that

$$\|z^u\|_{p,0,T} \le 2|z_0(u)| e^{C(T^p + |w|_{p,0,T}^p)}.$$

Proof. Fix $L \ge 0$ sufficiently large such that f is bounded by L and L is a Lipschitz constant for f. By the Love–Young inequality in the form of Lemma 4.4, there is a universal constant C_p such that, for every $0 \le s \le t \le T$,

$$|z^u|_{p,s,t} \le C_p \|f(\xi_\cdot^u) z_\cdot^u\|_{p,s,t} |w|_{p,s,t}.$$

In view of Lemma 4.2,

$$\|f(\xi_\cdot^u) z_\cdot^u\|_{p,s,t} \le 2 \|f(\xi_\cdot^u)\|_{p,s,t} \|z_\cdot^u\|_{p,s,t}$$

$$\le 2L(1 + |\xi^u|_{p,s,t})(|z_s^u| + |z^u|_{p,s,t}).$$

Hence,

$$|z^u|_{p,s,t} \le 2LC_p(1 + |\xi^u|_{p,s,t})(|z_s^u| + |z^u|_{p,s,t})|w|_{p,s,t}. \quad (4.40)$$

By (4.37), there is a constant C_1 independent of u and w such that $|\xi^u|_{p,s,t} \le 1$, if $(t-s) + |w|_{p,s,t} \le C_1$. Let $C_2 := \min\{C_1, (8LC_p)^{-1}\}$. Then, if $(t-s) + |w|_{p,s,t} \le C_2$,

$$|z^u|_{p,s,t} \le 4LC_p(|z_s^u| + |z^u|_{p,s,t})|w|_{p,s,t} \le \frac{1}{2}(|z_s^u| + |z^u|_{p,s,t}),$$

i.e., $|z^u|_{p,s,t} \leq |z^u_s|$. Hence, Gronwall's inequality (Lemma 4.5) yields

$$\|z^u\|_{p,0,T} \leq 2|z_0(u)|e^{C(T^p+|w|^p_{p,0,T})}$$

for $C = 2^p C_2^{-p}$. □

An important observation of the proof is that p-variation estimates for z^u depend on the p-variation regularity of ξ^u via the coefficient $f(\xi^u)$. The boundedness assumptions on b and σ allow to control $|\xi^u|_{p,s,t}$ via Lemma 4.6 and lead to an exponential bound for the p-variation norm of z^u in terms of $|w|_{p,s,t}$. In view of the exponential moment bound (4.6) and taking the boundedness of z_0 as a function in u into account, we conclude that for every $l \geq 1$,

$$\sup_{u \in \mathcal{U}} \mathrm{E}[\|z^u\|^l_{\infty,0,T}] < \infty.$$

With a little extra effort (but essentially the same argument), the same type of estimate can be obtained for y^u in place z^u.

Having the results on the differentiability of ξ^u in the parameter at hand, one can proceed to study the second subsystem of (4.4) given by

$$x^u_t = x_0(u) + \int_0^t \hat{b}(r, \xi^u_r, x^u_r, u)\, dr + \sum_{j=1}^{m_2} \int_0^t \hat{\sigma}^j(r, \xi^u_r, x^u_r, u)\, d^-B^j_r,$$

where the forward integrals coincide with the Itô integrals by Theorem 4.7. Existence and uniqueness are standard results under the Lipschitz conditions implied by (B_1)–(B_3); see, e.g., Chapter 1.6 in Ref. [34]. The differentiability of SDEs with respect to a parameter is also classical in the semimartingale case; see, e.g., Theorem 39 in Ref. [44]. For the particular SDE satisfied by x^u, some technicalities related to the coupling of ξ^u with the equation must be taken into account, but the proofs follow routine argument. Of course, in contrast to the Young integration, Itô's stochastic calculus is tailor-made for obtaining the required $L^l(\Omega, P)$-bounds via the Burkholder–Davis–Gundy inequality.

4.4.2. *On the proof of Theorem 4.2*

In this section, we sketch the proof of Theorem 4.2. We first discuss the existence and uniqueness of the matrix-valued homogeneous equations (4.9)–(4.10). Given the specific form of these equations, it is straightforward to check that solution processes need to be of the form

$$\Phi_t^{s_0} = \begin{pmatrix} \phi_t^{s_0} & 0 \\ \tilde{\phi}_t^{s_0} & \hat{\phi}_t^{s_0} \end{pmatrix}, \quad \Psi_t^{s_0} = \begin{pmatrix} \psi_t^{s_0} & 0 \\ \tilde{\psi}_t^{s_0} & \hat{\psi}_t^{s_0} \end{pmatrix},$$

for every $t \in [s_0, T]$. Here, the "component" processes solve the lower-dimensional matrix-valued SDEs

$$\phi_t^{s_0} = I_{n_1} + \int_{s_0}^t b_\xi^u(r) \phi_r^{s_0}\, dr + \sum_{j=1}^m \int_{s_0}^t \sigma_\xi^{u,j}(r) \phi_r^{s_0}\, d^- w_r^j,$$

$$\hat{\phi}_t^{s_0} = I_{n_2} + \int_{s_0}^t \hat{b}_x^u(r) \hat{\phi}_r^{s_0}\, dr + \sum_{j=1}^{m_2} \int_{s_0}^t \hat{\sigma}_x^{u,j}(r) \hat{\phi}_r^u\, d^- B_r^j,$$

$$\tilde{\phi}_t^{s_0} = \int_{s_0}^t \hat{b}_x^u(r) \tilde{\phi}_r^{s_0} + \hat{b}_\xi^u(r) \phi_r^{s_0}\, dr + \sum_{j=1}^{m_2} \int_{s_0}^t \hat{\sigma}_x^{u,j}(r) \tilde{\phi}_r^{s_0} + \hat{\sigma}_\xi^{u,j}(r) \phi_r^{s_0} d^- B_r^j$$

and

$$\psi_t^{s_0} = I_{n_1} - \int_{s_0}^t \psi_r^{s_0} b_\xi^u(r)\, dr - \sum_{j=1}^m \int_{s_0}^t \psi_r^{s_0} \sigma_\xi^{u,j}(r)\, d^- w_r^j,$$

$$\hat{\psi}_t^{s_0} = I_{n_2} - \int_{s_0}^t \hat{\psi}_r^{s_0} \left(\hat{b}_x^u(r) - \sum_{j=1}^{m_2} \hat{\sigma}_x^{u,j}(r)^2 \right) dr - \sum_{j=1}^{m_2} \int_{s_0}^t \hat{\psi}_r^{s_0} \hat{\sigma}_x^{u,j}(r)\, d^- B_r^j,$$

$$\tilde{\psi}_t^{s_0} = -\int_{s_0}^t \tilde{\psi}_r^{s_0} b_\xi^u(r) + \hat{\psi}_r^{s_0} \left[\hat{b}_\xi^u(r) - \sum_{j=1}^{m_2} \hat{\sigma}_x^{u,j}(r) \hat{\sigma}_\xi^{u,j}(r) \right] dr$$

$$- \sum_{j=1}^{m_1} \int_{s_0}^t \tilde{\psi}_r^{s_0} \sigma_\xi^{u,j}(r)\, d^- w_r^j - \sum_{j=1}^{m_2} \int_{s_0}^t \hat{\psi}_r^{s_0} \hat{\sigma}_\xi^{u,j}(r) d^- B_r^j,$$

respectively. By Theorem 4.7, the first equation of each of the systems is, pathwise, a Young differential equation driven by the p-variation process w, for which existence and uniqueness can be reduced to Proposition 2.2 in Ref. [45]. A bound of the form $C \exp\{C|w|_{p,0,T}^p\}$ for the p-variation norm of the solutions can be derived using the techniques explained in Lemma 4.7, which, in view of the exponential moment bound (4.6), implies that ϕ^{s_0}, $\psi^{s_0} \in L_{\mathbb{F}}^l(\Omega, C^{p,0}[s_0, T], \mathbb{R}^{n_1 \times n_1})$. The second and third equations in the system for Φ^{s_0} and the second equation in the system for Ψ^{s_0} are linear matrix-valued SDEs driven by a Brownian motion in the sense of Itô integration (applying Theorem 4.7, again). Existence, uniqueness, and L^l-integrability are classical for these equations; see, e.g., Chapter 1.6 in Ref. [34]. Thus, the most interesting equation is the one for $\tilde{\psi}^{s_0}$, which features a linear term in $\tilde{\psi}^{s_0}$ inside the forward integral w.r.t. to the p-variation process w and an inhomogeneity in terms of the forward integral w.r.t to the Brownian motion B. A formal application of the variation-of-constants formula provides the candidate solution

$$\tilde{\psi}_t^{s_0} := \left[-\int_{s_0}^t \hat{\psi}_r^{s_0} \left[\hat{b}_\xi^u(r) - \sum_{j=1}^{m_2} \hat{\sigma}_x^{u,j}(r) \hat{\sigma}_\xi^{u,j}(r) \right] (\psi_r^{s_0})^{-1} dr \right.$$
$$\left. - \sum_{j=1}^{m_2} \int_{s_0}^t \hat{\psi}_r^{s_0} \hat{\sigma}_\xi^{u,j}(r) (\psi_r^{s_0})^{-1} dB_r^j \right] \psi_t^{s_0} =: X_t \psi_t^{s_0}.$$

Now, the integration-by-parts formula in Theorem 4.8 yields

$$X_t \psi_t^{s_0} = -\int_{s_0}^t \left(\hat{\psi}_r^{s_0} \left[\hat{b}_\xi^u(r) - \sum_{j=1}^{m_2} \hat{\sigma}_x^{u,j}(r) \hat{\sigma}_\xi^{u,j}(r) \right] + (X_r \psi_r^{s_0}) b_\xi^u(r) \right) dr$$
$$- \sum_{j=1}^m \int_{s_0}^t (X_r \psi_r^{s_0}) \sigma_\xi^{u,j}(r) d^- w_r^j - \sum_{j=1}^{m_2} \int_{s_0}^t \hat{\psi}_r^{s_0} \hat{\sigma}_\xi^{u,j}(r) dB_r^j,$$

i.e., $\tilde{\psi}^{s_0} = X \psi^{s_0}$ is a solution to the last SDE in the Ψ^{s_0} system. Note that ψ^{s_0} is indeed invertible and $\phi^{s_0} = (\psi^{s_0})^{-1}$, which can again be verified through integration by parts. The L^l-integrability of $\tilde{\psi}^{s_0}$ is a simple consequence of Hölder's inequality, the Burkholder–Davis–Gundy inequality and the already established integrability properties of $\hat{\psi}^{s_0}$, ψ^{s_0}, and ϕ^{s_0}. Uniqueness can be derived by computing $\tilde{\psi}^{s_0} \phi^{s_0}$ in the same way (where $\tilde{\psi}^{s_0}$ is an arbitrary solution to the last SDE in the Ψ^{s_0} system) and using once more that $\phi^{s_0} = (\psi^{s_0})^{-1}$.

With existence, uniqueness, and the required integrability properties of Φ^{s_0} and Ψ^{s_0} at hand, a direct computation using Theorem 4.8

once again shows $\Psi^{so}\Phi^{so} = I_{n_1+n_2}$, i.e., Φ^{so} and Ψ^{so} are the inverses to each other. Write $\Phi = \Phi^0$, and define

$$Y_t := \begin{pmatrix} D\xi_0(u) \\ Dx_0(u) \end{pmatrix} + \int_0^t (\Phi_r)^{-1} \left[\begin{pmatrix} b_u^u(r) \\ \hat{b}_u^u(r) \end{pmatrix} - \sum_{j=1}^{m_2} \begin{pmatrix} 0 \\ \hat{\sigma}_x^{u,j}(r)\hat{\sigma}_u^{u,j}(r) \end{pmatrix} \right] dr$$

$$+ \sum_{j=1}^{m_1} \int_0^t (\Phi_r)^{-1} \begin{pmatrix} \sigma_u^{u,j}(r) \\ 0 \end{pmatrix} d^- w_r^j$$

$$+ \sum_{j=1}^{m_2} \int_0^t (\Phi_r)^{-1} \begin{pmatrix} 0 \\ \hat{\sigma}_u^{u,j}(r) \end{pmatrix} d^- B_r^j.$$

Then, a final application of the integration-by-parts formula verifies that $\mathcal{Y}_t^u = \Phi_t Y_t$ solves (4.7).

Note that this argument also implies the existence of a solution for (4.7). Uniqueness can be derived in the same way by computing $\Psi_t^0 \mathcal{Y}_t^u$ for some arbitrary solution $\mathcal{Y}_t^u$ to (4.7).

4.4.3. On the proof of Theorem 4.3

Before we derive the adjoint equation for $(\Lambda_t)_{t \in [0,T]}$, we first prove the gradient representation (4.12).

Lemma 4.8. *The gradient of the cost functional J admits the representation (4.12).*

Proof. Inserting (4.11) into (4.8) and taking Remark 4.6 into account, we get

$$\nabla J(u) = E\left[\sum_{\mu=1}^M E[g_\mu(\mathcal{X}_{T_\mu}^u)]g_\mu'(\mathcal{X}_{T_\mu}^u)\Phi_{T_\mu} \begin{pmatrix} D\xi_0(u) \\ Dx_0(u) \end{pmatrix} \right.$$

$$+ \sum_{\mu=1}^M \int_0^{T_\mu} E[g_\mu(\mathcal{X}_{T_\mu}^u)]g_\mu'(\mathcal{X}_{T_\mu}^u)\Phi_{T_\mu}\Phi_r^{-1}$$

$$\cdot \left[\begin{pmatrix} b_u^u(r) \\ \hat{b}_u^u(r) \end{pmatrix} - \sum_{j=1}^{m_2} \begin{pmatrix} 0 \\ \hat{\sigma}_x^{u,j}(r)\hat{\sigma}_u^{u,j}(r) \end{pmatrix} \right] dr$$

$$+ \sum_{\mu=1}^{M} \sum_{j=1}^{m_1} \int_0^{T_\mu} E[g_\mu(\mathcal{X}^u_{T_\mu})] g'_\mu(\mathcal{X}^u_{T_\mu}) \Phi_{T_\mu} \Phi_r^{-1} \begin{pmatrix} \sigma_u^{u,j}(r) \\ 0 \end{pmatrix} d^- w_r^j$$

$$+ \sum_{\mu=1}^{M} \sum_{j=1}^{m_2} \int_0^{T_\mu} E[g_\mu(\mathcal{X}^u_{T_\mu})] g'_\mu(\mathcal{X}^u_{T_\mu}) \Phi_{T_\mu} \Phi_r^{-1} \begin{pmatrix} 0 \\ \hat{\sigma}_u^{u,j}(r) \end{pmatrix} d^- B_r^j \Bigg].$$

By interchanging summation and integration, we then obtain

$$\nabla J(u)$$

$$= \mathrm{E}\Bigg[\sum_{\mu=1}^{M} E[g_\mu(\mathcal{X}^u_{T_\mu})] g'_\mu(\mathcal{X}^u_{T_\mu}) \Phi_{T_\mu} \begin{pmatrix} D\xi_0(u) \\ Dx_0(u) \end{pmatrix}$$

$$+ \int_0^T \sum_{\mu; T_\mu \geq r} E[g_\mu(\mathcal{X}^u_{T_\mu})] g'_\mu(\mathcal{X}^u_{T_\mu}) \Phi_{T_\mu} \Phi_r^{-1}$$

$$\cdot \Bigg[\begin{pmatrix} b_u^u(r) \\ \hat{b}_u^u(r) \end{pmatrix} - \sum_{j=1}^{m_2} \begin{pmatrix} 0 \\ \hat{\sigma}_x^{u,j}(r) \hat{\sigma}_u^{u,j}(r) \end{pmatrix} \Bigg] dr$$

$$+ \sum_{j=1}^{m_1} \int_0^T \sum_{\mu; T_\mu \geq r} E[g_\mu(\mathcal{X}^u_{T_\mu})] g'_\mu(\mathcal{X}^u_{T_\mu}) \Phi_{T_\mu} \Phi_r^{-1} \begin{pmatrix} \sigma_u^{u,j}(r) \\ 0 \end{pmatrix} d^- w_r^j$$

$$+ \sum_{j=1}^{m_2} \int_0^T \sum_{\mu; T_\mu \geq r} E[g_\mu(\mathcal{X}^u_{T_\mu})]^\top g'_\mu(\mathcal{X}^u_{T_\mu}) \Phi_{T_\mu} \Phi_r^{-1} \begin{pmatrix} 0 \\ \hat{\sigma}_u^{u,j}(r) \end{pmatrix} d^- B_r^j \Bigg].$$

Substituting the definition (4.13) of Λ into this expression finally yields (4.12). $\square$

Proof of Theorem 4.3. The integrability of Λ is inherited from Φ and Ψ. Recall that $\Phi_t^{-1} = \Psi_t$ by Theorem 4.2. Inserting the expression (4.10) for $\Psi_t - \Psi_{T_\mu}$ into the definition of Λ and interchanging summation and integration again, we obtain, thanks to Remark 4.6,

$$\Lambda_t = \sum_{T_\mu \geq t} E[g_\mu(\mathcal{X}^u_{T_\mu})] g'_\mu(\mathcal{X}^u_{T_\mu}) \Phi_{T_\mu} \Phi_t^{-1}$$

$$= \sum_{T_\mu \geq t} E[g_\mu(\mathcal{X}^u_{T_\mu})] g'_\mu(\mathcal{X}^u_{T_\mu}) \Phi_{T_\mu} (\Phi_{T_\mu}^{-1} + \Psi_t - \Psi_{T_\mu})$$

$$= \sum_{T_\mu \geq t} E[g_\mu(\mathcal{X}_{T_\mu}^u)]g_\mu'(\mathcal{X}_{T_\mu}^u) + \int_t^T \sum_{T_\mu \geq r} E[g_\mu(\mathcal{X}_{T_\mu}^u)]g_\mu'(\mathcal{X}_{T_\mu}^u)\Phi_{T_\mu}\Phi_r^{-1}$$

$$\cdot \left[\begin{pmatrix} b_\xi^u(r) & 0 \\ \hat{b}_\xi^u(r) & \hat{b}_x^u(r) \end{pmatrix} - \sum_{j=1}^{m_2} \begin{pmatrix} 0 & 0 \\ \hat{\sigma}_\xi^u(r) & \hat{\sigma}_x^{u,j}(r) \end{pmatrix}^2 \right] dr$$

$$+ \sum_{j=1}^{m_1} \int_t^T \sum_{T_\mu \geq r} E[g_\mu(\mathcal{X}_{T_\mu}^u)]g_\mu'(\mathcal{X}_{T_\mu}^u)\Phi_{T_\mu}\Phi_r^{-1} \begin{pmatrix} \sigma_\xi^{u,j}(r) & 0 \\ 0 & 0 \end{pmatrix} d^-w_r^j$$

$$+ \sum_{j=1}^{m_2} \int_t^T \sum_{T_\mu \geq r} E[g_\mu(\mathcal{X}_{T_\mu}^u)]g_\mu'(\mathcal{X}_{T_\mu}^u)\Phi_{T_\mu}\Phi_r^{-1}$$

$$\times \begin{pmatrix} 0 & 0 \\ \hat{\sigma}_\xi^{u,j}(r) & \hat{\sigma}_x^{u,j}(r) \end{pmatrix} d^-B_r^j.$$

Recalling the definition of Λ_r, the proof is completed. □

4.4.4. On the proof of Theorem 4.4

As in Section 4.4.1, we place emphasis on the techniques of proof for the Young SDEs. The key difficulty is to control the dependence of the constants on the driving path w in order to come up with $L^l(\Omega, P)$-estimates. For the sake of illustration, we consider the following simplified variant of the linear equation for y^u in (4.39):

$$z_t^u = z_0(u) + \int_0^t f(\xi_s^u) z_s^u dw_s, \quad 0 \leq t \leq T, \tag{4.41}$$

under the assumptions of Lemma 4.7. Given a partition $\Pi^E = (t_i)_{i=0,\dots,n}$ of $[0, T]$, we consider the Euler scheme

$$z_t^{n,u} = z_{t_i}^{n,u} + f(\xi_{t_i}^u) z_{t_i}^{n,u}(w_t - w_{t_i}), \; t \in (t_i, t_{i+1}], \; z_0^{n,u} = z_0(u). \tag{4.42}$$

Note that the Euler scheme should actually depend on the Euler approximation $\xi^{n,u}$ for ξ^u via the coefficient $f(\xi_{t_i}^{n,u})$, leading to an extra error term; however, the scheme in (4.42) already contains all essential difficulties, including the dependence of the coefficient on the path of ξ^u in (4.41).

Theorem 4.9. *Under the standing assumptions and the assumptions specified in Lemma 4.7, there is a constant C independent of $\Pi^E = (t_i)_{i=0,\ldots,n}$, u, and (the realization of) w such that*

$$\|z^u - z^{n,u}\|_{p,0,T} \leq \left(\max_{i=0,\ldots,n-1} |t_{i+1} - t_i| + |w|_{p,t_i,t_{i+1}} \right)^{2-p} C e^{C|w|_{p,0,T}^p}.$$

In view of the exponential moment bound (4.6), Theorem 4.9 and Hölder's inequality imply the existence of a constant $C_{z,l}$ such that

$$E[\|z^u - z^{n,u}\|_{\infty,0,T}^l]^{1/l} \leq C_{z,l}(\delta_{2l}(\Pi^E))^{2-p}$$

for every $l \geq 1$ (cp. Theorem 4.4), where δ_l is defined in (4.21).

The exponential bound on $|w|_{p,0,T}^p$ in Theorem 4.9 suggests an application of Gronwall's lemma. However, Lemma 4.5 is not well suited for the Euler approximation. Indeed, for $t_i < s < t < t_{i+1}$, any estimate for $|z^u - z^{n,u}|_{p,s,t}$ will depend on $z_{t_i}^{n,u}$ (where t_i is outside the interval $[s,t]$), while Lemma 4.5 requires an estimate in terms of $|z_s^u - z_s^{n,u}|$. The following variant of Gronwall's lemma is tailor-made to deal with such a situation and will be applied in the proof of Theorem 4.9.

Lemma 4.9 (Gronwall-type lemma on the Euler partition). *Let $\Pi^E = (t_i)_{i=0,\ldots,n}$ be a partition of $[0,T]$, and let $x \in W^p([0,T], \mathbb{R}^{n \times d})$, where $p \in (1,2)$. Furthermore, let $w : [0,T] \to \mathbb{R}^m$ ($m = d$ or $m = 1$) be a continuous function of finite p-variation and $K_1, a > 0$ be constants. If for every $t_i \in \Pi^E$, $i \in \{0,\ldots,n-1\}$, we have*

$$|x|_{p,t_i,t_{i+1}} \leq a(K_1 + |x_{t_i}|)(|t_{i+1} - t_i| + |w|_{p,t_i,t_{i+1}}), \tag{4.43}$$

and if there exists a constant $K_2 \leq \frac{1}{a}$ such that for every $t_l, t_k \in \Pi^E$ with $0 \leq t_l < t_{l+1} < t_k \leq T$,

$$|t_k - t_l| + |w|_{p,t_l,t_k} \leq K_2 \quad \Rightarrow \quad |x|_{p,t_l,t_k} \leq K_1 + |x_{t_l}|, \tag{4.44}$$

then

$$|x|_{p,0,T} \leq \frac{1}{2}(K_1 + |x_0|)\left(2^p K_2^{-p}\left(T^p + |w|_{p,0,T}^p\right) + 1\right)$$
$$\cdot \exp\left(2^p 3 K_2^{-p}\left(T^p + |w|_{p,0,T}^p\right) + 2\right).$$

Compared to Lemma 4.5, the estimate (4.44) only needs to hold for intervals whose boundary points are from the Euler partition Π^E, while the corresponding estimate (4.26) in Lemma 4.5 must be verified for all subintervals $[s,t] \subset [0,T]$. The price to pay is the extra condition (4.43) on the p-variation of x on the small subintervals $[t_i, t_{i+1}]$.

For the proof, we first need to introduce some notation. Write $\Pi^g = (\tau_i)_{i=0,\ldots,N}$ for the greedy sequence defined via (4.30) with $\mu = K_2$. As the points in the greedy sequence, in general, are not included in the Euler partition Π^E, we approximate them using neighboring points in the Euler partition. This leads to the subpartition Π^c of Π^E consisting of the time points $t \in \Pi^E$ satisfying

$$\exists \tau \in \Pi^g \text{ such that } t = t_{\underline{n}(\tau)} \text{ or } t = t_{\overline{n}(\tau)},$$

where

$$\underline{n} : [0,T] \to \mathbb{N}, s \mapsto \min\{i \in \{0,\ldots,n\} \mid t_i \in \Pi^E \text{ and } t_i \geq s\},$$

$$\overline{n} : [0,T] \to \mathbb{N}, s \mapsto \max\{i \in \{0,\ldots,n\} \mid t_i \in \Pi^E \text{ and } t_i \leq s\}.$$

Generic points in Π^c will be denoted by θ_j (with the convention $\theta_j < \theta_{j+1}$). The construction of Π^c is illustrated in Figure 4.1.

We mention the following properties of Π^c:

i) If $\tau = t$ for a $\tau \in \Pi^g$ and $t \in \Pi^E$, then there exists $\theta \in \Pi^c$ such that $\theta = t = t_{\underline{n}(\tau)} = t_{\overline{n}(\tau)}$.
ii) There can be multiple partition points $\tau \in \Pi^E$ such that $\theta_j = t_{\overline{n}(\tau)}$ and $\theta_{j+1} = t_{\underline{n}(\tau)}$, e.g., τ_1, τ_2 in Figure 4.1.

In the situation mentioned in (ii), let $\tau_{j-1} < \theta_i \leq \tau_j < \cdots < \tau_{j+m} \leq \theta_{i+1} < \tau_{j+m+1}$. Then, $m = N(\theta_i, \theta_{i+1})$, where $N(s,t)$ has been defined immediately before (4.31). Moreover, by Lemma 4.3

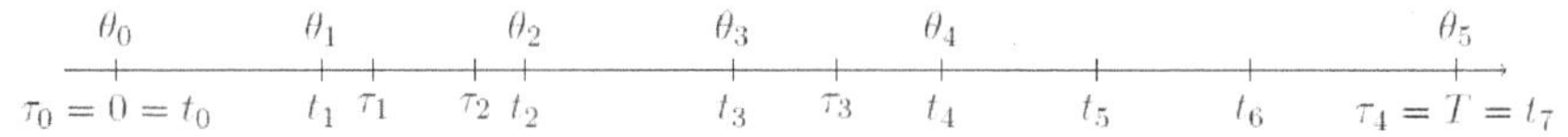

Fig. 4.1. Graphical illustration of the construction of the partition Π^c.

and the defining property (4.30) of the greedy sequence,

$$|\theta_{i+1} - \theta_i| + |w|_{p,\theta_i,\theta_{i+1}}$$
$$\leq |\tau_{j+m+1} - \tau_{j-1}| + |w|_{p,\tau_{j-1},\tau_{j+m+1}}$$
$$\leq \sum_{i=0}^{m+1} |\tau_{j+i} - \tau_{j-1+i}| + \left((m+2)^{p-1} \sum_{i=0}^{m+1} |w|_{p,\tau_{j-1+i},\tau_{j+i}}^p\right)^{\frac{1}{p}}$$
$$\leq \sum_{i=0}^{m+1} |\tau_{j+i} - \tau_{j-1+i}| + (m+2)^{1-\frac{1}{p}} \sum_{i=0}^{m+1} |w|_{p,\tau_{j-1+i},\tau_{j+i}}$$
$$\leq (N(\theta_i,\theta_{i+1}) + 2)^{1-\frac{1}{p}} \sum_{i=0}^{N(\theta_i,\theta_{i+1})+1} (|\tau_{j+i} - \tau_{j-1+i}| + |w|_{p,\tau_{j-1+i},\tau_{j+i}})$$
$$\leq (N(\theta_i,\theta_{i+1}) + 2)^{2-\frac{1}{p}} K_2. \tag{4.45}$$

Concerning Π^c, we also introduce the notation

$$\underline{\mathcal{N}} : [0,T] \to \mathbb{N}, s \mapsto \min\{i \in \mathbb{N}_0 | \theta_i \in \Pi^c \text{ and } \theta_i \geq s\},$$
$$\overline{\mathcal{N}} : [0,T] \to \mathbb{N}, s \mapsto \max\{i \in \mathbb{N}_0 | \theta_i \in \Pi^c \text{ and } \theta_i \leq s\}.$$

We also define $\mathcal{N}(s,t) := \overline{\mathcal{N}}(t) - \underline{\mathcal{N}}(s)$. Then, by construction, $\mathcal{N}(s,t) \leq 2N(s,t) + 1$ for all $(s,t) \in \Delta([0,T])$.

We note that an alternative way to transfer the greedy sequence technique to partitions has been recently suggested in Ref. [46], which can be used to bound the discrete-time p-variation norm (i.e., for functions restricted to the grid only) of solutions to stochastic difference equations.

Proof of Lemma 4.9. Recall that the greedy sequence is constructed for the constant $\mu = K_2$. The number of subintervals defined by the partitions Π^g and Π^c is denoted by $N = N(0,T)$ and $\mathcal{N} = \mathcal{N}(0,T)$, respectively.

We consider the p-variation of x on the subintervals $[\theta_i, \theta_{i+1}]$ of the partition Π^c for $i \in \{0, \ldots, \mathcal{N} - 1\}$, and we distinguish the following cases.

Case 1: There exist $\tau_l \in \Pi^g$ and $i \in \{0, \ldots, \mathcal{N} - 1\}$ such that $\theta_i = t_{\overline{n}(\tau_l)}$ and $\theta_{i+1} = t_{\underline{n}(\tau_l)}$ (e.g., $[\theta_1, \theta_2]$, $[\theta_3, \theta_4]$ in Figure 4.1).

By construction, it follows that there exists $j \in \{0, \ldots, n\}$ such that $\theta_i = t_j$ and $\theta_{i+1} = t_{j+1}$. We estimate using (4.43), (4.45), and $aK_2 \leq 1$,

$$|x|_{p,\theta_i,\theta_{i+1}} \leq a(K_1 + |x_{\theta_i}|)(|\theta_{i+1} - \theta_i| + |w|_{p,\theta_i,\theta_{i+1}})$$
$$\leq (K_1 + |x_{\theta_i}|)(N(\theta_i, \theta_{i+1}) + 2)^{2-\frac{1}{p}}. \quad (4.46)$$

Case 2: There exist $\tau_j, \tau_{j+1} \in \Pi^g$ and $i \in \{0, \ldots, \mathcal{N} - 1\}$ such that $\theta_i = t_{\underline{n}(\tau_j)}$ and $\theta_{i+1} = t_{\overline{n}(\tau_{j+1})}$ (e.g., $[\theta_i, \theta_{i+1}]$ for $i \in \{0, 2, 4\}$ in Figure 4.1). Then, there exists a finite number $k - l = m \geq 1$ of subintervals of Π^E in the interval $[\theta_i, \theta_{i+1}]$. Let $\theta_i = t_l < t_{l+1} < \cdots < t_{l+m} = t_k = \theta_{i+1}$; if $m = 1$, we have, by (4.43),

$$|x|_{p,\theta_i,\theta_{i+1}} = |x|_{p,t_l,t_{l+1}} \leq a(K_1 + |x_{t_l}|)(|t_{l+1} - t_l| + |w|_{p,t_l,t_{l+1}}).$$

By assumption on the form of $[\theta_i, \theta_{i+1}]$, we have

$$|t_{l+1} - t_l| + |w|_{p,t_l,t_{l+1}} \leq |\tau_{j+1} - \tau_j| + |w|_{p,\tau_j,\tau_{j+1}} \leq K_2 \leq \frac{1}{a},$$

which yields

$$|x|_{p,\theta_i,\theta_{i+1}} = |x|_{p,t_l,t_{l+1}} \leq K_1 + |x_{\theta_i}|. \quad (4.47)$$

Now, let $m \geq 2$; since

$$|t_k - t_l| + |w|_{p,t_l,t_k} \leq K_2,$$

we have by (4.44) that

$$|x|_{p,\theta_i,\theta_{i+1}} = |x|_{p,t_l,t_k} \leq K_1 + |x_{t_l}| = K_1 + |x_{\theta_i}|. \quad (4.48)$$

Summarizing, by taking (4.46), (4.47), and (4.48) into account, we have

$$|x|_{p,\theta_i,\theta_{i+1}} \leq (N(\theta_i, \theta_{i+1}) + 2)^{2-\frac{1}{p}}(K_1 + |x_{\theta_i}|) \quad (4.49)$$

for every $i \in \{0, \ldots, \mathcal{N} - 1\}$. We show inductively that

$$|x_{\theta_i}| + K_1 \leq e^{2(N(0,\theta_i) + \mathcal{N}(0,\theta_i))}(K_1 + |x_0|) \quad (4.50)$$

for every $i \in \{0, \ldots, \mathcal{N}\}$, noting that the base case $i = 0$ is trivial. Now, assume that (4.50) holds for some $i \in \{0, \ldots, \mathcal{N} - 1\}$, then, by (4.49),

$$|x_{\theta_{i+1}}| + K_1 \leq |x_{\theta_i}| + |x|_{p,\theta_i,\theta_{i+1}} + K_1$$
$$\leq (|x_{\theta_i}| + K_1)((N(\theta_i, \theta_{i+1}) + 2)^{2-\frac{1}{p}} + 1).$$

Noting that

$$(x+2)^{2-\frac{1}{p}} \leq (x+2)^{\frac{3}{2}} \leq \frac{1}{2}e^{2(x+1)}$$

for every $x \geq 0$, we obtain

$$|x_{\theta_{i+1}}| + K_1 \leq (|x_{\theta_i}| + K_1)e^{2(N(\theta_i,\theta_{i+1})+1)}.$$

Since $\mathcal{N}(\theta_i, \theta_{i+1}) = 1$, the induction hypothesis yields

$$|x_{\theta_{i+1}}| + K_1 \leq (|x_0| + K_1)e^{2(N(0,\theta_i)+\mathcal{N}(0,\theta_i)+N(\theta_i,\theta_{i+1})+\mathcal{N}(\theta_i,\theta_{i+1}))}$$
$$\leq (|x_0| + K_1)e^{2(N(0,\theta_{i+1})+\mathcal{N}(0,\theta_{i+1}))},$$

which completes the proof of (4.50).

Combining (4.49) and (4.50), we have, for $0 \leq i \leq \mathcal{N} - 1$,

$$|x|_{p,\theta_i,\theta_{i+1}} \leq (N(\theta_i,\theta_{i+1})+2)^{2-\frac{1}{p}}(K_1 + |x_{\theta_i}|)$$
$$\leq \frac{1}{2}e^{2(N(\theta_i,\theta_{i+1})+1)}e^{2(N(0,\theta_i)+\mathcal{N}(0,\theta_i))}(K_1 + |x_0|)$$
$$\leq \frac{1}{2}e^{2(N(0,\theta_{i+1})+\mathcal{N}(0,\theta_{i+1}))}(K_1 + |x_0|).$$

These considerations enable us to complete the proof. We have

$$|x|_{p,0,T} \leq \left(\mathcal{N}(0,T)^{p-1} \sum_{i=0}^{\mathcal{N}(0,T)-1} |x|_{p,\theta_i,\theta_{i+1}}^p\right)^{\frac{1}{p}}$$
$$\leq \mathcal{N}(0,T)^{1-\frac{1}{p}}\frac{1}{2}(K_1 + |x_0|)\left(\sum_{i=0}^{\mathcal{N}(0,T)-1} e^{2p(N(0,\theta_{i+1})+\mathcal{N}(0,\theta_{i+1}))}\right)^{\frac{1}{p}}$$
$$\leq \mathcal{N}(0,T)\frac{1}{2}(K_1 + |x_0|)e^{2(N(0,T)+\mathcal{N}(0,T))}.$$

Now, keep in mind that $\mathcal{N}(0,T) \leq 2N(0,T) + 1$ by the construction of Π^c. This implies

$$|x|_{p,0,T} \leq (2N(0,T)+1)\frac{1}{2}(K_1 + |x_0|)e^{6N(0,T)+2}.$$

Taking (4.31) into account, we know that

$$N(0,T) \leq 2^{p-1}K_2^{-p}\left(T^p + |w|_{p,0,T}^p\right),$$

and we conclude that

$$|x|_{p,0,T} \leq \frac{1}{2}(K_1+|x_0|)\left(2^p K_2^{-p}\left(T^p + |w|_{p,0,T}^p\right)+1\right)$$
$$\cdot \exp\left(2^p 3 K_2^{-p}\left(T^p + |w|_{p,0,T}^p\right)+2\right). \qquad \square$$

We are now in a position to present the proof of Theorem 4.9.

Proof of Theorem 4.9. *Step 1: Preliminary estimates and some notation.*

We fix $u \in \mathcal{U}$ and the partition $\Pi^E = (t_i)_{i=0,\ldots,n}$, and we choose $L \geq 1$ sufficiently large such that L is a Lipschitz constant for f and an upper bound for $|f|$ and $|\xi_0|$. Write $A := f(\xi^u)$, $z^n := z^{n,u}$, and $z := z^u$, and let $\delta_i := |t_{i+1} - t_i| + |w|_{p,t_i,t_{i+1}}$ and $\delta := \max_{i=0,\ldots,n-1} \delta_i$.

We first derive bounds for the p-variation seminorm of A and Az. By Lemma 4.2, $|A|_{p,s,t} \leq L|\xi|_{p,s,t}$ for every $0 \leq s \leq t \leq T$. Hence, by (4.37), there is a constant $C_1 > 0$ (independent of u, Π^E, and w) such that

$$(t-s) + |w|_{p,s,t} \leq C_1 \quad \Rightarrow \quad |A|_{p,s,t} \leq L. \qquad (4.51)$$

Moreover, by Lemma 4.6 and (4.38),

$$|A|_{p,t_i,t_{i+1}} \leq \frac{L}{2C_1}(1 + L + 2^{p-1}C_1^{-p}(T^p + |w|_{p,0,T}^p))\delta_i.$$

Similarly, by Lemma 4.7, (4.38), and (4.40), there is a constant C_2 (independent of u, Π^E, and w) such that

$$|z|_{p,t_i,t_{i+1}} \leq C_2(1 + |w|_{p,0,T}^p)e^{C_2(T^p + |w|_{p,0,T}^p)}\delta_i.$$

By Lemma 4.2,

$$|Az|_{p,t_i,t_{i+1}} \leq \|A\|_{\infty,t_i,t_{i+1}}|z|_{p,t_i,t_{i+1}} + \|z\|_{\infty,t_i,t_{i+1}}|A|_{p,t_i,t_{i+1}}.$$

Combining the previous estimates and bounding $\|z\|_{\infty,t_i,t_{i+1}}$ by Lemma 4.7, we find a constant C' (independent of u, Π^E, and w) such that

$$|Az|_{p,t_i,t_{i+1}} \leq C' e^{C'|w|_{p,0,T}^p} \delta_i, \qquad (4.52)$$

for every $i = 0, \ldots, n-1$. We define

$$C_1' := C_1'(w) := C' e^{C'|w|_{p,0,T}^p},$$
$$K_1 := K_1(w) := (C_p + 1)C_1'(w)(T + |w|_{p,0,T})^{p-1}\delta^{2-p},$$
$$K_2 = \left(C_1^{-1} + 4L + 8C_p L\right)^{-1}, \qquad (4.53)$$

where $C_p \geq 1$ is the constant from the Love–Young inequality for $q = p$.

Step 2: Verification of (4.43) for $z - z^n$.

Define $\Delta_t := z_t - z_t^n$. We fix a grid point t_i, and let $t_i \leq s \leq t \leq t_{i+1}$. Then,

$$\Delta_t - \Delta_s = \int_s^t A_r z_r - A_s z_s dw_r + (A_s z_s - A_{t_i} z_{t_i})(w_t - w_s)$$
$$+ A_{t_i}(z_{t_i} - z_{t_i}^n)(w_t - w_s).$$

Hence, by the Love–Young inequality (4.23), (4.52), and (4.53),

$$|\Delta_t - \Delta_s| \leq C_p |Az|_{p,s,t}|w|_{p,s,t} + |Az|_{p,t_i,s}|w|_{p,s,t} + L|\Delta_{t_i}||w|_{p,s,t}$$
$$\leq L\left(|\Delta_{t_i}| + (C_p + 1)|Az|_{p,t_i,t_{i+1}}\right)|w|_{p,s,t}$$
$$\leq L\left(|\Delta_{t_i}| + (C_p + 1)C_1'\delta^{2-p}\delta^{p-1}\right)(|t - s| + |w|_{p,s,t})$$
$$\leq L\left(|\Delta_{t_i}| + K_1\right)(|t - s| + |w|_{p,s,t}), \qquad (4.54)$$

noting that $L \geq 1$ and $\delta \leq (T + |w|_{p,0,T})$. Then, by Lemma 4.1,

$$|\Delta|_{p,t_i,t_{i+1}} \leq L\left(|\Delta_{t_i}| + K_1\right)(|t_{i+1} - t_i| + |w|_{p,t_i,t_{i+1}}). \qquad (4.55)$$

Step 3: Estimates for $z - z^n$ on the grid.
We now fix $t_\lambda, t_\kappa \in \Pi^E$ such that $t_\lambda \leq t_\kappa$. Then,

$$\Delta_{t_\kappa} - \Delta_{t_\lambda} = \sum_{i=\lambda}^{\kappa-1} \int_{t_i}^{t_{i+1}} (A_r z_r - A_{t_i} z_{t_i}) dw_r + A_{t_i}(z_{t_i} - z_{t_i}^n)(w_{t_{i+1}} - w_{t_i}).$$

The first term can be estimated (summand by summand) using the Love–Young inequality (4.23), while for the second term the variant of the Love–Young estimate (4.24) for Riemann sums (see Corollary 3.87 in Ref. [29]) applies. We, thus, obtain

$$|\Delta_{t_\kappa} - \Delta_{t_\lambda}| \leq C_p \sum_{i=\lambda}^{\kappa-1} |Az|_{p,t_i,t_{i+1}} |w|_{p,t_i,t_{i+1}}$$
$$+ C_p \|A(z - z^n)\|_{p,t_\lambda,t_\kappa} |w|_{p,t_\lambda,t_\kappa}$$
$$=: (I) + (II).$$

For the first term, we note that, by (4.52) and Lemma 4.3,

$$(I) \leq C_p C_1' \sum_{i=\lambda}^{\kappa-1} (|t_{i+1} - t_i| + |w|_{p,t_i,t_{i+1}}) |w|_{p,t_i,t_{i+1}}$$
$$\leq C_p C_1' (t_\kappa - t_\lambda) \delta + C_p C_1' \delta^{2-p} \sum_{i=\lambda}^{\kappa-1} |w|_{p,t_i,t_{i+1}}^p$$
$$\leq C_p C_1' (t_\kappa - t_\lambda) \delta^{2-p} (T + |w|_{p,0,T})^{p-1} + C_p C_1' \delta^{2-p} |w|_{p,t_\lambda,t_\kappa} |w|_{p,0,T}^{p-1}$$
$$\leq K_1 (|t_\kappa - t_\lambda| + |w|_{p,t_\lambda,t_\kappa}).$$

For the second term, Lemma 4.2 yields

$$(II) \leq 2C_p \|A\|_{p,t_\lambda,t_\kappa} \|\Delta\|_{p,t_\lambda,t_\kappa} |w|_{p,t_\lambda,t_\kappa}$$
$$\leq 2C_p (L + |A|_{p,t_\lambda,t_\kappa}) \|\Delta\|_{p,t_\lambda,t_\kappa} (|t_\kappa - t_\lambda| + |w|_{p,t_\lambda,t_\kappa}).$$

Gathering the terms, we obtain

$$|\Delta_{t_\kappa} - \Delta_{t_\lambda}|$$
$$\leq (K_1 + 2C_p(L + |A|_{p,t_\lambda,t_\kappa}) \|\Delta\|_{p,t_\lambda,t_\kappa}) (|t_\kappa - t_\lambda| + |w|_{p,t_\lambda,t_\kappa}).$$
(4.56)

Step 4: Verification of (4.44) *for* $z - z^n$.

Fix $t_l, t_k \in \Pi^E$ such that $t_l < t_{l+1} < t_k$ and $|t_k - t_l| + |w|_{p,t_l,t_k} \leq K_2$ for the constant K_2 defined in (4.53). Let $t_l \leq s \leq t \leq t_k$. We distinguish two cases:

(a) $s, t \in [t_i, t_{i+1}]$ for some t_i, i.e., s and t are in the same subinterval.
(b) $s \in [t_{\lambda-1}, t_\lambda)$ and $t \in (t_\kappa, t_{\kappa+1}]$ for $t_\lambda \leq t_\kappa$, i.e., s and t are in different subintervals.

In case (a), we obtain, thanks to (4.54) and recalling (4.25),

$$|\Delta_t - \Delta_s| \leq L \left(|\Delta_{t_l}| + |\Delta|_{p,t_l,t_k} + K_1\right)\left(|t-s| + |w|_{p,s,t}\right). \qquad (4.57)$$

In case (b), we decompose

$$|\Delta_t - \Delta_s| \leq |\Delta_t - \Delta_{t_\kappa}| + |\Delta_{t_\kappa} - \Delta_{t_\lambda}| + |\Delta_{t_\lambda} - \Delta_s|.$$

Applying (4.57) to the first and third terms and using (4.56) for the second one, we get, in view of (4.51) and since $s \leq t_\lambda \leq t_\kappa \leq t$,

$$|\Delta_t - \Delta_s| \leq 2L\left(|\Delta_{t_l}| + |\Delta|_{p,t_l,t_k} + K_1\right)\left(|t-s| + |w|_{p,s,t}\right)$$
$$+ \left(K_1 + 4C_p L \|\Delta\|_{p,t_l,t_k}\right)\left(|t-s| + |w|_{p,s,t}\right).$$

In view of (4.57), this estimate is valid for every $t_l \leq s \leq t \leq t_k$ (and not just in case (b)). Hence, by Lemma 4.1 and the definition K_2,

$$|\Delta|_{p,t_l,t_k} \leq ((2L + 4C_p L)(|\Delta_{t_l}| + |\Delta|_{p,t_l,t_k}) + 3LK_1)$$
$$\times (|t_l - t_k| + |w|_{p,t_l,t_k})$$
$$\leq \frac{1}{2}(|\Delta|_{p,t_l,t_k} + |\Delta_{t_l}| + K_1).$$

Thus,

$$|\Delta|_{p,t_l,t_k} \leq |\Delta_{t_l}| + K_1. \qquad (4.58)$$

Step 5: Application of Gronwall's lemma for Euler partitions.
By (4.55) and (4.58), we may apply Gronwall's inequality in the form of Lemma 4.9. Taking into account that $\Delta_0 = z_0 - z_0^n = 0$, we obtain

$$\|z - z^n\|_{p,0,T} = |z - z^n|_{p,0,T}$$
$$\leq \frac{K_1}{2}\left(2^p K_2^{-p}\left(T^p + |w|_{p,0,T}^p\right) + 1\right)$$
$$\times \exp\left(2^p 3 K_2^{-p}\left(T^p + |w|_{p,0,T}^p\right) + 2\right).$$

Inserting the definition of K_1 and K_2, we observe that the right-hand side can be bounded by $\delta^{2-p} C e^{C|w|_{p,0,T}^p}$ for some constant C independent of Π^E, u, and w. □

The Euler schemes for the Young SDEs in the first n_1 lines of $\mathcal{X}^u$ and $\mathcal{Y}^u$ can be analyzed through the same techniques, which we illustrated in the proof of Theorem 4.9. Due to the boundedness of the coefficient functions, the Young SDEs in $\mathcal{X}^u$ are actually much simpler and do not require Gronwall's lemma on Euler partitions in the form of Lemma 4.9.

The Euler schemes for the component processes in the last n_2 lines of $\mathcal{X}^u$ and $\mathcal{Y}^u$ are driven by a Brownian motion. They can be analyzed through standard methods (see, e.g., Ref. [40]). There is a little extra work because the coefficients in the Euler scheme of these equations depend on the Euler approximations of the Young SDEs. In particular, the error estimates for the Young SDEs enter the error analysis of the SDEs driven by the Brownian motion. For this reason, the strong convergence rates deteriorate from $1/2$ to $\min(2 - p, 1/2)$.

We finally turn to the Euler approximation (4.20) to the adjoint equation. For fixed $t_k \in \Pi^E$ and recalling (4.18), we define

$$\Phi^{t_k,u,n}_{t_{i+1}} = \Phi^{t_k,u,n}_{t_i} + \eta^{n,u}_{t_i,t_{i+1}} \Phi^{t_k,u,n}_{t_i}, \; i > k, \quad \Phi^{t_k,u,n}_{t_k} = I_{n_1+n_2}.$$

A direct computation shows

$$\Lambda^{n,u}_{t_i} = \sum_{\mu;\, T_\mu \geq t_i} E[g_\mu(\mathcal{X}^{n,u}_{T_\mu})] g'_\mu(\mathcal{X}^{n,u}_{T_\mu}) \Phi^{t_i,u,n}_{T_\mu}, \tag{4.59}$$

while, by the definition of Λ^u in (4.13),

$$\Lambda_t^u = \sum_{\mu;\, T_\mu \geq t} E[g_\mu(\mathcal{X}_{T_\mu}^u)] g_\mu'(\mathcal{X}_{T_\mu}^u) \Phi_{T_\mu}^{t,u}. \tag{4.60}$$

Now, fix $s \in [0,T]$, and let t_k be the smallest grid point larger or equal to s. Then, $\Phi^{t_k,u,n}$ can be considered the Euler scheme for $\Phi^{s,u}$. The same convergence rates as for the Euler approximation to $\mathcal{Y}^u$ can be derived using the same techniques, with the constants being independent of s. In view of (4.59)–(4.60), these rates carry over to the approximation of Λ^u by $\Lambda^{n,u}$.

4.5. Case Study and Numerical Experiments

4.5.1. *Monte Carlo implementation*

As demonstrated in Section 4.2.3, we can approximate the cost function and its gradient with respect to the parameter via Euler schemes. Since for the calculation of the discretized cost function and the discretized gradient we need to evaluate expected values, we apply the Monte Carlo method to come up with an implementable scheme. A comprehensive introduction to Monte Carlo methods is given in Ref. [47]. Using A independent copies $(\mathcal{X}^{n,u,a}, \mathcal{Y}^{n,u,a})_{a=1,\ldots,A}$ of the Euler schemes in (4.15), with (4.16) restricted to the discrete-time grid Π^E, we can approximate the discretized cost function and the discretized gradient using the Monte Carlo estimators

$$J^{n,A}(u) = \frac{1}{2} \sum_{\mu=1}^M \left(\frac{1}{A} \sum_{a=1}^A g_\mu(\mathcal{X}_{T_\mu}^{n,u,a}) \right)^2$$

$$(\nabla J)^{n,A}(u) = \sum_{\mu=1}^M \left(\frac{1}{A} \sum_{a=1}^A g_\mu(\mathcal{X}_{T_\mu}^{n,u,a}) \right) \left(\frac{1}{A} \sum_{a=1}^A g_\mu'(\mathcal{X}_{T_\mu}^{n,u,a}) \mathcal{Y}_{T_\mu}^{n,u,a} \right). \tag{4.61}$$

Using the central limit theorem, it is well established that the corresponding approximation error behaves asymptotically like $\mathcal{O}(A^{-\frac{1}{2}})$; see Ref. [47].

Accordingly, we replace the expectations by the sample mean in the Euler scheme with the adjoint equation, leading to the backward recursion

$$\Lambda_{t_i}^{n,u,a} = \Lambda_{t_{i+1}}^{n,u,a} + \Lambda_{t_{i+1}}^{n,u,a} \eta_{t_i,t_{i+1}}^{n,u,a}$$

$$+ \sum_{\mu; T_\mu = t_i} \left(\frac{1}{A} \sum_{\alpha=1}^{A} g_\mu(\mathcal{X}_{t_i}^{n,u,\alpha}) \right) g'_\mu(\mathcal{X}_{t_i}^{n,u,a}),$$

initialized at

$$\Lambda_{t_n}^{n,u,a} = \sum_{\mu; T_\mu = T} \left(\frac{1}{A} \sum_{\alpha=1}^{A} g_\mu(\mathcal{X}_T^{n,u,\alpha}) \right) g'_\mu(\mathcal{X}_T^{n,u,a}),$$

where

$$\eta_{t_i,t_{i+1}}^{n,u,a} = \begin{pmatrix} b_\xi(t_i, \xi_{t_i}^{n,u,a}, u) & 0 \\ \hat{b}_\xi(t_i, \xi_{t_i}^{n,u,a}, x_{t_i}^{n,u,a}, u) & \hat{b}_x(t_i, \xi_{t_i}^{n,u,a}, x_{t_i}^{n,u,a}, u) \end{pmatrix} (t_{i+1} - t_i)$$

$$+ \sum_{j=1}^{m_1} \begin{pmatrix} \sigma_\xi^j(t_i, \xi_{t_i}^{n,u,a}, u) & 0 \\ 0 & 0 \end{pmatrix} (w_{t_{i+1}}^{j,a} - w_{t_i}^{j,a})$$

$$+ \sum_{j=1}^{m_2} \begin{pmatrix} 0 & 0 \\ \hat{\sigma}_\xi^j(t_i, \xi_{t_i}^{n,u,a}, x_{t_i}^{n,u,a}, u) & \hat{\sigma}_x^j(t_i, \xi_{t_i}^{n,u,a}, x_{t_i}^{n,u,a}, u) \end{pmatrix}$$

$$\times (B_{t_{i+1}}^{j,a} - B_{t_i}^{j,a}).$$

Note that the realizations $(\Lambda^{n,u,a})_{a=1,\ldots,A}$ are not independent due to the presence of the sample means.

The following proposition is the analog of Theorem 4.5 in the Monte Carlo setup. Its proof remains unchanged, except replacing the mean with the sample mean in the last step of the proof.

Proposition 4.2. *For every* $u \in \mathcal{U}$, *we have*

$$(\nabla J)^{n,A}(u) = \frac{1}{A} \sum_{a=1}^{A} \left(\Lambda_{t_0}^{n,u,a} D\mathcal{X}_0^u + \sum_{i=0}^{n-1} \Lambda_{t_{i+1}}^{n,u,a} \eta_{t_i,t_{i+1}}^{n,u,a} \right). \tag{4.62}$$

This proposition states that we get exactly the same result when realizing the Monte Carlo paths and calculating the gradient via the discrete sensitivity equation (4.61) or the adjoint method (4.62). Since we are now able to approximate the value of the cost function, as well as its gradient with respect to the parameter (using two different methods), we can apply smooth gradient-based optimization algorithms to find the minimum of the cost function. As already mentioned, the adjoint method has the advantage that, instead of $(n_1 + n_2) \cdot d$ forward solves of the recursion for $\mathcal{Y}^{n,u}$, we only have to perform $n_1 + n_2$ backward solves. Hence, the computational cost of a gradient evaluation does not depend on the number of parameters in the adjoint approach. In particular, in the case of time-dependent parameters, this reduces the numerical effort substantially in comparison to the sensitivity method.

4.5.2. Case study: Calibrating a fractional Heston-type model

In this section, we illustrate how the results from the previous sections can be applied to calibrate a financial model with a volatility process driven by the process of finite p-variation for $p \in (1, 2)$. There are several models which incorporate the long memory phenomenon of volatility by using a fractional Brownian motion with Hurst parameter $H \in (0.5, 1)$ as the driving process for the volatility; see, e.g., Refs. [6–10]. We choose a fractional version of the Cox–Ingersoll–Ross (CIR) process given by

$$v_t = v_0 + \int_0^t \kappa(\theta - v_r) \, dr + \int_0^t \zeta \sqrt{v_r} \, dB_r^H,$$

where B^H is a fractional Brownian motion with the Hurst parameter $H \in (0.5, 1)$. It has been shown in Refs. [8] and [10] that this equation has a unique positive solution when the integral $\int_0^t \zeta \sqrt{v_r} \, dB_r^H$ is interpreted as a pathwise Young integral. Furthermore, in Ref. [8], the authors show that the process v_t is mean-reverting to the parameter θ, and hence the parameters can be interpreted similarly to the standard CIR model. Another feature, which we want to incorporate, is the correlation between the volatility process and the asset price process. To this end, fix $T > 0$, and let $(\Omega, \mathcal{F}, P)$ be a probability space carrying a two-sided Brownian motion $(B_t^1)_{t \in \mathbb{R}}$ and a Brownian motion $(B_t^2)_{t \in [0,T]}$ independent of B^1. Then, a fractional

Brownian motion B_t^H with the Hurst parameter $H \in (0.5, 1)$ can be constructed via the following integral transformation of B^1 (see Ref. [48]):

$$B_t^H = C_H \left(\int_0^t (t-u)^{H-\frac{1}{2}} dB_u^1 + \int_{-\infty}^0 (t-u)^{H-\frac{1}{2}} - (-u)^{H-\frac{1}{2}} dB_u^1 \right),$$

where

$$C_H = \sqrt{\frac{2H\Gamma(\frac{3}{2} - H)}{\Gamma(H + \frac{1}{2})\Gamma(2 - 2H)}}.$$

By defining $B_t = \rho B_t^1 + \sqrt{1-\rho^2} B_t^2$, we obtain a standard Brownian motion B_t, which is correlated with B^1 via $Corr(B_t, B_t^1) = \rho$ for all $t \in [0, T]$. We write $\mathbb{F}$ for the augmentation of the filtration generated by $(B_t^H, B_t)_{t \in [0, T]}$. Note that ρ is not the correlation between B^H and the Brownian motion driving the asset price process B, but between B and the Brownian motion B_1 from which B^H has been constructed. Thus, we generate the desired correlation between the volatility process v and the asset price S in the following model, in a similar way as in Ref. [10]:

$$v_t = v_0 + \int_0^t \kappa(\theta - v_s) \, ds + \int_0^t \zeta \sqrt{v_s} \, dB_s^H,$$

$$S_t = S_0 + \int_0^t (r - d) S_s \, ds + \int_0^t \sqrt{v_s} S_s \, d(\rho B_s^1 + \sqrt{1-\rho^2} B_s^2).$$

(4.63)

Here, the spot price S_0, the riskless rate r, and the dividend yield d are given. We assume that the market we are trading in only consists of the asset S and a riskless bond e^{-rt} for $t \in [0, T]$.

We aim at calibrating the model with respect to the parameters $u = (v_0, \kappa, \theta, \zeta, \rho)$ to a set of market-observed European call option prices. In order to apply the results derived in the theoretical part of this chapter, we must smoothen the coefficient functions of the SDE system (4.63). We basically follow the approach in Ref. [16] when using a piecewise polynomial error function π_1 to smoothen out the positive part $\pi(x) := (x)_+ := \max(x, 0)$ at 0 and a similar construction for the function π_2, which smoothens out the square-root function at 0. These functions are given by

$$\pi_1(x) = \begin{cases} 0, & x < -\varepsilon_1, \\ -\dfrac{1}{16(\varepsilon_1)^3}x^4 + \dfrac{3}{8\varepsilon_1}x^2 + \dfrac{1}{2}x + \dfrac{3\varepsilon_1}{16}, & -\varepsilon_1 \leq x \leq \varepsilon_1, \\ x, & x > \varepsilon_1, \end{cases}$$

for $x \in \mathbb{R}$ and an error parameter $\varepsilon_1 > 0$, which we choose to set as 0.01 for all calculations. The second function is given by

$$\pi_2(x) = \begin{cases} 0, & x < -\varepsilon_2, \\ -\dfrac{1}{256\varepsilon_2^{6,5}}(-15x^7 + 7\varepsilon_2 x^6 + 65\varepsilon_2^2 x^5 - 33\varepsilon_2^3 x^4 \\ \quad -117\varepsilon_2^4 x^3 + 77\varepsilon_2^5 x^2 + 195\varepsilon_2^6 x + 77\varepsilon_2^7), & -\varepsilon_2 \leq x \leq \varepsilon_2, \\ \sqrt{x}, & x > \varepsilon_2, \end{cases}$$

for $x \in \mathbb{R}$ and an error parameter $\varepsilon_2 > 0$, which we choose to set as 0.001 for all calculations. To achieve boundedness of these two functions, we theoretically compose them with a smooth truncation function. By choosing the truncation level sufficiently large, this truncation can be ignored in practice since v_t is mean-reverting, and for a moderate time horizon, we do not expect the log-price of the asset in our model to explode along typical realizations. This reasoning is justified by our numerical findings. The dynamics of the adjusted fractional Heston-type model are then given by

$$v_t = v_0 + \int_0^t \kappa(\theta - \pi_1(v_s))\,ds + \int_0^t \zeta\pi_2(v_s)\,dB_s^H,$$

$$S_t = S_0 + \int_0^t (r-d)S_s\,ds + \int_0^t \pi_2(v_s)S_s\,d(\rho B_s^1 + \sqrt{1-\rho^2}B_s^2),$$

(4.64)

and after a log-transformation $\hat{S}_t = \log(S_t)$ in the asset equation, this yields

$$v_t = v_0 + \int_0^t \kappa(\theta - \pi_1(v_s))\,ds + \int_0^t \zeta\pi_2(v_s)\,dB_s^H,$$

$$\hat{S}_t = \hat{S}_0 + \int_0^t (r-d) - \frac{1}{2}\pi_2(v_s)^2\,ds + \int_0^t \pi_2(v_s)\,d(\rho B_s^1 + \sqrt{1-\rho^2}B_s^2).$$

Note that under these adjustments, (the truncated version of) $\pi_2(v_r)$ is bounded, and hence the SDE (4.64) has the explicit solution

$$S_t = S_0 e^{\left((r-d)t - \frac{1}{2}\int_0^t \pi_2(v_s)^2\, ds + \int_0^t \pi_2(v_s)\, dB_s\right)}.$$

The dividend-adjusted discounted price process $e^{-(r-d)t} S_t$ is then a martingale with respect to P, and the price for a call option with maturity T_μ and strike K_μ at time 0 in this model (taking P as a pricing measure) is given by

$$e^{-rT_\mu} \mathrm{E}\left[\left(e^{\hat{S}^u_{T_\mu}} - K_\mu\right)_+\right]$$

using the risk-neutral pricing formula. We approximate this value by

$$C^{\mathrm{mod}}_\mu(u) = e^{-rT_\mu} \mathrm{E}\left[\pi_1\left(e^{\hat{S}^u_{T_\mu}} - K_\mu\right)\right],$$

and the cost function, thus, translates to

$$J(u) = \frac{1}{2}\sum_{\mu=1}^M \mathrm{E}\left[g_\mu\begin{pmatrix} v^u_{T_\mu} \\ \hat{S}^u_{T_\mu} \end{pmatrix}\right]^2 = \frac{1}{2}\sum_{\mu=1}^M \left(C^{\mathrm{mod}}_\mu(u) - C^{\mathrm{obs}}_\mu\right)^2,$$

where $g_\mu(x_1, x_2) = e^{-rT_\mu}\pi_1(e^{x_2} - K_\mu) - C^{\mathrm{obs}}_\mu$ and C^{obs}_μ is the observed market price for a call option with maturity T_μ struck at K_μ. As the parameter set for the calibration problem, we choose

$$\mathcal{U} := \{(v_0, \kappa, \theta, \zeta, \rho) \in \mathbb{R}^5|\, v_0 \in (0.0001, 1),\, \kappa \in (0.0001, 2),$$
$$\theta \in (0.0001, 2),\, \zeta \in (0.0001, 4)\, \rho \in (-0.99, 0.99)\}.$$

Note that a fractional Brownian motion with the Hurst parameter $H > \frac{1}{2}$ has Hölder continuous paths with Hölder index H' for every $1/2 < H' < H$; see, e.g., Ref. [49, p. 274]. Hence, in view of Remark 4.2, it satisfies assumption (W) for every $p \in (1/H, 2)$. Moreover, the Hölder assumption in Remark 4.4 can be verified for every Hölder index $H' < H$. Finally, the assumptions $(H_1), (H_2)$, $(H_3), (B_1), (B_2), (B_3), (E_1), (E_2)$, and (G) are also fulfilled (taking the choice of $\mathcal{U}$ into account).

We calibrate the model to market prices for call options on the EUROSTOXX 50 as of 7 October 2003. The data set is reported in Ref. [50] and consists of the prices for 144 call options (in total) with six different maturities: 0.0361, 0.2000, 1.1944, 2.1916, 4.2056, and 5.1639 (in years). We exclude the call option data for the strikes 2499.76 and 4990.91 in order to remove static arbitrage opportunities from the data set. After this modification of the data set, it still consists of 136 call option prices. Following Ref. [50], we set $S_0 = 2461.44$, $r = 0.03$, and $d = 0$.

For the numerical calibration, we minimize the Monte Carlo estimate $J^{n,A}$ of the discretized cost functional using the MATLAB® fmincon function, with the trust region reflective algorithm and a function tolerance of 10^{-6} feeding in the gradient approximation $(\nabla J)^{n,A}$, which is computed via the adjoint representation in Proposition 4.2. We initialize the parameter values at

$$v_0 = 0.1, \quad \kappa = 1, \quad \theta = 0.05, \quad \zeta = 0.3, \quad \rho = -0.7$$

and first run the calibration with 10,000 Monte Carlo samples and a time grid which divides the time between each of the neighboring maturities into 40 subintervals (leading to $n = 240$ and a mesh size of 0.0504). The resulting parameters are stored and taken as input for a second optimization stage with 100,000 Monte Carlo samples and a time grid consisting of 480 subintervals, halving each of the 240 intervals of the first stage. The empirical mean and the empirical standard deviation of the parameters found in the second stage over 25 independent repetitions of the algorithm are reported in Table 4.1 for various choices of the Hurst parameter: $H \in \{0.5, 0.55, 0.6, 0.65, 0.7, 0.75, 0.8, 0.85, 0.9, 0.95\}$.

Table 4.1. Calibrated parameters for different values of $H \in (\frac{1}{2}, 1)$.

H	v_0 μ	Sd	κ μ	Sd	θ μ	Sd	ζ μ	Sd	ρ μ	Sd
0.5	0.072	0.0030	0.809	0.0990	0.071	0.0022	0.438	0.0428	−0.657	0.0208
0.55	0.070	0.0016	0.838	0.0662	0.054	0.0008	0.413	0.0365	−0.657	0.0115
0.6	0.070	0.0018	0.971	0.1197	0.045	0.0013	0.424	0.0400	−0.667	0.0156
0.65	0.070	0.0017	1.030	0.0733	0.042	0.0016	0.409	0.0287	−0.690	0.0155
0.7	0.069	0.0018	1.055	0.0669	0.043	0.0018	0.383	0.0221	−0.724	0.0189
0.75	0.069	0.0016	1.085	0.0609	0.043	0.0012	0.362	0.0182	−0.759	0.0152
0.8	0.067	0.0011	1.163	0.0920	0.045	0.0012	0.360	0.0164	−0.822	0.0235
0.85	0.067	0.0014	1.327	0.1149	0.047	0.0010	0.369	0.0208	−0.925	0.0233
0.9	0.065	0.0012	1.199	0.0651	0.046	0.0004	0.354	0.0141	−0.990	0.0001
0.95	0.066	0.0011	1.076	0.0537	0.039	0.0005	0.368	0.0168	−0.990	0.0000

Table 4.2. Summary statistics for the calibration results.

H	AvgErr (sample mean)	AvgErr (emp. standard dev.)	Avg runtime in sec
0.5	$8.788 \cdot 10^{-4}$	$7.229 \cdot 10^{-6}$	973.30
0.55	$6.913 \cdot 10^{-4}$	$7.014 \cdot 10^{-6}$	654.87
0.6	$6.890 \cdot 10^{-4}$	$7.628 \cdot 10^{-6}$	553.73
0.65	$6.577 \cdot 10^{-4}$	$7.598 \cdot 10^{-6}$	453.74
0.7	$6.987 \cdot 10^{-4}$	$7.501 \cdot 10^{-6}$	390.58
0.75	$8.059 \cdot 10^{-4}$	$7.659 \cdot 10^{-6}$	374.23
0.8	$8.821 \cdot 10^{-4}$	$6.467 \cdot 10^{-6}$	422.38
0.85	$9.869 \cdot 10^{-4}$	$6.725 \cdot 10^{-6}$	538.93
0.9	$2.156 \cdot 10^{-3}$	$9.565 \cdot 10^{-6}$	541.30
0.95	$3.276 \cdot 10^{-3}$	$7.675 \cdot 10^{-6}$	977.54

In order to evaluate the fit of the calibration procedure, we simulate, for each choice of the Hurst parameter H, a new independent Monte Carlo sample of size 100.000 and compute the Monte Carlo estimator $\hat{C}_\mu^{\mathrm{mod}}(u^*)$ for the model price $C_\mu^{\mathrm{mod}}(u^*)$ along the finer time grid, where u^* is the optimal parameter vector found in the calibration routine (see Table 4.1). Table 4.2 contains some summary statistics for the estimated average error avgErr $= \frac{1}{136 S_0} \sum_{\mu=1}^{136} |\hat{C}_\mu^{\mathrm{mod}}(u^*) - C_\mu^{\mathrm{obs}}|$ over the 136 option prices. Precisely, we report the sample mean and the empirical standard deviation of avgErr over 100 independent repetitions as well as the run time for the calibration step (depending on the Hurst parameter). Our results suggest that the best fit to the data can be achieved by adding a moderate long-range dependence into the stochastic volatility process corresponding to a Hurst parameter of about $H = 0.65$.

The option price function of the calibrated model with the Hurst parameter $H = 0.65$ is plotted in Figures 4.2–4.4 for the six maturities, for which price data are available (marked by "*"). The figures illustrate the excellent fit of the calibrated model across all maturities and strikes.

4.5.3. *Additional numerical experiments*

We finally perform some numerical experiments in order to illustrate the rates of convergence derived in Theorem 4.4 and Remark 4.4 and the computational benefit from simulating the gradient of the cost

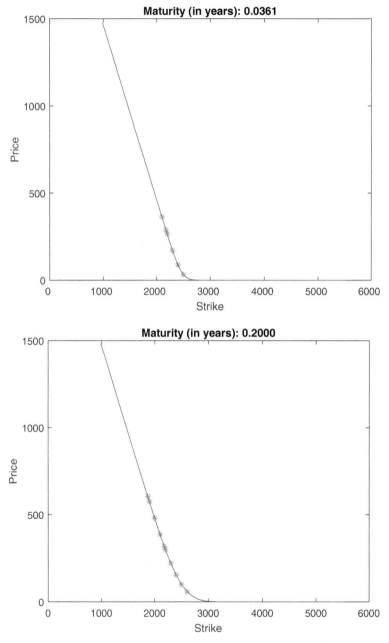

Fig. 4.2. Call price function of the calibrated model (solid line) with $H = 0.65$ and observed option prices (marked by "*") for several maturities.

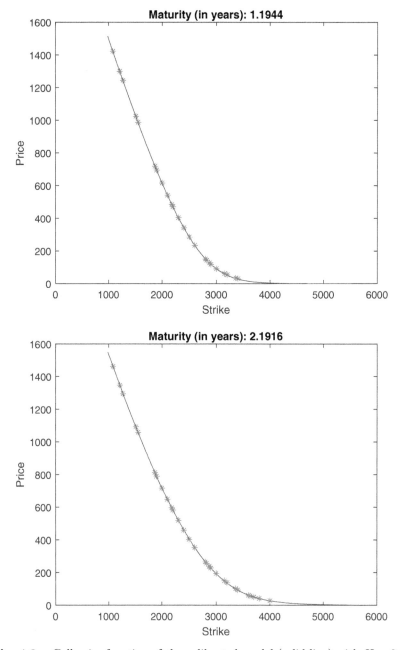

Fig. 4.3. Call price function of the calibrated model (solid line) with $H = 0.65$ and observed option prices (marked by "*") for several maturities.

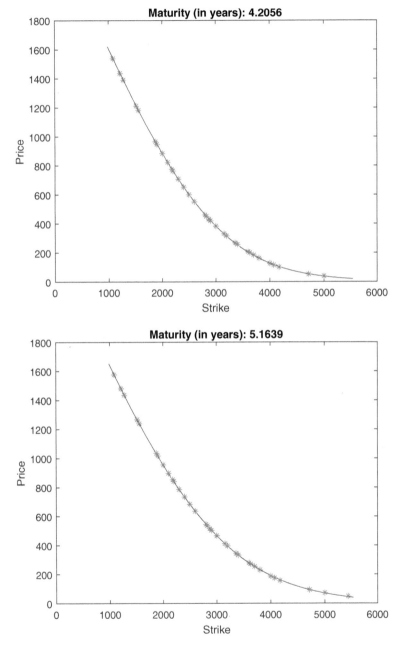

Fig. 4.4. Call price function of the calibrated model (solid line) with $H = 0.65$ and observed option prices (marked by "*") for several maturities.

functional via the adjoint equation Λ^u as compared to the sensitivity equation $\mathcal{Y}^u$.

Note that, in the fractional Brownian motion case, by (4.22) and Remark 4.4, for every $H' < H$,

$$E[|(\nabla J)^{n,A}(u) - (\nabla J)(u)|] = \mathcal{O}\left(A^{-\frac{1}{2}} + |\Pi^E|^{(2H'-1)\wedge\frac{1}{2}}\right).$$

In the following experiment, we fix the sample size as $A = 100{,}000$ but refine the time partition by decomposing the time between two maturities into 2^i subintervals for $i = 4, \ldots, 9$, leading to a partition into $n_i = 6 \cdot 2^i$ subintervals in total. This corresponds to a mesh size of $0.126 \cdot 2^{-(i-4)}$. Note that, by the triangle inequality,

$$E\left[|(\nabla J)^{n_i,A}(u) - (\nabla J)^{n_{i-1},A}(u)|\right] = \mathcal{O}\left(A^{-\frac{1}{2}} + 2^{-i((2H'-1)\wedge\frac{1}{2})}\right).$$

We choose $H = 0.8$, leading to $(2H'-1) \wedge \frac{1}{2} = \frac{1}{2}$ for sufficiently large $H' < H$, and fix

$$u = (v_0, \kappa, \theta, \zeta, \rho)^\top = (0.016, 1, 0.02, 0.3, -0.7)^\top.$$

In this setting, we sample 20 independent copies,

$$((\nabla J)^{n_4,A,j}(u), \ldots, (\nabla J)^{n_9,A,j}(u))_{j=1,\ldots,20},$$

of $((\nabla J)^{n_4,A}(u), \ldots, (\nabla J)^{n_9,A}(u))$ and consider the error

$$Err_i = \frac{1}{20}\sum_{j=1}^{20}|(\nabla J)^{n_i,A,j}(u) - (\nabla J)^{n_{i-1},A,j}(u)|$$

for $i = 5, \ldots, 9$. The theoretical considerations above suggest that Err_i decays as $2^{-i/2}$ provided the time-discretization error dominates the Monte Carlo error. This is confirmed by the log-log plot of the mesh size $0.126 \cdot 2^{-(i-5)}$ of the $(i-1)$th partition versus Err_i in Figure 4.5, where the dashed reference line exhibits a slope of 0.5.

We finally compare the run time for computing $(\nabla J)^{n,A}$ based on the representation (4.61), for which we employ the Euler approximation of the sensitivity equation, and based on the discretization of the adjoint equation; see (4.62). To this end, we replace the constants κ, θ, ζ, and ρ with piecewise constant functions in time, where each of these functions can take I values. Hence, the total number of

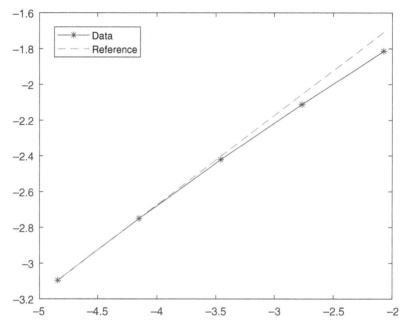

Fig. 4.5. Log-log plot of the mesh size $0.126 \cdot 2^{-(i-5)}$ against Err_i for $i = 5, \ldots, 9$ and $H = 0.8$.

Table 4.3. Runtime (RT; in s) for the computation of the gradient of the cost function with the two different methods.

Number of parameters	5	9	13	17	21	25
RT adjoint	19.0	18.0	18.0	18.0	18.0	18.0
RT sensitivity	22.0	26.0	30.0	35.0	39.0	43.0

parameters then becomes $4I + 1$. The run times reported in Table 4.3 correspond to a single evaluation of the gradient $(\nabla J)^{n,A}$ based on a Monte Carlo sample of size $A = 100{,}000$ and time discretization into $n = 480$ subintervals. In line with the theoretical considerations, the run time for computing the gradient via the adjoint method does not depend on the number of model parameters, while the computational time for simulating the sensitivity equation linearly increases with the number of parameters. The reported run times, thus, demonstrate the computational benefits of applying the adjoint method based on the new type of anticipating BSDE (4.14) for model calibration, in particular when the parameter vector is high dimensional.

References

[1] R. Cont, Empirical properties of asset returns: Stylized facts and statistical issues. *Quantitative Finance*. **1**, 223–236 (2001). https://doi.org/10.1080/713665670.

[2] S. Heston, A closed-form solution for options with stochastic volatility with applications to bond and currency options. *Review of Financial Studies*. **6**, 327–343 (1993). https://doi.org/10.1093/rfs/6.2.327.

[3] J. Hull and A. White, The pricing of options on assets with stochastic volatilities. *The Journal of Finance*. **42**(2), 281–300 (1987). https://doi.org/10.1111/j.1540-6261.1987.tb02568.x.

[4] M. Chesney and L. Scott, Pricing European currency options: A comparison of the modified Black-Scholes model and a random variance model. *The Journal of Financial and Quantitative Analysis*. **24**(3), 267–284 (1989). https://doi.org/10.2307/2330812.

[5] E. M. Stein and J. C. Stein, Stock price distributions with stochastic volatility: An analytic approach. *The Review of Financial Studies*. **4**(4), 727–752 (2015). https://doi.org/10.1093/rfs/4.4.727.

[6] F. Comte and E. Renault, Long memory in continuous-time stochastic volatility models. *Mathematical Finance*. **8**(4), 291–323 (1998). https://doi.org/10.1111/1467-9965.00057.

[7] A. Chronopoulou and F. G. Viens, Estimation and pricing under long-memory stochastic volatility. *Annals of Finance*. **8**(2–3), 379–403 (2012). https://doi.org/10.1007/s10436-010-0156-4.

[8] E. Lépinette and F. Mehrdoust, A fractional version of the Heston model with Hurst parameter $H \in (1/2, 1)$. *Dynamic Systems and Applications*. **26**(3&4), 535–548 (2017).

[9] V. Bezborodov, L. Di Persio, and Y. Mishura, Option pricing with fractional stochastic volatility and discontinuous payoff function of polynomial growth. *Methodology and Computing in Applied Probability*. **21**(1), 331–366 (2019). https://doi.org/10.1007/s11009-018-9650-3.

[10] Y. Mishura and A. Yurchenko-Tytarenko, Approximating expected value of an option with non-Lipschitz payoff in fractional Heston-type model. *International Journal of Theoretical and Applied Finance*. **23**(5), 2050031, 36 (2020). https://doi.org/10.1142/S0219024920500314.

[11] M. Fukasawa, Asymptotic analysis for stochastic volatility: Martingale expansion. *Finance and Stochastics.* **15**(4), 635–654 (2011). https://doi.org/10.1007/s00780-010-0136-6.

[12] J. Gatheral, T. Jaisson, and M. Rosenbaum, Volatility is rough. *Quantitative Finance.* **18**(6), 933–949 (2018). https://doi.org/10.1080/14697688.2017.1393551.

[13] C. Bayer, P. Friz, and J. Gatheral, Pricing under rough volatility. *Quantitative Finance.* **16**(6), 887–904 (2016). https://doi.org/10.1080/14697688.2015.1099717.

[14] M. Fukasawa, Short-time at-the-money skew and rough fractional volatility. *Quantitative Finance.* **17**(2), 189–198 (2017). https://doi.org/10.1080/14697688.2016.1197410.

[15] O. El Euch and M. Rosenbaum, The characteristic function of rough Heston models. *Mathematical Finance.* **29**(1), 3–38 (2019). https://doi.org/10.1111/mafi.12173.

[16] C. Kaebe, J. H. Maruhn, and E. W. Sachs, Adjoint-based Monte Carlo calibration of financial market models. *Finance and Stochastics.* **13**(3), 351–379 (2009). https://doi.org/10.1007/s00780-009-0097-9.

[17] L. C. Young, An inequality of the Hölder type, connected with Stieltjes integration. *Acta Mathematica.* **67**(1), 251–282 (1936). https://doi.org/10.1007/BF02401743.

[18] A. Shapiro. Monte Carlo sampling methods. In *Handbooks in Operations Research and Management Science. Volume 10: Stochastic Programming.* Elsevier Science B. V., Amsterdam (2003), pp. 353–425. https://doi.org/10.1016/S0927-0507(03)10006-0.

[19] F. Russo and P. Vallois, Forward, backward and symmetric stochastic integration. *Probability Theory and Related Fields.* **97**(3), 403–421 (1993). https://doi.org/10.1007/BF01195073.

[20] M. B. Giles and N. A. Pierce, An introduction to the adjoint approach to design. *Flow, Turbulence and Combustion.* **65**, 393–415 (2000). https://doi.org/10.1023/A:1011430410075.

[21] M. Giles and P. Glasserman, Smoking adjoints: Fast evaluation of Greeks in Monte Carlo calculations. *Risk.* **19**, 88–92 (2006).

[22] N. Nikolova, R. Safian, E. Soliman, M. Bakr, and J. Bandler, Accelerated gradient based optimization using adjoint sensitivities. *IEEE Transactions on Antennas and Propagation.* **52**, 2147–2157 (2004). https://doi.org/10.1109/TAP.2004.832313.

[23] A. Lejay, Controlled differential equations as Young integrals: A simple approach. *Journal of Differential Equations.* **249**(8), 1777–1798 (2010). https://doi.org/10.1016/j.jde.2010.05.006.

[24] T. Cass, C. Litterer, and T. Lyons, Integrability and tail estimates for Gaussian rough differential equations. *Annals of Probability.* **41**(4), 3026–3050 (2013). https://doi.org/10.1214/12-AOP821.

[25] N. D. Cong, L. H. Duc, and P. T. Hong, Nonautonomous Young differential equations revisited. *Journal of Dynamics and Differential Equations.* **30**(4), 1921–1943 (2018). https://doi.org/10.1007/s10884-017-9634-y.

[26] M. Thiel. Calibration of Non-Semimartingale Models - An Adjoint Approach. PhD Thesis, Saarland University (2023). http://dx.doi.org/10.22028/D291-40950.

[27] K. French, G. Schwert, and R. Stambaugh, Expected stock returns and volatility. *Journal of Financial Economics.* **19**(1), 3–29 (1987). https://doi.org/10.1016/0304-405X(87)90026-2.

[28] R. M. Dudley and R. Norvaiša. *Differentiability of Six Operators on Nonsmooth Functions and p-Variation.* Lecture Notes in Mathematics, Vol. 1703. Springer-Verlag, Berlin (1999). https://doi.org/10.1007/BFb0100744.

[29] R. M. Dudley and R. Norvaiša. *Concrete Functional Calculus.* Springer Monographs in Mathematics. Springer, New York (2011). https://doi.org/10.1007/978-1-4419-6950-7.

[30] I. Karatzas and S. E. Shreve. *Brownian Motion and Stochastic Calculus*, 2nd edn. Graduate Texts in Mathematics, Vol. 113. Springer-Verlag, New York (1991). https://doi.org/10.1007/978-1-4612-0949-2.

[31] F. Russo and P. Vallois. Elements of stochastic calculus via regularization. In *Séminaire de Probabilités XL*. Lecture Notes in Mathematics, Vol. 1899. Springer, Berlin (2007), pp. 147–185. https://doi.org/10.1007/978-3-540-71189-6_7.

[32] N. C. Jain and D. Monrad, Gaussian measures in B_p. *Annals of Probability.* **11**(1), 46–57 (1983). https://doi.org/10.1214/aop/1176993659.

[33] A. Ambrosetti and G. Prodi. *A Primer of Nonlinear Analysis.* Cambridge Studies in Advanced Mathematics, Vol. 34. Cambridge University Press, Cambridge (1995).

[34] J. Yong and X. Y. Zhou. *Stochastic Controls: Hamiltonian Systems and HJB Equations.* Applications of Mathematics (New York), Vol. 43. Springer-Verlag, New York (1999). https://doi.org/10.1007/978-1-4612-1466-3.

[35] Y. Hu, Y. Liu, and D. Nualart, Rate of convergence and asymptotic error distribution of Euler approximation schemes for fractional diffusions. *Annals of Applied Probability.* **26**(2), 1147–1207 (2016). https://doi.org/10.1214/15-AAP1114.

[36] Y. Mishura and G. Shevchenko, The rate of convergence for Euler approximations of solutions of stochastic differential equations driven by fractional Brownian motion. *Stochastics.* **80**(5), 489–511 (2008). https://doi.org/10.1080/17442500802024892.

[37] A. Neuenkirch, Optimal approximation of SDE's with additive fractional noise. *Journal of Complexity.* **22**(4), 459–474 (2006). https://doi.org/10.1016/j.jco.2006.02.001.

[38] I. Nourdin, Schémas d'approximation associés à une équation différentielle dirigée par une fonction höldérienne; cas du mouvement brownien fractionnaire. *Comptes Rendus de l'Académie des Sciences Paris.* **340**(8), 611–614 (2005). https://doi.org/10.1016/j.crma.2005.03.013.

[39] A. Neuenkirch and I. Nourdin, Exact rate of convergence of some approximation schemes associated to SDEs driven by a fractional Brownian motion. *Journal of Theoretical Probability.* **20**(4), 871–899 (2007). https://doi.org/10.1007/s10959-007-0083-0.

[40] P. E. Kloeden and E. Platen, *Numerical Solution of Stochastic Differential Equations.* Applications of Mathematics (New York), Vol. 23. Springer-Verlag, Berlin (1992). https://doi.org/10.1007/978-3-662-12616-5.

[41] A. Chronopoulou and S. Tindel, On inference for fractional differential equations. *Statistical Inference for Stochastic Processes.* **16**(1), 29–61 (2013). https://doi.org/10.1007/s11203-013-9076-z.

[42] P. K. Friz and N. B. Victoir, *Multidimensional Stochastic Processes as Rough Paths.* Cambridge Studies in Advanced Mathematics, Vol. 120. Cambridge University Press, Cambridge (2010). https://doi.org/10.1017/CBO9780511845079.

[43] Y. Han, Y. Hu, and J. Song, Maximum principle for general controlled systems driven by fractional Brownian motions. *Applied Mathematics & Optimization.* **67**(2), 279–322 (2013). https://doi.org/10.1007/s00245-012-9188-7.

[44] P. E. Protter. *Stochastic Integration and Differential Equations*, 2nd edn. Applications of Mathematics (New York), Vol. 21. Springer-Verlag, Berlin (2004).

[45] N. D. Cong, L. H. Duc, and P. T. Hong, Lyapunov spectrum of nonautonomous linear Young differential equations. *Journal of Dynamics and Differential Equations.* **32**(4), 1749–1777 (2020). https://doi.org/10.1007/s10884-019-09780-z.

[46] N. D. Cong, L. H. Duc, and P. T. Hong, Numerical attractors via discrete rough paths. *Journal of Dynamics and Differential Equations.* (2023). https://doi.org/10.1007/s10884-023-10280-4.

[47] P. Glasserman, *Monte Carlo Methods in Financial Engineering.* Applications of Mathematics (New York), Vol. 53. Springer-Verlag, New York (2004).

[48] B. B. Mandelbrot and J. W. Van Ness, Fractional Brownian motions, fractional noises and applications. *SIAM Review*. **10**, 422–437 (1968). https://doi.org/10.1137/1010093.
[49] D. Nualart. *The Malliavin Calculus and Related Topics*, 2nd edn. Probability and Its Applications (New York), Springer-Verlag, Berlin (2006).
[50] W. Schoutens, E. Simons, and J. Tistaert, A perfect calibration! now what? *Wilmott Magazine*. 66–78 (2004).

Chapter 5

Strong Convergence Analysis of a Fractional Exponential Integrator and Finite Element Method for Time-Fractional SPDEs Driven by Gaussian and Non-Gaussian Noises

Aurelien Junior Noupelah[*,‡] and Antoine Tambue[†,§]

[*]*The African Institute for Mathematical Sciences (AIMS) of Cameroon, P.O. Box 608, Crystal Gardens, Limbe, Cameroon*

[†]*Department of Computer Science, Electrical Engineering and Mathematical Sciences, Western Norway University of Applied Sciences, Inndalsveien 28, 5063 Bergen, Norway*

[‡]*aurelien.noupelah@aims-cameroon.org, noupsjunior@yahoo.fr*
[§]*antonio@aims.ac.za, tambuea@gmail.com*

In this chapter, we provide the first strong convergence result of a numerical approximation of the general second-order semilinear stochastic fractional order evolution equation involving a Caputo derivative in time of order $\alpha \in (\frac{1}{2}, 1)$ and driven simultaneously by Gaussian and non-Gaussian noises. The Gaussian noise considered here is a Hilbert space valued Q-Wiener process, and the non-Gaussian noise is defined through a compensated Poisson random measure associated with a Lévy process. The linear operator is not necessary self-adjoint.

The fractional stochastic partial differential equation is discretized in space by the finite element method and in time by a variant of the exponential integrator scheme. We investigate the mean square error estimate of our fully discrete scheme, and the result shows how the convergence orders depend on the regularity of the initial data and the power of the fractional derivative.

5.1. Introduction

We consider the following stochastic partial differential equation (SPDE) with initial value

$$\begin{cases} \partial_t^\alpha X(t) = AX(t) + F(X(t)) + B(X(t))\dfrac{dW(t)}{dt} \\ \qquad + \dfrac{\int_{\mathcal{X}} G(z, X(t))\widetilde{N}(dz, dt)}{dt}, \\ X(0) = X_0, \quad t \in [0, T], \end{cases} \quad (5.1)$$

on the Hilbert space $H = \left(L^2(\Lambda), \langle \cdot, \cdot \rangle_H, \|\cdot\|\right)$, $\Lambda \subset \mathbb{R}^d$, $d = 1, 2, 3$, where $T > 0$ is the final time and A is a linear operator which is unbounded, not necessarily self-adjoint, and is assumed to generate an analytic semigroup $S(t) := e^{tA}$. Note that ∂_t^α denotes the Caputo fractional derivative with $\alpha \in (\frac{1}{2}, 1)$, $W(t) = W(x, t)$ is a H-valued Q-Wiener process defined in a filtered probability space $(\Omega, \mathcal{F}, \mathbb{P}, \{\mathcal{F}_t\}_{t \geq 0})$, where the covariance operator $Q : H \to H$ is a positive and linear self-adjoint operator. The filtration is assumed to fulfill the usual assumptions (see Ref. [29, Def 2.1.11]). The Q-Wiener process $W(t)$ can be represented as follows [29]:

$$W(x, t) = \sum_{i \in \mathbb{N}^d} \beta_i(t) Q^{\frac{1}{2}} e_i(x) = \sum_{i \in \mathbb{N}^d} \sqrt{q_i} \beta_i(t) e_i(x),$$

where q_i, e_i, $i \in \mathbb{N}^d$ are, respectively, the eigenvalues and eigenfunctions of the covariance operator Q and β_i are mutually independent and identically distributed standard normal distributions. The mark set $\mathcal{X}$ is defined by $\mathcal{X} := H - \{0\}$. For a given set Γ, we denote by $\mathcal{B}(\Gamma)$ the smallest σ-algebra containing all open sets of Γ. Let $(\mathcal{X}, \mathcal{B}(\mathcal{X}), v)$ be a σ-finite measurable space and v (with $v \neq 0$)

be a Levy measurable on $\mathcal{B}(\mathcal{X})$ such that

$$v(\{0\}) = 0, \quad \int_{\mathcal{X}} \min(\|z\|^2, 1) v(dz) < \infty.$$

Let $N(dz, dt)$ be the H-valued Poisson distributed σ-finite measure on the product σ-algebra $\mathcal{B}(\mathcal{X})$ and $\mathcal{B}(\mathbb{R}_+)$ with intensity $v(dz)dt$, where dt is the Lebesgue measure on $\mathcal{B}(\mathbb{R}_+)$. In our model problem (5.1), $\widetilde{N}(dz, dt)$ stands for the compensated Poisson random measure defined by

$$\widetilde{N}(dz, dt) := N(dz, dt) - v(dz)dt.$$

Note that $\widetilde{N}(dz, dt)$ is a noncontinuous martingale with mean 0 (see, e.g., Ref. [19]). The Wiener process W and the compensated Poisson measure $\widetilde{N}$ are supposed to be independent. Precise assumptions on the nonlinear functions F, B, and G to ensure the existence of a mild solution for (5.1) will be given in the following section.

In the past few decades, fractional calculus has become a subject of increasing interest to researchers in various fields of science and technology. In particular, the theory of fractional partial differential equations (PDEs) has been the focus of studies over time, and since most of these equations have no analytical solutions, numerical schemes are the only tools to provide good approximations. For a deterministic equation ($G = B = 0$) and a self-adjoint operator A, Lin and Xu [15] have considered the numerical approximation of the time-fractional diffusion equation and proposed an algorithm based on the finite difference scheme in time and the Legendre spectral method in space. In the same context, high-order finite element method and mixed finite element scheme have been studied in Refs. [12, 16]. The numerical methods to solve the fractional heat equation with Dirichlet condition, involving a Riemann–Liouville fractional derivative in time, have been presented in Refs. [4, 30]. Gao et al. [6] presented a novel fractional numerical method (called L_{1-2} formula) to approximate the Caputo fractional derivative order α ($0 < \alpha < 1$), with a modification of the classical L_1 formula, and proved that the computational efficiency and the numerical accuracy of the new formula are superior to the standard L_1 formula. The Galerkin finite element approximation for time-fractional Navier–Stokes and the semilinear time-fractional subdiffusion problem are studied in Refs. [1, 14]. In Ref. [24], the authors developed an alternative numerical method based on the Keller box method for the subdiffusion equation, and in Ref. [3], Elzaki *et al.* used the decomposition method coupled with the Elzaki transform to construct appropriate solutions for multi-dimensional waves.[a] The authors in Ref. [32] developed and used a new type of discrete fractional Gronwall inequality to analyze the stability and convergence of the Galerkin spectral method for a linear time-fractional subdiffusion equation. Note that the time-stepping methods used in all the works mentioned until now are based on finite difference methods. However, these schemes are explicit but unstable, unless the time step size is very small. To address that drawback, numerical methods based on exponential integrators of the Adams type have been proposed in Ref. [7]. The price to pay is the computation of Mittag–Leffler (ML) matrix functions. As the ML matrix function is the generalized form of the exponential of a matrix function, the studies in Refs. [8, 20, 27] have extended some exponential computational techniques to ML. Note that up to now all the numerical algorithms presented are for time-fractional deterministic PDEs with self-adjoint linear operators.

However, in order to represent real-world physical phenomena more accurately, it is necessary to take into account stochastic disturbances from uncertain input data. The uncertainty is usually modeled by including the standard Brownian motion (Gaussian noise), and the corresponding model equation is given by (5.1) with $G = 0$. Few works have been done for numerical methods for Gaussian noise and time-fractional SPDE (5.1) with $G = 0$, even when the linear operator A is self-adjoint. To the best of our knowledge, Ref. [35] is the first of the basic theories and numerical methods for a class of these fractional SPDEs. In Ref. [36], the authors developed the fully discrete Galerkin finite element method for solving the time-fractional stochastic diffusion[b] equations based on the approximations of the ML function. Indeed, the temporal integration is similar to the deterministic exponential scheme in Ref. [7]. In Ref. [9], the authors provided rigorous convergence of numerical methods for solving stochastic time-fractional PDEs where the temporal discretization is done using the backward-Euler convolution quadrature. Note that all of the above works have been done for a self-adjoint

---

[a]Burger and Klein–Gordon equations of fractional order.
[b]So, the corresponding linear operator is self-adjoint.

linear operator A, so the numerical study of (5.1) with $G = 0$ and a non-self-adjoint operator A is still an open problem in the field to the best of our knowledge.

Furthermore, in finance, for example, the unpredictable nature of many events such as market crashes, announcements made by the central banks, changing credit risk, insurance for a changing risk, and changing face of operational risk [2,26] might have sudden and significant impacts on the stock price. In such situations, the more realistic model is built by incorporating non-Gaussian noise, such as Lévy process or Poisson random measure, to model such events. The corresponding equation is our model equation given in (5.1). As we have mentioned, studies on numerical schemes for such SPDEs of type (5.1) driven by Gaussian and non-Gaussian noises are lacking in the scientific literature; therefore, our goal is to fill that gap by extending the exponential scheme [17,22] to time-fractional SPDEs of type (5.1). The extension is extremely complicated since the ML function is more challenging than the exponential function. Using novel technical results that we have developed here, we have achieved strong convergence of our full discrete scheme for (5.1). Our strong convergence results examine how the convergence orders depend on the regularity of the initial data and the power of fractional derivatives.

The rest of the chapter is structured as follows. In Section 5.2, mathematical settings for cylindrical Brownian motion, random Poisson measure, Caputo-type fractional derivative, Laplace transform, and Mainardi's Wright-type function are presented, along with the well-posedness and regularity results of the mild solution of the SPDE (5.1). In Section 5.3, numerical schemes based on the stochastic exponential integrator scheme for the SPDE (5.1) are presented. We give some regularity estimates of the semidiscrete problem and analyze the spatial error in Section 5.4. We end the chapter in Section 5.5 by presenting the strong convergence proof of the full scheme for (5.1) based on finite element for spatial discretization and exponential integrator for temporal discretization.

5.2. Mathematical Setting, Main Assumptions, and Well-Posedness Problem

In this section, some notations and preliminary results needed throughout this work are provided. Let $(K, \langle ., . \rangle_K, \|.\|)$ be a separable

Hilbert space. For $p \geq 2$ and for a Banach space U, we denote by $L^p(\Omega, U)$ the Banach space of p-integrable U-valued random variables. We denote by $L(U, K)$ the space of bounded linear mapping from U to K endowed with the usual operator norm $\|.\|_{L(U,K)}$ and by $\mathcal{L}_2(U, K) = HS(U, K)$ the space of Hilbert–Schmidt operators from U to K equipped with the following norm:

$$\|l\|_{\mathcal{L}_2(U,K)} := \left(\sum_{i \in \mathbb{N}^d} \|l\psi_i\|^2 \right)^{\frac{1}{2}}, \quad l \in \mathcal{L}_2(U, K), \tag{5.2}$$

where $(\psi_i)_{i \in \mathbb{N}^d}$ is an orthonormal basis on U. The sum in (5.2) is independent of the choice of the orthonormal basis of U. We use the notation $L(U, U) =: L(U)$ and $\mathcal{L}_2(U, U) =: \mathcal{L}_2(U)$. It is well known that for all $l \in L(U, K)$ and $l_1 \in \mathcal{L}_2(U)$, $ll_1 \in \mathcal{L}_2(U, K)$ and

$$\|ll_1\|_{\mathcal{L}_2(U,K)} \leq \|l\|_{L(U,K)} \|l_1\|_{\mathcal{L}_2(U)}.$$

We denote by $L_2^0 := HS(Q^{\frac{1}{2}}(H), H)$ the space of Hilbert–Schmidt operators from $Q^{\frac{1}{2}}(H)$ to H, with the corresponding norm $\|.\|_{L_2^0}$ defined by

$$\|l\|_{L_2^0} := \left\| lQ^{\frac{1}{2}} \right\|_{HS} = \left(\sum_{i \in \mathbb{N}^d} \|lQ^{\frac{1}{2}} e_i\|^2 \right)^{\frac{1}{2}}, \quad l \in L_2^0, \tag{5.3}$$

where $(e_i)_{i \in \mathbb{N}^d}$ is an orthonormal basis of H. The sum in (5.3) is also independent of the choice of the orthonormal basis of H. Let $L_\nu^2(\chi \times [0, T]; H)$ be the space of all mappings $\theta : \chi \times [0, T] \times \Omega \to H$ such that θ is jointly measurable and $\mathcal{F}_t$-adapted for all $z \in \chi$, $0 \leq s \leq T$ satisfying

$$\int_0^T \int_\chi \|\theta(z, s)\|^2 \nu(dz) ds < \infty.$$

The following lemma is a result that will be used throughout this chapter.

Lemma 5.1 (Itô Isometry: [28, (4.30)], [19, (3.56)]).
(i) Let $\phi \in L^2([0,T]; L_2^0)$. Then, the following holds:
$$\mathbb{E}\left[\left\|\int_0^T \phi(s)dW(s)\right\|^2\right] = \mathbb{E}\left[\int_0^T \|\phi(s)\|_{L_2^0}^2 ds\right]. \qquad (5.4)$$

(ii) Let $\theta \in L_\nu^2(\chi \times [0,T]; \mathcal{H})$. Then, the following holds:
$$\mathbb{E}\left[\left\|\int_0^T \int_\chi \theta(z,s)\tilde{N}(dz,ds)\right\|^2\right] = \mathbb{E}\left[\int_0^T \int_\chi \|\theta(z,s)\|^2 \nu(dz)ds\right]. \qquad (5.5)$$

Definition 5.1. The Caputo-type derivative of order α with respect to t is defined by
$$\partial_t^\alpha X(t) = \begin{cases} \dfrac{1}{\Gamma(1-\alpha)} \displaystyle\int_0^t \dfrac{\partial X(s)}{\partial s} \dfrac{ds}{(t-s)^\alpha}, & 0 < \alpha < 1, \\ \dfrac{\partial X}{\partial t}, & \alpha = 1, \end{cases}$$
where $\Gamma(\cdot)$ is the Gamma function.

Let's introduce the generalized ML function $E_{\alpha,\beta}(t)$ defined as follows:
$$E_{\alpha,\beta}(t) = \sum_{k=0}^\infty \frac{t^k}{\Gamma(\alpha k + \beta)},$$
and its Laplace transform is given by (see [10])
$$\mathcal{L}(t^{\beta-1}E_{\alpha,\beta}(\lambda t^\alpha)) = \int_0^\infty e^{-\varsigma t}t^{\beta-1}E_{\alpha,\beta}(\lambda t^\alpha)dt = \frac{\varsigma^{\alpha-\beta}}{\varsigma^\alpha - \lambda}.$$

Remember that the Laplace transform is defined by
$$\tilde{f}(\varsigma) = \mathcal{L}(f(t)) = \int_0^\infty e^{-\varsigma t}f(t)dt.$$

Now, we give the definition of the mild solution to (5.1).

Definition 5.2 ([35, Definition 2.2]). For any $0 < \alpha < 1$, a stochastic process $\{X(t), t \in [0,T]\}$, is called the mild solution of (5.1) if:

1. $X(t)$ is $\mathcal{F}_t$-adapted on the filtration $(\Omega, \mathcal{F}, \mathbb{P}, \{\mathcal{F}_t\}_{t\geq 0})$;
2. $\{X(t), t \in [0,T]\}$ is measurable and $\mathbb{E}\left[\int_0^T \|X(t)\|^2 dt\right] < \infty$;
3. for all $t \in [0,T]$,

$$X(t) = S_1(t)X_0 + \int_0^t (t-s)^{\alpha-1} S_2(t-s) F(X(s)) ds$$

$$+ \int_0^t (t-s)^{\alpha-1} S_2(t-s) B(X(s)) dW(s)$$

$$+ \int_0^t \int_{\mathcal{X}} (t-s)^{\alpha-1} S_2(t-s) G(z, X(s)) \widetilde{N}(dz, ds) \quad (5.6)$$

hold a.s., where $S_1(t) = E_{\alpha,1}(At^\alpha)$ and $S_2(t) = E_{\alpha,\alpha}(At^\alpha)$.

Remark 5.1. Considering Mainardi's Wright-type function (see Ref. [18]),

$$M_\alpha(\theta) = \sum_{n=0}^{\infty} \frac{(-1)^n \theta^n}{n!\Gamma(1-\alpha(1+\theta))}, \quad 0 < \alpha < 1, \quad \theta > 0,$$

then the following results hold:

$$M_\alpha(\theta) \geq 0, \quad \int_0^\infty \theta^\mu M_\alpha(\theta) d\theta = \frac{\Gamma(1+\mu)}{\Gamma(1+\alpha\mu)}, \quad -1 < \mu < \infty, \quad \theta > 0,$$
$$(5.7)$$

and

$$E_{\alpha,1}(t) = \int_0^\infty M_\alpha(\theta) e^{t\theta} d\theta, \quad E_{\alpha,\alpha}(t) = \int_0^\infty \alpha\theta M_\alpha(\theta) e^{t\theta} d\theta. \quad (5.8)$$

Using (5.8), we rewrite the operators $S_1(t)$ and $S_2(t)$ as follows:

$$S_1(t) = E_{\alpha,1}(At^\alpha) = \int_0^\infty M_\alpha(\theta) e^{A\theta t^\alpha} d\theta = \int_0^\infty M_\alpha(\theta) S(\theta t^\alpha) d\theta,$$
$$(5.9)$$

and

$$S_2(t) = E_{\alpha,\alpha}(At^\alpha) = \int_0^\infty \alpha\theta M_\alpha(\theta)e^{A\theta t^\alpha}d\theta = \int_0^\infty \alpha\theta M_\alpha(\theta)S(\theta t^\alpha)d\theta.$$
(5.10)

Combining (5.6), (5.7), and (5.8), we obtain the result presented in Ref. [34, Definition 2.5], and we have the following lemma.

Lemma 5.2 ([34, Lemma 2.8]). *The operators $\{S_1(t)\}_{t\geq 0}$ and $\{S_2(t)\}_{t\geq 0}$ depending on $\alpha \in (\frac{3}{4}, 1)$ are bounded linear operator, and the following estimates hold:*

$$\|S_1(t)v\| \leq C_1 e^{-\gamma t}\|v\|, \quad \|S_2(t)v\| \leq \frac{C_1\alpha}{\Gamma(1+\alpha)}e^{-\gamma t}\|v\|, \quad v \in H,$$
(5.11)

and for some constants $C_1, \gamma > 0$.

In order to ensure the existence and uniqueness of the mild solution for the SPDE (5.1) and for the purpose of convergence analysis, we make the following assumptions.

Assumption 5.1 (Initial value). We assume that the initial data $X_0 : \Omega \to H$ to be a $\mathcal{F}_0$-measurable mapping and $X_0 \in L^2(\Omega, D((-A)^{\frac{\beta}{2}}))$ with $0 \leq \beta < 2$.

Assumption 5.2 (Nonlinearity term F). We assume the non-linear mapping $F: H \to H$ to be linear growth and Lipschitz continuous, i.e., there exists a constant $L > 0$ such that

$$\|F(u) - F(v)\|^2 \leq L\|u - v\|^2, \quad \|F(v)\|^2 \leq L(1 + \|v\|^2), \quad u, v \in H.$$

Assumption 5.3 (Lipschitz condition). We assume that the diffusion and jump coefficients, $B: H \to L_2^0$ and $G: \mathcal{X} \times H \to H$, respectively, satisfy the global Lipschitz condition, i.e., there exists a positive constant $L > 0$ such that

$$\|B(u) - B(v)\|_{L_2^0}^2 \leq L\|u - v\|^2,$$

$$\int_{\mathcal{X}} \|G(z, u) - G(z, v)\|^2 v(dz) \leq L\|u - v\|^2, \quad u, v \in H.$$

Assumption 5.4 (Linear growth). For $\tau \in [0, 1)$ and some constant $L > 0$, the following bound holds:

$$\|(-A)^\tau B(u)\|_{L_2^0}^2 \leq L(1 + \|(-A)^\tau u\|^2),$$

$$\int_{\mathcal{X}} \|(-A)^\tau G(z, u)\|^2 v(dz) \leq L(1 + \|(-A)^\tau u\|^2), \quad u \in H.$$

Theorem 5.1. *Under Assumptions 5.1–5.4, the SPDE (5.1) admits a unique mild solution $X(t) \in (D[0, T], H)$ asymptotically stable in mean square, that is,*

$$\mathbb{E}\left[\sup_{0 \leq t \leq T} \|X(t)\|^2\right] < \infty, \tag{5.12}$$

where, by $(D[0, T], H)$, we denote the space of all adapted càdlàg processes defined on $[0, T]$ with values in H.

Proof. See Ref. [34, Corollary 3.2] and the work done in Ref. [23, Theorem 1] for a proof of the result.

In all that follows, C denotes a positive constant that may change from line to line. In the Banach space $D((-A)^{\frac{\alpha}{2}})$, $\alpha \in \mathbb{R}$, we use the notation $\|(-A)^{\frac{\alpha}{2}} \cdot \| = \|\cdot\|_\alpha$, and we now present the following regularity results. □

5.3. Regularity of the Mild Solution

We discuss the space and regularity of the mild solution $X(t)$ of (5.1) given by (5.6) in this section. In the rest of this chapter, to simplify the presentation, we assume the SPDE (5.1) to be second order of the following type:

$$\partial_t^\alpha X(t, x) = [\nabla \cdot (D \nabla X(t, x)) - q \cdot \nabla X(t, x) + f(x, X(t, x))]$$

$$+ b(x, X(t, x)) \frac{dW(t, x)}{dt} + \frac{\int_{\mathcal{X}} g(z, x, X(t, x)) \widetilde{N}(dz, dt)}{dt},$$

where $f : \Lambda \times \mathbb{R} \to \mathbb{R}$ is globally continuous, $b : \Lambda \times \mathbb{R} \to \mathbb{R}$ is continuously differentiable with globally bounded derivatives and $g : \mathcal{X} \times \Lambda \times \mathbb{R} \to \mathbb{R}$ is globally Lipschitz continuous. In the abstract framework (5.1), the linear operator A is the $L^2(\Lambda)$ realization (see Ref. [5, p. 812]) of the following differential operator:

$$\mathcal{A}u = \sum_{i,j=1}^{d} \frac{\partial}{\partial x_i}\left(D_{i,j}(x)\frac{\partial u}{\partial x_j}\right) - \sum_{i=1}^{d} q_i(x)\frac{\partial u}{\partial x_i},$$

$$D = (D_{i,j})_{1\leq i,j \leq d}, \quad q = (q_i)_{1\leq i \leq d},$$

where $D_{i,j} \in L^\infty(\Lambda)$, $q_i \in L^\infty(\Lambda)$. We assume that there exists a positive constant $c_1 > 0$ such that

$$\sum_{i,j=1}^{d} D_{i,j}(x)\xi_i\xi_j \geq c_1|\xi|^2, \quad \xi \in \mathbb{R}^d, \ x \in \bar{\Lambda}.$$

The functions $F : H \to H$, $B : H \to L_2^0$, and $G : \mathcal{X} \times H \to H$ are defined by

$$(F(v))(x) = f(x, v(x)), \quad (B(v)u)(x) = b(x, v(x)) \cdot u(x),$$
$$G(z, v)(x) = g(z, x, v(x)),$$

for all $x \in \Lambda$, $v \in H$, $u \in Q^{1/2}(H)$, and $z \in \mathcal{X}$. As in Refs. [5, 17], we introduce two spaces $\mathbb{H}$ and V such that $\mathbb{H} \subset V$; the two spaces depend on the boundary conditions of Λ and the domain of the operator A. For Dirichlet (or first-type) boundary conditions, we take

$$V = \mathbb{H} = H_0^1(\Lambda) = \{v \in H^1(\Lambda) : v = 0 \text{ on } \partial\Lambda\}.$$

For the Robin (third-type) boundary condition and the Neumann (second-type) boundary condition, which is a special case of the Robin boundary condition, we take $V = H^1(\Omega)$:

$$\mathbb{H} = \{v \in H^2(\Lambda) : \partial v/\partial v_\mathcal{A} + \alpha_0 v = 0 \text{ on } \partial\Lambda\}, \quad \alpha_0 \in \mathbb{R},$$

where $\partial v/\partial v_\mathcal{A}$ is the normal derivative of v and $v_\mathcal{A}$ is the exterior-pointing normal $n = (n_i)$ to the boundary of $\mathcal{A}$ given by

$$\partial v/\partial v_\mathcal{A} = \sum_{i,j=1}^{d} n_i(x) D_{i,j}(x)\frac{\partial v}{\partial x_j}, \quad x \in \partial\Lambda.$$

Using Gårding's inequality (see, e.g., Ref. [31]), it holds that there exist two constants c_0 and $\lambda_0 > 0$ such that the bilinear form $a(.,.)$ associated with $-A$ satisfies

$$a(v,v) \geq \lambda_0 \|v\|^2_{H^1(\Lambda)} - c_0 \|v\|^2, \quad v \in V.$$

By adding and subtracting $c_0 X dt$ from both sides of (5.13), we have a new linear operator, still denoted by A, and the corresponding bilinear form is also still denoted by a. Therefore, the following coercivity property holds:

$$a(v,v) \geq \lambda_0 \|v\|^2_1, \quad v \in V. \qquad (5.13)$$

Note that the expression of the nonlinear term F has changed as we included the term $c_0 X$ in the new nonlinear term that we still denote by F.

The coercivity property (5.13) implies that A is the infinitesimal generator of a contraction semigroup $S(t) = e^{tA}$ on $L^2(\Lambda)$. Note that the coercivity property (5.13) also implies that the real part of the eigenvalues of $-A$ are positive; therefore, its fractional powers are well defined for any $\alpha > 0$ by

$$\begin{cases} (-A)^{-\alpha} = \dfrac{1}{\Gamma(\alpha)} \displaystyle\int_0^\infty t^{\alpha-1} e^{tA} dt, \\ (-A)^{\alpha} = (-A^{-\alpha})^{-1}, \end{cases}$$

where $\Gamma(\alpha)$ is the Gamma function. As the real part of the eigenvalues of A is negative, the generalized ML operator $E_{\alpha,\beta}(tA)$ is therefore well defined by

$$\|(\lambda I + A)^{-1}\|_{L(L^2(\Lambda))} \leq \dfrac{C_1}{|\lambda|}, \quad \lambda \in S_\theta,$$

where $S_\theta := \{\lambda \in \mathbb{C} : \lambda = \rho e^{i\phi},\ \rho > 0,\ 0 \leq |\phi| \leq \theta\}$ (see, e.g., Ref. [11]).

We recall the following properties of the semigroup $S(t)$ generated by $-A$, which will be useful throughout this chapter.

Proposition 5.1 (Smoothing properties of the semigroup [25]). *Let $\alpha > 0$, $\delta \geq 0$, and $0 \leq \gamma \leq 1$. Then, there exists a* constant $C > 0$ such that

$$\|(-A)^\delta S(t)\|_{L(H)} \leq Ct^{-\delta}, \quad \|(-A)^{-\gamma}(I - S(t))\|_{L(H)} \leq Ct^\gamma, \quad t > 0,$$

$$(-A)^\delta S(t) = S(t)(-A)^\delta \quad \text{on } D((-A)^\delta), \tag{5.14}$$

$$\|D_t^l S(t)v\| \leq Ct^{-l-(\gamma-\alpha)/2}\|v\|_\alpha, \quad v \in D((-A)^\alpha),$$

where $l = 0, 1$ and $D^l = \frac{d^l}{dt^l}$. If $\delta > \gamma$, then $D((-A)^\delta) \supset D((-A)^\gamma)$.

Firstly, we have the following useful lemma.

Lemma 5.3. Let $0 \leq t_1 < t_2 \leq T$ and $0 < a < 1$. Then, we have

$$t_2^a - t_1^a \leq (t_2 - t_1)^a.$$

Proof. Using the integral form and the variable change $t = t_2 u + t_1(1-u)$ yields

$$t_2^a - t_1^a = \int_{t_1}^{t_2} a\, t^{a-1} dt$$

$$= a(t_2 - t_1) \int_0^1 [t_1 + (t_2 - t_1)u]^{a-1}\, du$$

$$\leq a(t_2 - t_1) \int_0^1 (t_2 - t_1)^{a-1} u^{a-1}\, du$$

$$\leq (t_2 - t_1)^a \int_0^1 a\, u^{a-1}\, du$$

$$\leq (t_2 - t_1)^a. \qquad \square$$

The following lemma is an extension of the smoothing properties of the semigroup $S(t)$, Proposition 5.1 to the operators $\{S_1(t)\}_{t\geq 0}$ and $\{S_2(t)\}_{t\geq 0}$.

Lemma 5.4. Let $t \in (0,T)$, $0 < t_1 < t_2 \leq T$, $T < \infty$, $\frac{1}{2} < \alpha < 1$, $\rho \geq 0$, $0 \leq \eta < 1$, $0 \leq \sigma \leq \nu \leq 1$, and $\delta \geq 0$. Then, there exists a constant $C > 0$ such that for all $i = 1, 2$ and $u \in H$,

$$\|(-A)^\rho S_i(t)\|_{L(H)} \leq Ct^{-\alpha\rho}, \tag{5.15}$$

$$\|(-A)^{-\eta}(S_1(t_2) - S_1(t_1))\|_{L(H)} \leq Ct^{\alpha\eta}, \tag{5.16}$$

$$\|(-A)^\nu \left[t_1^{\alpha-1} S_2(t_1) - t_2^{\alpha-1} S_2(t_2)\right] u\|$$
$$\leq C(t_2 - t_1)^{1-[1-(\nu-\sigma)]\alpha} \|(-A)^\sigma u\|, \qquad (5.17)$$

and

$$(-A)^\delta S_i(t) = S_i(t)(-A)^\delta \quad \text{on } D((-A)^\delta). \qquad (5.18)$$

Proof. See Ref. [35, Lemma 3.3] for the proof of (5.15) and Ref. [36, Lemma 3.3] for that of (5.17). (5.18) is simply a consequence of (5.14) using (5.7) and (5.8). Concerning (5.16), using triangle inequality, Proposition 5.1, (5.7), and Lemma 5.3, we have

$$\|(-A)^{-\eta}(S_1(t_2) - S_1(t_1))\|_{L(H)}$$
$$= \left\|\int_0^\infty (-A)^{-\eta} M_\alpha(\theta) \left(S(\theta t_2^\alpha) - S(\theta t_1^\alpha)\right) d\theta\right\|_{L(H)}$$
$$\leq \int_0^\infty M_\alpha(\theta) \|S(\theta t_1^\alpha)\|_{L(H)} \left\|(-A)^{-\eta}\left(e^{A\theta(t_2^\alpha - t_1^\alpha)} - I\right)\right\|_{L(H)} d\theta$$
$$\leq C \int_0^\infty [\theta(t_2^\alpha - t_1^\alpha)]^\eta M_\alpha(\theta) d\theta$$
$$\leq C(t_2^\alpha - t_1^\alpha)^\eta \int_0^\infty \theta^\eta M_\alpha(\theta) d\theta$$
$$\leq C \frac{\Gamma(1+\eta)}{\Gamma(1+\alpha\eta)} (t_2 - t_1)^{\alpha\eta} \leq C(t_2 - t_1) t^{\alpha\eta}. \qquad \square$$

Moreover, (5.15)–(5.18) hold if A, S_1, and S_2 are replaced by their discrete versions A_h, S_{1h}, and S_{2h}, respectively, defined in Section 5.4.

Now, we give a spatial regularity result for the solution $X(t)$ in the following lemma.

Lemma 5.5. *Let Assumptions 5.1–5.4 be fulfilled. Then, the following space regularity holds:*

$$\|(-A)^{\beta/2} X(t)\|_{L^2(\Omega, H)} \leq C(1 + \|(-A)^{\beta/2} X_0\|_{L^2(\Omega, H)}), \quad 0 \leq t \leq T. \qquad (5.19)$$

Moreover, (5.19) holds if A and X are replaced by their discrete versions A_h and X^h, respectively, defined in Section 5.4.

Proof. By the definition of the mild solution (5.6),

$$X(t) = S_1(t)X_0 + \int_0^t (t-s)^{\alpha-1} S_2(t-s) F(X(s)) ds$$

$$+ \int_0^t (t-s)^{\alpha-1} S_2(t-s) B(X(s)) dW(s)$$

$$+ \int_0^t \int_{\mathcal{X}} (t-s)^{\alpha-1} S_2(t-s) G(z, X(s)) \widetilde{N}(dz, ds).$$

Then, taking the L^2-norm, using triangle inequality, the classical estimate $(\sum_{i=1}^n a_i)^2 \leq n \sum_{i=1}^n a_i^2$, the Itô isometry (5.4), and (5.5) for the last two terms yield

$$\left\| (-A)^{\beta/2} X(t) \right\|^2_{L^2(\Omega, H)}$$

$$\leq 4 \left\| (-A)^{\beta/2} S_1(t) X_0 \right\|^2_{L^2(\Omega, H)}$$

$$+ 4 \left(\int_0^t (t-s)^{\alpha-1} \left\| (-A)^{\beta/2} S_2(t-s) F(X(s)) \right\|^2_{L^2(\Omega, H)} ds \right)^2$$

$$+ 4 \int_0^t (t-s)^{2\alpha-2} \mathbb{E}\left(\left\| (-A)^{\beta/2} S_2(t-s) B(X(s)) \right\|^2_{L_2^0} \right) ds$$

$$+ 4 \int_0^t \int_{\mathcal{X}} (t-s)^{2\alpha-2} \left\| (-A)^{\beta/2} S_2(t-s) G(z, X(s)) \right\|^2_{L^2(\Omega, H)} v(dz) ds$$

$$=: 4 \sum_{i=1}^4 I_i^2. \tag{5.20}$$

We bound I_i^2, $i = 1, 2, 3, 4$ one by one. First, by the boundedness of $S_1(t)$ in Lemma 5.2 and (5.18), we have

$$I_1^2 := \left\| (-A)^{\beta/2} S_1(t) X_0 \right\|^2_{L^2(\Omega, H)} \leq C \left\| (-A)^{\beta/2} X_0 \right\|^2_{L^2(\Omega, H)}. \tag{5.21}$$

For I_2, by the semigroup property (5.15) with $\rho = \frac{\beta}{2}$, Assumption 5.2, and (5.12), we get

$$I_2^2 := \left(\int_0^t (t-s)^{\alpha-1} \left\|(-A)^{\beta/2} S_2(t-s) F(X(s))\right\|_{L^2(\Omega,H)} ds\right)^2$$

$$\leq C \left(\int_0^t (t-s)^{\alpha-1}(t-s)^{-\frac{\alpha\beta}{2}} \left(1 + \|X(s)\|_{L^2(\Omega,H)}\right) ds\right)^2$$

$$\leq C \left(\int_0^t (t-s)^{\alpha-\frac{\alpha\beta}{2}-1} ds\right)^2 \left(1 + \mathbb{E}\left[\sup_{0\leq s\leq t} \|X(s)\|^2\right]\right)$$

$$\leq Ct^{\alpha(2-\beta)} \left(1 + \mathbb{E}\left[\sup_{0\leq t\leq T} \|X(t)\|^2\right]\right) \leq C. \tag{5.22}$$

Applying (5.14), the boundedness of $S_2(t)$ in Lemma 5.2, and Assumption 5.4 with $\tau = \frac{\beta}{2}$, we deduce

$$I_3^2 := \int_0^t (t-s)^{2\alpha-2} \mathbb{E}\left[\left\|(-A)^{\beta/2} S_2(t-s) B(X(s))\right\|_{L_2^0}^2\right] ds$$

$$\leq C \int_0^t (t-s)^{2\alpha-2} \|S_2(t-s)\|_{L(H)}^2 \left(1 + \mathbb{E}\left[\|(-A)^{\beta/2} X(s)\|^2\right]\right) ds$$

$$\leq C \int_0^t (t-s)^{2\alpha-2} \left(1 + \|(-A)^{\beta/2} X(s)\|_{L^2(\Omega;H)}^2\right) ds$$

$$\leq C + C \int_0^t (t-s)^{2\alpha-2} \|(-A)^{\beta/2} X(s)\|_{L^2(\Omega;H)}^2 ds. \tag{5.23}$$

For I_4, analogous to I_3, we use also (5.14), the boundedness of $S_2(t)$ in Lemma 5.2, and Assumption 5.4 with $\tau = \frac{\beta}{2}$ to obtain

$$I_4^2 := \mathbb{E}\left[\int_0^t \int_\mathcal{X} (t-s)^{2\alpha-2} \left\|(-A)^{\beta/2} S_2(t-s) G(z, X(s))\right\|^2 v(dz) ds\right]$$

$$= \mathbb{E}\left[\int_0^t \int_\mathcal{X} (t-s)^{2\alpha-2} \|S_2(t-s)\|_{L(H)}^2\right.$$

$$\left.\times \left\|(-A)^{\beta/2} G(z, X(s))\right\|^2 v(dz) ds\right]$$

$$\leq C\mathbb{E}\left[\int_0^t (t-s)^{2\alpha-2} \int_\mathcal{X} \left\|(-A)^{\beta/2}G(z,X(s))\right\|^2 v(dz)ds\right]$$

$$\leq C\int_0^t (t-s)^{2\alpha-2}\left(1+\|(-A)^{\beta/2}X(s)\|_{L^2(\Omega;H)}^2\right)ds$$

$$\leq C + C\int_0^t (t-s)^{2\alpha-2}\|(-A)^{\beta/2}X(s)\|_{L^2(\Omega;H)}^2 ds. \tag{5.24}$$

Putting (5.21), (5.22), (5.23), and (5.24) in (5.20) hence yields

$$\|(-A)^{\beta/2}X(t)\|_{L^2(\Omega,H)}^2 \leq C(1+\|(-A)^{\beta/2}X_0\|_{L^2(\Omega,H)}^2)$$
$$+ C\int_0^t (t-s)^{2\alpha-2}\|(-A)^{\beta/2}X(s)\|_{L^2(\Omega;H)}^2 ds,$$

and applying fractional Gronwall's lemma [13,33, Lemma A.2] proves (5.19). □

Now, by the following theorem, we provide a temporal regularity of the solution process of (5.1).

Theorem 5.2. *Suppose that Assumptions 5.1–5.4 are fulfilled. Then, the following estimate holds:*

$$\|X(t_2) - X(t_1)\|_{L^2(\Omega;H)} \leq C(t_2-t_1)^{\frac{\min(\alpha\beta,2\alpha-1)}{2}}, \quad 0 \leq t_1 < t_2 \leq T. \tag{5.25}$$

Moreover, (5.25) holds when A and X are replaced by their semidiscrete versions A_h and X^h. respectively, defined in Section 5.4.

Proof. For $0 \leq t_1 < t_2 \leq T$, we rewrite the mild solution (5.6) at times $t = t_2$ and $t = t_1$, and we subtract $X(t_2)$ by $X(t_1)$ as

$$X(t_2) - X(t_1)$$
$$= (S_1(t_2) - S_1(t_1))X_0$$
$$+ \int_0^{t_1}\left[(t_2-s)^{\alpha-1}S_2(t_2-s) - (t_1-s)^{\alpha-1}S_2(t_1-s)\right]F(X(s))ds$$

$$+ \int_0^{t_1} [(t_2-s)^{\alpha-1} S_2(t_2-s) - (t_1-s)^{\alpha-1} S_2(t_1-s)]$$
$$\times B(X(s)) dW(s)$$
$$+ \int_0^{t_1} \int_{\mathcal{X}} [(t_2-s)^{\alpha-1} S_2(t_2-s) - (t_1-s)^{\alpha-1} S_2(t_1-s)]$$
$$\times G(z, X(s)) \widetilde{N}(dz, ds)$$
$$+ \int_{t_1}^{t_2} (t_2-s)^{\alpha-1} S_2(t_2-s) F(X(s)) ds$$
$$+ \int_{t_1}^{t_2} (t_2-s)^{\alpha-1} S_2(t_2-s) B(X(s)) dW(s)$$
$$+ \int_{t_1}^{t_2} \int_{\mathcal{X}} (t_2-s)^{\alpha-1} S_2(t_2-s) G(z, X(s)) \widetilde{N}(dz, ds).$$

Taking the L^2 norm in both sides and using triangle inequality yields

$$\|X(t_2) - X(t_1)\|_{L^2(\Omega; H)} \leq \sum_{i=1}^{7} J_i. \tag{5.26}$$

Inserting an appropriate power of $(-A)$, using Lemma 5.4 with $\eta = \frac{\beta}{2}$, and Assumption 5.1 implies

$$J_1 := \|(S_1(t_2) - S_1(t_1)) X_0\|_{L^2(\Omega; H)}$$
$$= \|(-A)^{-\frac{\beta}{2}} (S_1(t_2) - S_1(t_1)) (-A)^{\frac{\beta}{2}} X_0\|_{L^2(\Omega; H)}$$
$$\leq C(t_2 - t_1)^{\frac{\alpha\beta}{2}} \|(-A)^{\frac{\beta}{2}} X_0\|_{L^2(\Omega; H)}$$
$$\leq C(t_2 - t_1)^{\frac{\alpha\beta}{2}}. \tag{5.27}$$

The estimate of J_2 is already obtained in Ref. [23, (45)], and we have

$$J_2 \leq C(t_2 - t_1)^{\frac{2-(2-\beta)\alpha}{2}}. \tag{5.28}$$

Using the Itô isometry (5.5), inserting an appropriate power of $(-A)$, (5.17) with $\nu = \frac{\beta}{2}$ and $\sigma = 0$, Assumption 5.4 with $r = 0$, (5.12), and the Cauchy–Schwartz inequality, we obtain

$$J_4^2 = \left\| \int_0^{t_1} \int_{\mathcal{X}} [(t_2 - s)^{\alpha-1} S_2(t_2 - s) - (t_1 - s)^{\alpha-1} S_2(t_1 - s)] \right.$$

$$\left. \times G(z, X(s)) \widetilde{N}(dz, ds) \right\|^2_{L^2(\Omega; H)}$$

$$= \mathbb{E}\left[\int_0^{t_1} \int_{\mathcal{X}} \left\| (-A)^{\frac{\beta}{2}} [(t_2 - s)^{\alpha-1} S_2(t_2 - s) - (t_1 - s)^{\alpha-1} S_2(t_1 - s)] \right. \right.$$

$$\left. \left. \times (-A)^{-\frac{\beta}{2}} G(z, X(s)) \right\|^2 \nu(dz) ds \right]$$

$$\leq C \int_0^{t_1} (t_2 - t_1)^{2-(2-\beta)\alpha} \mathbb{E}\left[\int_{\mathcal{X}} \left\| (-A)^{-\frac{\beta}{2}} G(z, X(s)) \right\|^2 \nu(dz) \right] ds$$

$$\leq C(t_2 - t_1)^{2-(2-\beta)\alpha} t_1 \mathbb{E}\left[\int_{\mathcal{X}} \| G(z, X(s)) \|^2 \nu(dz) \right]$$

$$\leq C(t_2 - t_1)^{2-(2-\beta)\alpha} t_1 \left(1 + \mathbb{E}\left[\sup_{0 \leq s \leq T} \| X(s) \|^2 \right] \right)$$

$$\leq C(t_2 - t_1)^{2-(2-\beta)\alpha}. \tag{5.29}$$

By using a similar procedure as bounding J_4, using the Itô isometry (5.4), inserting an appropriate power of $(-A)$, (5.17) with $\nu = \frac{\beta}{2}$ and $\sigma = 0$, Assumption 5.4 with $r = 0$, (5.12), and the Cauchy–Schwartz inequality, we have

$$J_3^2 \leq \left\| \int_0^{t_1} [(t_2 - s)^{\alpha-1} S_2(t_2 - s) \right.$$

$$\left. - (t_1 - s)^{\alpha-1} S_2(t_1 - s)] B(X(s)) dW(s) \right\|^2_{L^2(\Omega; H)}$$

$$\leq C(t_2 - t_1)^{2-(2-\beta)\alpha}. \tag{5.30}$$

By the triangle inequality, the boundedness of operator $S_2(t)$ (5.11), Assumption 5.2, and (5.12), we get

$$J_5 := \left\| \int_{t_1}^{t_2} (t_2-s)^{\alpha-1} S_2(t_2-s) F(X(s)) ds \right\|_{L^2(\Omega;H)}$$

$$\leq \int_{t_1}^{t_2} (t_2-s)^{\alpha-1} \| S_2(t_2-s) F(X(s)) \|_{L^2(\Omega;H)} ds$$

$$\leq C \int_{t_1}^{t_2} (t_2-s)^{\alpha-1} \| F(X(s)) \|_{L^2(\Omega;H)} ds$$

$$\leq C \left(\int_{t_1}^{t_2} (t_2-s)^{\alpha-1} ds \right) \left(1 + \mathbb{E}[\sup_{0 \leq s \leq T} \|X(s)\|^2] \right)^{\frac{1}{2}}$$

$$\leq C(t_2-t_1)^{\alpha}. \tag{5.31}$$

Now, to bound the sixth term, using the Itô isometry property (5.4) gives

$$J_6^2 := \left\| \int_{t_1}^{t_2} (t_2-s)^{\alpha-1} S_2(t_2-s) B(X(s)) dW(s) \right\|_{L^2(\Omega;H)}^2$$

$$= \mathbb{E} \left[\int_{t_1}^{t_2} \| (t_2-s)^{\alpha-1} S_2(t_2-s) B(X(s)) \|_{L_2^0}^2 ds \right].$$

By the boundedness of the operator $S_2(t)$ (5.11), Assumption 5.4 with $\tau = 0$, and (5.12), we have

$$J_6^2 \leq \int_{t_1}^{t_2} (t_2-s)^{2\alpha-2} \| S_2(t_2-s) \|_{L(H)}^2 \mathbb{E} \|B(X(s))\|_{L_2^0}^2 ds$$

$$\leq C \left(\int_{t_1}^{t_2} (t_2-s)^{2\alpha-2} ds \right) \left(1 + \mathbb{E}[\sup_{0 \leq s \leq T} \|X(s)\|^2] \right)$$

$$\leq C(t_2-t_1)^{2\alpha-1}. \tag{5.32}$$

By using a similar procedure as bounding J_6, using the Itô isometry (5.5), the boundedness of operator $S_2(t)$ (5.11), Assumption 5.4 with

$r = 0$, and (5.12), we easily have

$$J_7^2 = \left\| \int_{t_1}^{t_2} \int_{\mathcal{X}} (t_2 - s)^{\alpha-1} S_2(t_2 - s) G(z, X(s)) \widetilde{N}(dz, ds) \right\|_{L^2(\Omega;H)}^2$$
$$\leq C(t_2 - t_1)^{2\alpha - 1}. \tag{5.33}$$

We substitute (5.27), (5.28), (5.30), (5.29), (5.31), (5.32), and (5.33) in (5.26); moreover, we note that for $\alpha \leq 1$, $2 - (2-\beta)\alpha \geq \alpha\beta$. Then,

$$\|X(t_2) - X(t_1)\|_{L^2(\Omega;H)} \leq C(t_2 - t_1)^{\frac{\min(\alpha\beta, 2\alpha-1)}{2}},$$

where $0 < t_1 < t_2 \leq T$. The proof of Theorem 5.2 is thus completed. □

5.4. Space Approximation and Error Estimates

We consider the discretization of the spatial domain using finite element triangulation with a maximal length of h satisfying the usual regularity assumptions. Let $V_h \subset V$ denote the space of continuous functions that are piecewise linear over triangulation J_h. To discretize in space, we introduce P_h from $L^2(\Omega)$ to V_h, defined for $u \in L^2(\Omega)$ by

$$(P_h u, \xi) = (u, \xi), \quad \forall \xi \in V_h.$$

The discrete operator $A_h : V_h \to V_h$ is defined by

$$(A_h \rho, \xi) = -a(\rho, \xi), \quad \forall \rho, \xi \in V_h,$$

where a is the corresponding bilinear form of A. Like the operator A, the discrete operator A_h is also the generator of an analytic semigroup $S_h(t) := e^{tA_h}$. The semidiscrete space version of problem (5.1) is to find $X^h(t) = X^h(\cdot, t)$ such that for $t \in [0, T]$,

$$\begin{cases} \partial_t^\alpha X^h(t) = A_h X^h(t) + P_h F(X^h(t)) + P_h B(X^h(t)) \dfrac{dW(t)}{dt} \\ \qquad + \dfrac{\int_{\mathcal{X}} P_h G(z, X^h(t)) \widetilde{N}(dz, dt)}{dt}, \\ X^h(0) = P_h X_0, \quad t \in [0, T]. \end{cases} \tag{5.34}$$

Note that A_h, P_hF, P_hB, and P_hG satisfy the same assumptions as A, F, B, and G, respectively. The mild solution of (5.34) can be represented as follows:

$$X^h(t) = S_{1h}(t)X_0^h + \int_0^t (t-s)^{\alpha-1} S_{2h}(t-s) P_h F(X^h(s)) ds$$

$$+ \int_0^t (t-s)^{\alpha-1} S_{2h}(t-s) P_h B(X^h(s)) dW(s)$$

$$+ \int_0^t \int_{\mathcal{X}} (t-s)^{\alpha-1} S_{2h}(t-s) P_h G(z, X^h(s)) \widetilde{N}(dz, ds), \tag{5.35}$$

where S_{1h} and S_{2h} are the semidiscrete version of S_1 and S_2, respectively, defined by (5.9) and (5.10). Let us define the error operators

$$T_h(t) := S(t) - S_h(t)P_h, \quad T_{1h}(t) := S(t) - S_{1h}(t)P_h,$$

$$T_{2h}(t) := S(t) - S_{2h}(t)P_h.$$

Then, we have the following lemma.

Lemma 5.6.

(i) *Let $r \in [0, 2]$, $\rho \leq r$, $t \in (0, T]$, $v \in D((-A)^\rho)$. Then, there exists a positive constant C such that*

$$\|T_h(t)v\| \leq Ch^r t^{-(r-\rho)/2} \|v\|_\rho, \tag{5.36}$$

and

$$\|T_{ih}(t)v\| \leq Ch^r t^{-\alpha(r-\rho)/2} \|v\|_\rho, \quad \forall i \in \{1, 2\}. \tag{5.37}$$

(ii) *Let $0 \leq \gamma \leq 1$. Then, there exists a constant C such that*

$$\left\| \int_0^t s^{\alpha-1} T_{2h}(s) v ds \right\| \leq Ch^{2-\gamma} \|v\|_{-\gamma}, \quad v \in D((-A)^{-\gamma}), \ t > 0. \tag{5.38}$$

Proof.

(i) See Ref. [21, Lemma 3.1] for the proof of (5.36) for the proof of (5.37), using (5.9), (5.10), their semidiscrete forms, (5.36), (5.7), and (5.8), we get

$$\|T_{1h}(t)v\| = \|(S_1(t) - S_{1h}(t)P_h)v\|$$
$$= \left\|\int_0^\infty M_\alpha(\theta)(S(\theta t^\alpha) - S_h(\theta t^\alpha)P_h)vd\theta\right\|$$
$$\leq \int_0^\infty M_\alpha(\theta)\|T_h(\theta t^\alpha)v\|d\theta$$
$$\leq C\int_0^\infty M_\alpha(\theta)h^r(\theta t^\alpha)^{-(r-\rho)/2}\|v\|_\rho d\theta$$
$$\leq Ch^r t^{-\alpha(r-\rho)/2}\int_0^\infty M_\alpha(\theta)\theta^{-(r-\rho)/2}d\theta\|v\|_\rho$$
$$\leq C\frac{\Gamma\left(1 - \frac{r-\rho}{2}\right)}{\Gamma\left(1 - \frac{\alpha(r-\rho)}{2}\right)}h^r t^{-\alpha(r-\rho)/2}\|v\|_\rho$$
$$\leq Ch^r t^{-\alpha(r-\rho)/2}\|v\|_\rho,$$

and

$$\|T_{2h}(t)v\| = \|(S_2(t) - S_{2h}(t)P_h)v\|$$
$$= \left\|\int_0^\infty \alpha\theta M_\alpha(\theta)(S(\theta t^\alpha) - S_h(\theta t^\alpha)P_h)vd\theta\right\|$$
$$\leq \int_0^\infty \alpha\theta M_\alpha(\theta)\|T_h(\theta t^\alpha)v\|d\theta$$
$$\leq C\int_0^\infty \alpha\theta M_\alpha(\theta)h^r(\theta t^\alpha)^{-(r-\rho)/2}\|v\|_\rho d\theta$$
$$\leq C\alpha h^r t^{-\alpha(r-\rho)/2}\int_0^\infty M_\alpha(\theta)\theta^{1-(r-\rho)/2}d\theta\|v\|_\rho$$
$$\leq C\frac{\Gamma\left(2 - \frac{r-\rho}{2}\right)}{\Gamma\left(1 + \alpha\left(1 - \frac{r-\rho}{2}\right)\right)}h^r t^{-\alpha(r-\rho)/2}\|v\|_\rho$$
$$\leq Ch^r t^{-\alpha(r-\rho)/2}\|v\|_\rho.$$

(ii) Using (5.10) and its semidiscrete form, (5.7), we obtain

$$\int_0^t s^{\alpha-1} T_{2h}(s) v ds$$

$$= \int_0^t s^{\alpha-1} S_2(s) v ds - \int_0^t s^{\alpha-1} S_{2h}(s) P_h v ds$$

$$= \int_0^t s^{\alpha-1} \int_0^\infty \alpha\theta M_\alpha(\theta) S(\theta s^\alpha) v d\theta ds$$

$$- \int_0^t s^{\alpha-1} \int_0^\infty \alpha\theta M_\alpha(\theta) S_h(\theta s^\alpha) P_h v d\theta ds$$

$$= \int_0^\infty A^{-1} M_\alpha(\theta) \int_0^t A\alpha\theta s^{\alpha-1} S(\theta s^\alpha) v d\theta ds$$

$$- \int_0^\infty A_h^{-1} M_\alpha(\theta) \int_0^t A_h \alpha\theta s^{\alpha-1} S_h(\theta s^\alpha) P_h v d\theta ds$$

$$= \int_0^\infty A^{-1} M_\alpha(\theta) [S(\theta s^\alpha)]_0^t v d\theta ds$$

$$- \int_0^\infty A_h^{-1} M_\alpha(\theta) [S_h(\theta s^\alpha)]_0^t P_h v d\theta ds$$

$$= \int_0^\infty A^{-1} M_\alpha(\theta) (S(\theta s^\alpha) - I) v d\theta ds$$

$$- \int_0^\infty A_h^{-1} M_\alpha(\theta) (S_h(\theta s^\alpha) - I) P_h v d\theta ds$$

$$= \int_0^\infty M_\alpha(\theta)(A_h^{-1} P_h - A^{-1}) v d\theta$$

$$+ \int_0^\infty M_\alpha(\theta)(A^{-1} S(\theta t^\alpha) - A_h^{-1} S_h(\theta t^\alpha) P_h) v d\theta$$

$$= (A_h^{-1} P_h - A^{-1}) v + \int_0^\infty M_\alpha(\theta)(A^{-1} S(\theta t^\alpha)$$

$$- A_h^{-1} S_h(\theta t^\alpha) P_h) v d\theta,$$

and from Ref. [21, (65), (69)], we have

$$\left\| \int_0^t s^{\alpha-1} T_{2h}(s) v \, ds \right\|$$

$$\leq \left\| (A_h^{-1} P_h - A^{-1}) v \right\|$$

$$+ \left\| \int_0^\infty M_\alpha(\theta)(A^{-1} S(\theta t^\alpha) - A_h^{-1} S_h(\theta t^\alpha) P_h) v \, d\theta \right\|$$

$$\leq \left\| (A_h^{-1} P_h - A^{-1}) v \right\|$$

$$+ \int_0^\infty M_\alpha(\theta) \left\| (A^{-1} S(\theta t^\alpha) - A_h^{-1} S_h(\theta t^\alpha) P_h) v \right\| d\theta$$

$$\leq Ch^{2-\gamma} \|v\|_{-\gamma} + C \int_0^\infty M_\alpha(\theta) h^{2-\gamma} \|v\|_{-\gamma} \, d\theta$$

$$\leq Ch^{2-\gamma} \|v\|_{-\gamma} + Ch^{2-\gamma} \|v\|_{-\gamma} \int_0^\infty M_\alpha(\theta) \, d\theta$$

$$\leq Ch^{2-\gamma} \|v\|_{-\gamma}.$$

This completes the proof of Lemma 5.6. □

We are now in a position to prove the following theorem, which provides an estimate in mean square sense of the error between the solution of SPDE (5.1) and the spatially semidiscrete approximation (5.34).

Theorem 5.3 (Space error). *Let X and X^h be the mild solutions of (5.1) and (5.34), respectively. Suppose that Assumptions 5.1–5.4 hold. We have the following estimates depending on the regularity parameter β of the initial solution X_0:*

$$\left\| X(t) - X^h(t) \right\|_{L^2(\Omega; H)} \leq Ch^\beta, \quad 0 \leq t \leq T.$$

Proof. Define $e(t) := X(t) - X^h(t)$. By (5.6) and (5.35), we deduce

$$e(t) = S_1(t) X_0 - S_{1h}(t) P_h X_0$$

$$+ \int_0^t (t-s)^{\alpha-1} S_2(t-s) F(X(s)) \, ds$$

$$- \int_0^t (t-s)^{\alpha-1} S_{2h}(t-s) P_h F(X^h(s)) ds$$

$$+ \int_0^t (t-s)^{\alpha-1} S_2(t-s) B(X(s)) dW(s)$$

$$- \int_0^t (t-s)^{\alpha-1} S_{2h}(t-s) P_h B(X^h(s)) dW(s)$$

$$+ \int_0^t \int_{\mathcal{X}} (t-s)^{\alpha-1} S_2(t-s) G(z, X(s)) \widetilde{N}(dz, ds)$$

$$- \int_0^t \int_{\mathcal{X}} (t-s)^{\alpha-1} S_{2h}(t-s) P_h G(z, X(s)) \widetilde{N}(dz, ds)$$

$$=: I + II + III + IV.$$

Thus, taking the L^2 norm and using the triangle inequality, we have

$$\|e(t)\|_{L^2(\Omega;H)} \leq \|I\|_{L^2(\Omega;H)} + \|II\|_{L^2(\Omega;H)} + \|III\|_{L^2(\Omega;H)}$$
$$+ \|IV\|_{L^2(\Omega;H)}.$$

We bound the above terms one by one. For the first term $\|I\|_{L^2(\Omega;H)}$, using (5.37) with $r = \rho = \beta$ and Assumption 5.1 yields

$$\|I\|_{L^2(\Omega;H)} := \|S_1(t) X_0 - S_{1h}(t) P_h X_0\|_{L^2(\Omega;H)}$$
$$= \|T_{1h}(t) X_0\|_{L^2(\Omega;H)}$$
$$\leq Ch^\beta \left\|(-A)^{\beta/2} X_0\right\|_{L^2(\Omega;H)} \leq Ch^\beta. \qquad (5.39)$$

For the second term $\|II\|_{L^2(\Omega;H)}$, adding and subtracting a term and using the triangle inequality gives

$$\|II\|_{L^2(\Omega;H)}$$
$$= \left\| \int_0^t (t-s)^{\alpha-1} S_2(t-s) F(X(s)) ds \right.$$
$$\left. - \int_0^t (t-s)^{\alpha-1} S_{2h}(t-s) P_h F(X^h(s)) ds \right\|_{L^2(\Omega;H)}$$

$$\leq \left\| \int_0^t (t-s)^{\alpha-1}(S_2(t-s) - S_{2h}(t-s)P_h)F(X(s))ds \right\|_{L^2(\Omega;H)}$$

$$+ \left\| \int_0^t (t-s)^{\alpha-1} S_{2h}(t-s)P_h(F(X(s)) - F(X^h(s)))ds \right\|_{L^2(\Omega;H)}$$

$$=: \|II_1\|_{L^2(\Omega;H)} + \|II_2\|_{L^2(\Omega;H)}. \tag{5.40}$$

To estimate the term $\|II_1\|_{L^2(\Omega;H)}$, we also add and subtract a term. Using the triangle inequality yields

$$\|II_1\|_{L^2(\Omega;H)}$$

$$:= \left\| \int_0^t (t-s)^{\alpha-1}(S_2(t-s) - S_{2h}(t-s)P_h)F(X(s))ds \right\|_{L^2(\Omega;H)}$$

$$\leq \left\| \int_0^t (t-s)^{\alpha-1}(S_2(t-s) - S_{2h}(t-s)P_h) \right.$$

$$\left. \times (F(X(s)) - F(X(t)))ds \right\|_{L^2(\Omega;H)}$$

$$+ \left\| \int_0^t (t-s)^{\alpha-1}(S_2(t-s) - S_{2h}(t-s)P_h)F(X(t))ds \right\|_{L^2(\Omega;H)}$$

$$=: \|II_{11}\|_{L^2(\Omega;H)} + \|II_{12}\|_{L^2(\Omega;H)}. \tag{5.41}$$

We estimate these two terms separately. Using the Cauchy–Schwartz inequality, (5.37) with $r = \beta$, $\rho = 0$, Assumption 5.2, and Theorem 5.2 leads to

$$\|II_{11}\|_{L^2(\Omega;H)}$$

$$:= \left\| \int_0^t (t-s)^{\alpha-1} T_{2h}(t-s)(F(X(s)) - F(X(t)))ds \right\|_{L^2(\Omega;H)}$$

$$\leq \int_0^t (t-s)^{\alpha-1} \|T_{2h}(t-s)(F(X(s)) - F(X(t)))\|_{L^2(\Omega;H)} ds$$

$$\leq Ch^\beta \int_0^t (t-s)^{\alpha-1}(t-s)^{-\frac{\alpha\beta}{2}}(t-s)^{\frac{\min(\alpha\beta, 2\alpha-1)}{2}} ds$$

$$\leq Ch^\beta \int_0^t (t-s)^{\min\left(\alpha-1, \frac{\alpha(4-\beta)-3}{2}\right)} ds$$

$$\leq Ch^\beta t^{\min\left(\alpha, \frac{\alpha(4-\beta)-1}{2}\right)} \leq Ch^\beta. \tag{5.42}$$

As with the term $\|II_{11}\|_{L^2(\Omega;H)}$, by applying (5.38) with $\gamma = 0$, Assumption 5.2, and Theorem 5.1, we get

$$\|II_{12}\|_{L^2(\Omega;H)} = \left\| \int_0^t (t-s)^{\alpha-1} T_{2h}(t-s) F(X(t)) ds \right\|_{L^2(\Omega;H)}$$

$$\leq Ch^2 \|F(X(t))\|_{L^2(\Omega;H)}$$

$$\leq Ch^2 \left(1 + \mathbb{E}\left[\sup_{0\leq t\leq T} \|X(t)\|^2 \right] \right)^{\frac{1}{2}} \leq Ch^2. \quad (5.43)$$

For the term $\|II_2\|_{L^2(\Omega;H)}$, using the Cauchy–Schwartz inequality, the fact that $S_{2h}(t-s)$ and P_h are bounded, and Assumption 5.2, we have

$$\|II_2\|^2_{L^2(\Omega;H)}$$

$$:= \left\| \int_0^t (t-s)^{\alpha-1} S_{2h}(t-s) P_h (F(X(s)) - F(X^h(s))) ds \right\|^2_{L^2(\Omega;H)}$$

$$\leq C \int_0^t (t-s)^{2\alpha-2} \|S_{2h}(t-s) P_h\|^2_{L(H)}$$

$$\times \left\| F(X(s)) - F(X^h(s)) \right\|^2_{L^2(\Omega;H)} ds$$

$$\leq C \int_0^t (t-s)^{2\alpha-2} \|e(s)\|^2_{L^2(\Omega;H)} ds. \quad (5.44)$$

Substituting (5.42) and (5.43) in (5.41) and hence putting (5.41) and (5.44) in (5.40) give

$$\|II\|^2_{L^2(\Omega;H)} \leq Ch^{2\beta} + Ch^4 + C \int_0^t (t-s)^{2\alpha-2} \|e(s)\|^2_{L^2(\Omega;H)} ds$$

$$\leq Ch^{2\beta} + C \int_0^t (t-s)^{2\alpha-2} \|e(s)\|^2_{L^2(\Omega;H)} ds. \quad (5.45)$$

For the fourth term $\|IV\|_{L^2(\Omega;H)}$, by adding and subtracting a term and using the triangle inequality yield

$$\|IV\|_{L^2(\Omega;H)}$$
$$= \left\| \int_0^t \int_{\mathcal{X}} (t-s)^{\alpha-1} S_2(t-s) G(z, X(s)) \widetilde{N}(dz, ds) \right.$$
$$\left. - \int_0^t \int_{\mathcal{X}} (t-s)^{\alpha-1} S_{2h}(t-s) P_h G(z, X(s)) \widetilde{N}(dz, ds) \right\|_{L^2(\Omega;H)}$$
$$\leq \left\| \int_0^t \int_{\mathcal{X}} (t-s)^{\alpha-1} \left(S_2(t-s) - S_{2h}(t-s) P_h \right) \right.$$
$$\left. \times G(z, X(s)) \widetilde{N}(dz, ds) \right\|_{L^2(\Omega;H)}$$
$$+ \left\| \int_0^t \int_{\mathcal{X}} (t-s)^{\alpha-1} S_{2h}(t-s) \right.$$
$$\left. \times P_h \left(G(z, X(s)) - G(z, X^h(s)) \right) \widetilde{N}(dz, ds) \right\|_{L^2(\Omega;H)}$$
$$=: \|IV_1\|_{L^2(\Omega;H)} + \|IV_2\|_{L^2(\Omega;H)}.$$

In a similar way as for $\|II\|^2_{L^2(\Omega;H)}$, using the Itô isometry (5.5), the boundedness of the operators $S_{2h}(t-s)$ and P_h, and Assumption 5.3, we deduce

$$\|IV_2\|^2_{L^2(\Omega;H)}$$
$$= \left\| \int_0^t \int_{\mathcal{X}} (t-s)^{\alpha-1} S_{2h}(t-s) \right.$$
$$\left. \times P_h \left(G(z, X(s)) - G(z, X^h(s)) \right) \widetilde{N}(dz, ds) \right\|^2_{L^2(\Omega;H)}$$
$$= \mathbb{E} \left[\int_0^t \int_{\mathcal{X}} (t-s)^{2\alpha-2} \|S_{2h}(t-s) P_h\|^2_{L(H)} \right.$$
$$\left. \times \left\| G(z, X(s)) - G(z, X^h(s)) \right\|^2 v(dz) ds \right]$$

$$\leq C\mathbb{E}\left[\int_0^t (t-s)^{2\alpha-2}\left(\int_{\mathcal{X}}\left\|G(z,X(s))-G(z,X^h(s))\right\|^2 v(dz)\right)ds\right]$$

$$\leq C\mathbb{E}\left[\int_0^t (t-s)^{2\alpha-2}\left\|X(s)-X^h(s)\right\|^2 ds\right]$$

$$\leq C\int_0^t (t-s)^{2\alpha-2}\|e(s)\|^2_{L^2(\Omega;H)}\,ds. \tag{5.46}$$

For the estimate $\|IV_1\|^2_{L^2(\Omega;H)}$, using the Itô isometry (5.5), (5.37) with $r=\rho=\beta$, Assumption 5.4 with $\tau=\frac{\beta}{2}$, and Lemma 5.5 leads to

$$\|IV_1\|^2_{L^2(\Omega;H)}$$

$$:=\left\|\int_0^t\int_{\mathcal{X}}(t-s)^{\alpha-1}T_{2h}(t-s)G(z,X(s))\tilde{N}(dz,ds)\right\|^2_{L^2(\Omega;H)}$$

$$=\mathbb{E}\left[\int_0^t\int_{\mathcal{X}}(t-s)^{2\alpha-2}\|T_{2h}(t-s)G(z,X(s))\|^2 v(dz)ds\right]$$

$$\leq Ch^{2\beta}\mathbb{E}\left[\int_0^t (t-s)^{2\alpha-2}\left(\int_{\mathcal{X}}\left\|(-A)^{\frac{\beta}{2}}G(z,X(s))\right\|^2 v(dz)\right)ds\right]$$

$$\leq Ch^{2\beta}\int_0^t (t-s)^{2\alpha-2}\left(1+\|(-A)^{\frac{\beta}{2}}X(s)\|^2_{L^2(\Omega,H)}\right)ds$$

$$\leq Ch^{2\beta}t^{2\alpha-1}\left(1+\|(-A)^{\frac{\beta}{2}}X_0\|^2_{L^2(\Omega,H)}\right)\leq Ch^{2\beta}. \tag{5.47}$$

Combining (5.46) and (5.47), it results that

$$\|IV\|^2_{L^2(\Omega;H)}\leq Ch^{2\beta}+C\int_0^t (t-s)^{2\alpha-2}\|e(s)\|^2_{L^2(\Omega;H)}\,ds. \tag{5.48}$$

Using the similar procedure as in $\|IV_1\|^2_{L^2(\Omega;H)}$, we have the following estimate:

$$\|III\|^2_{L^2(\Omega;H)}\leq Ch^{2\beta}+C\int_0^t (t-s)^{2\alpha-2}\|e(s)\|^2_{L^2(\Omega;H)}\,ds. \tag{5.49}$$

Combining (5.39), (5.45), (5.48), and (5.49) and applying the fractional Gronwall's lemma (see Refs. [13, 33]) completes the proof of Theorem 5.3. □

5.5. Fully Discrete Euler Scheme and Its Error Estimates

In this section, we consider a fully discrete approximation of SPDE (5.34). Before defining our numerical approximation of the mild solution of the semidiscrete problem (5.34), we present a useful way to appropriately rewrite this numerical approximation.

We recall that the mild solution at $t_m = m\Delta t$, $\Delta t > 0$, of the semidiscrete problem (5.34) is given by

$$X^h(t_m) = S_{1h}(t_m)X_0^h + \int_0^{t_m}(t_m-s)^{\alpha-1}S_{2h}(t_m-s)P_hF(X^h(s))ds$$

$$+ \int_0^{t_m}(t_m-s)^{\alpha-1}S_{2h}(t_m-s)P_hB(X^h(s))dW(s)$$

$$+ \int_0^{t_m}\int_{\mathcal{X}}(t_m-s)^{\alpha-1}S_{2h}(t_m-s)P_hG(z,X^h(s))\widetilde{N}(dz,ds).$$

By decomposing the integrals of the right-hand side of the previous equality using the Chasles relation, we obtain

$$X^h(t_m)$$

$$= S_{1h}(t_m)X_0^h + \sum_{j=0}^{m-1}\int_{t_j}^{t_{j+1}}(t_m-s)^{\alpha-1}S_{2h}(t_m-s)P_hF(X^h(s))ds$$

$$+ \sum_{j=0}^{m-1}\int_{t_j}^{t_{j+1}}(t_m-s)^{\alpha-1}S_{2h}(t_m-s)P_hB(X^h(s))dW(s)$$

$$+ \sum_{j=0}^{m-1}\int_{t_j}^{t_{j+1}}\int_{\mathcal{X}}(t_m-s)^{\alpha-1}S_{2h}(t_m-s)P_hG(z,X^h(s))\widetilde{N}(dz,ds).$$

$$(5.50)$$

To build our numerical scheme, we use the following approximations for all $s \in [t_j, t_{j+1})$, with $j \in \{0, 1, \ldots m-1\}$:

$$(t_m-s)^{\alpha-1}S_{2h}(t_{m+1}-s)P_hF(X^h(s))$$
$$\approx (t_m-t_j)^{\alpha-1}S_{2h}(t_m-t_j)P_hF(X^h(t_j)),$$

$$(t_{m+1} - s)^{\alpha-1} S_{2h}(t_{m+1} - s) P_h B(X^h(s))$$
$$\approx (t_m - t_j)^{\alpha-1} S_{2h}(t_m - t_j) P_h B(X^h(t_j)),$$
$$(t_{m+1} - s)^{\alpha-1} S_{2h}(t_{m+1} - s) P_h G(z, X^h(s))$$
$$\approx (t_m - t_j)^{\alpha-1} S_{2h}(t_m - t_j) P_h G(z, X^h(t_j)).$$

We can define our approximation X_m^h of $X^h(m\Delta t)$ by X_m^h

$$= S_{1h}(t_m) X_0^h + \sum_{j=0}^{m-1} \int_{t_j}^{t_{j+1}} (t_m - t_j)^{\alpha-1} S_{2h}(t_m - t_j) P_h F(X_j^h) ds$$

$$+ \sum_{j=0}^{m-1} \int_{t_j}^{t_{j+1}} (t_m - t_j)^{\alpha-1} S_{2h}(t_m - t_j) P_h B(X_j^h) dW(s)$$

$$+ \sum_{j=0}^{m-1} \int_{t_j}^{t_{j+1}} \int_{\mathcal{X}} (t_m - t_j)^{\alpha-1} S_{2h}(t_m - t_j) P_h G(z, X_j^h) \widetilde{N}(dz, ds).$$
(5.51)

Hence, using (5.9) and (5.10), it holds that
$$X_m^h = E_{\alpha,1}(A_h t_m^\alpha) X_0^h$$
$$+ \Delta t \sum_{j=0}^{m-1} (t_m - t_j)^{\alpha-1} E_{\alpha,\alpha}(A_h (t_m - t_j)^\alpha) S_{2h} P_h F(X_j^h)$$

$$+ \sum_{j=0}^{m-1} (t_m - t_j)^{\alpha-1} E_{\alpha,\alpha}(A_h (t_m - t_j)^\alpha) P_h B(X_j^h)(W_{t_{j+1}} - W_{t_j})$$

$$+ \sum_{j=0}^{m-1} (t_m - t_j)^{\alpha-1} E_{\alpha,\alpha}(A_h (t_m - t_j)^\alpha)$$

$$\times \int_{t_j}^{t_{j+1}} \int_{\mathcal{X}} P_h G(z, X_j^h) \widetilde{N}(dz, ds). \quad (5.52)$$

The strong convergence result of the fully discrete scheme is formulated in the following theorems.

Theorem 5.4 (Main result). *Let Assumptions 5.1–5.4 be fulfilled. Let X_m^h be the numerical approximation defined in (5.51). We have* the following estimates depending on the regularity parameter β of the initial solution X_0 and the power of the time-fractional derivative α:

$$\left\|X(t_m) - X_m^h\right\|_{L^2(\Omega;H)} \leq C\left(h^\beta + \Delta t^{\frac{\min(\alpha\beta, 2\alpha-1)}{2}}\right).$$

Proof. Using the triangle inequality yields

$$\left\|X(t_m) - X_m^h\right\|_{L^2(\Omega;H)}$$
$$\leq \left\|X(t_m) - X^h(t_m)\right\|_{L^2(\Omega;H)} + \left\|X^h(t_m) - X_m^h\right\|_{L^2(\Omega;H)}.$$

The space error is estimated in Theorem 5.3. It remains to estimate the time error. Recall the mild solution given by (5.50),

$$X^h(t_m)$$
$$= S_{1h}(t_m)X_0^h + \sum_{j=0}^{m-1}\int_{t_j}^{t_{j+1}}(t_m-s)^{\alpha-1}S_{2h}(t_m-s)P_hF(X^h(s))ds$$
$$+ \sum_{j=0}^{m-1}\int_{t_j}^{t_{j+1}}(t_m-s)^{\alpha-1}S_{2h}(t_m-s)P_hB(X^h(s))dW(s)$$
$$+ \sum_{j=0}^{m-1}\int_{t_j}^{t_{j+1}}\int_{\mathcal{X}}(t_m-s)^{\alpha-1}S_{2h}(t_m-s)P_hG(z,X^h(s))\widetilde{N}(dz,ds),$$

and the numerical solution X_m^h given by (5.51),

$$X_m^h = S_{1h}(t_m)X_0^h + \sum_{j=0}^{m-1}\int_{t_j}^{t_{j+1}}(t_m-t_j)^{\alpha-1}S_{2h}(t_m-t_j)P_hF(X_j^h)ds$$
$$+ \sum_{j=0}^{m-1}\int_{t_j}^{t_{j+1}}(t_m-t_j)^{\alpha-1}S_{2h}(t_m-t_j)P_hB(X_j^h)dW(s)$$
$$+ \sum_{j=0}^{m-1}\int_{t_j}^{t_{j+1}}\int_{\mathcal{X}}(t_m-t_j)^{\alpha-1}S_{2h}(t_m-t_j)P_hG(z,X_j^h)\widetilde{N}(dz,ds).$$

Subtracting these two previous equalities yields

$$X^h(t_m) - X_m^h$$
$$= \sum_{j=0}^{m-1} \int_{t_j}^{t_{j+1}} (t_m - s)^{\alpha-1} S_{2h}(t_m - s) P_h F(X^h(s))$$
$$- (t_m - t_j)^{\alpha-1} S_{2h}(t_m - t_j) P_h F(X_j^h) ds$$
$$+ \sum_{j=0}^{m-1} \int_{t_j}^{t_{j+1}} (t_m - s)^{\alpha-1} S_{2h}(t_m - s) P_h B(X^h(s))$$
$$- (t_m - t_j)^{\alpha-1} S_{2h}(t_m - t_j) P_h B(X_j^h) dW(s)$$
$$+ \sum_{j=0}^{m-1} \int_{t_j}^{t_{j+1}} \int_{\mathcal{X}} (t_m - s)^{\alpha-1} S_{2h}(t_m - s) P_h G(z, X^h(s))$$
$$- (t_m - t_j)^{\alpha-1} S_{2h}(t_m - t_j) P_h G(z, X_j^h) \widetilde{N}(dz, ds)$$
$$=: K_1 + K_2 + K_3.$$

Using the triangle inequality,

$$\left\| X^h(t_m) - X_m^h \right\|_{L^2(\Omega;H)}$$
$$\leq \|K_1\|_{L^2(\Omega;H)} + \|K_2\|_{L^2(\Omega;H)} + \|K_3\|_{L^2(\Omega;H)}. \quad (5.53)$$

By adding and subtracting a term, we recast K_1 as follows:

$$K_1 = \sum_{j=0}^{m-1} \int_{t_k}^{t_{k+1}} \left[(t_m - s)^{\alpha-1} S_{2h}(t_m - s) \right.$$
$$\left. - (t_m - t_j)^{\alpha-1} S_{2h}(t_m - t_j) \right] P_h F(X^h(s)) ds$$
$$+ \sum_{j=0}^{m-1} \int_{t_k}^{t_{k+1}} (t_m - t_j)^{\alpha-1} S_{2h}(t_m - t_j) P_h$$
$$\times \left[F(X^h(s)) - F(X^h(t_k)) \right] ds$$

$$+ \sum_{j=0}^{m-1} \int_{t_k}^{t_{k+1}} (t_m - t_j)^{\alpha-1} S_{2h}(t_m - t_j) P_h$$

$$\times \left[F(X^h(t_k)) - F(X_k^h) \right] ds$$

$$=: K_{11} + K_{12} + K_{13}.$$

The estimate of K_{11} is already obtained in Ref. [23, (89)], and we have

$$\|K_{11}\|_{L^2(\Omega;H)} \leq C \Delta t^{\frac{2-(2-\beta)\alpha}{2}}. \tag{5.54}$$

Let us recall the following estimate, for $\epsilon > 0$ small enough:

$$\sum_{k=1}^{m} t_k^{-1+\epsilon} \Delta t \leq C. \tag{5.55}$$

For the second estimate $\|K_{12}\|_{L^2(\Omega;H)}$, applying the triangle inequality, the boundedness of $S_{2h}(t)$ and P_h, Assumption 5.2, (5.25), the variable change $k = m - j$, and (5.55) with $\epsilon = \alpha$ yields

$$\|K_{12}\|_{L^2(\Omega;H)} \leq \sum_{j=0}^{m-1} \int_{t_j}^{t_{j+1}} \left\| (t_m - t_j)^{\alpha-1} S_{2h}(t_m - t_j) \right.$$

$$\left. \times P_h \left(F(X^h(s)) - F(X^h(t_j)) \right) \right\|_{L^2(\Omega;H)} ds$$

$$\leq \sum_{j=0}^{m-1} \int_{t_k}^{t_{k+1}} (t_m - t_j)^{\alpha-1} \|S_{2h}(t_m - t_j) P_h\|_{L(H)}$$

$$\times \left\| F(X^h(s)) - F(X^h(t_j)) \right\|_{L^2(\Omega;H)} ds$$

$$\leq \sum_{j=0}^{m-1} \int_{t_k}^{t_{k+1}} (t_m - t_j)^{\alpha-1} (s - t_j)^{\frac{\min(\alpha\beta, 2\alpha-1)}{2}} ds$$

$$\leq C\Delta t^{1+\frac{\min(\alpha\beta,2\alpha-1)}{2}} \sum_{j=0}^{m-1} (t_m - t_j)^{\alpha-1}$$

$$\leq C\Delta t^{\frac{\min(\alpha\beta,2\alpha-1)}{2}} \left(\sum_{j=0}^{m-1} (t_m - t_j)^{-1+\alpha} \Delta t \right)$$

$$\leq C\Delta t^{\frac{\min(\alpha\beta,2\alpha-1)}{2}}, \qquad (5.56)$$

and using again the triangle inequality, the boundedness of $S_{2h}(t)$ and P_h, and Assumption 5.2, we estimate $\|K_{13}\|_{L^2(\Omega;H)}$ as follows:

$$\|K_{13}\|_{L^2(\Omega;H)} \leq \sum_{j=0}^{m-1} \int_{t_j}^{t_{j+1}} \left\| (t_m - t_j)^{\alpha-1} S_{2h}(t_m - t_j) \right.$$

$$\left. \times P_h \left(F(X^h(t_j)) - F(X_j^h) \right) \right\|_{L^2(\Omega;H)} ds$$

$$\leq \sum_{j=0}^{m-1} \int_{t_j}^{t_{j+1}} (t_m - t_j)^{\alpha-1} \|S_{2h}(t_m - t_j) P_h\|_{L(H)}$$

$$\times \left\| F(X^h(t_j)) - F(X_j^h) \right\|_{L^2(\Omega;H)} ds$$

$$\leq C\Delta t \sum_{j=0}^{m-1} (t_m - t_j)^{\alpha-1} \left\| X^h(t_j) - X_j^h \right\|_{L^2(\Omega;H)}$$

$$\leq C\Delta t^\alpha \sum_{j=0}^{m-1} \left\| X^h(t_j) - X_j^h \right\|_{L^2(\Omega;H)}. \qquad (5.57)$$

Adding (5.54), (5.56), and (5.57), it holds that

$$\|K_1\|_{L^2(\Omega;H)} \leq C\Delta t^{\frac{\min(\alpha\beta,2\alpha-1)}{2}} + C\Delta t^\alpha \sum_{k=0}^{m-1} \left\| X^h(t_k) - X_k^h \right\|_{L^2(\Omega;H)}.$$

$$(5.58)$$

We will not give details of the estimate of K_2, as it is similar to that of K_3. Let us now estimate the norm of K_3. By adding and subtracting the same term, we rewrite it in three terms as follows:

$$K_3 = \sum_{j=0}^{m-1} \int_{t_j}^{t_{k+1}} \int_{\mathcal{X}} \Big[(t_m - s)^{\alpha-1} S_{2h}(t_m - s) P_h G(z, X^h(s))$$

$$- (t_m - t_j)^{\alpha-1} S_{2h}(t_m - t_j) \Big] P_h G(z, X^h(s)) \widetilde{N}(dz, ds)$$

$$+ \sum_{j=0}^{m-1} \int_{t_j}^{t_{j+1}} \int_{\mathcal{X}} (t_m - t_j)^{\alpha-1} S_{2h}(t_m - t_j)$$

$$\times P_h \Big[G(z, X^h(s)) - G(z, X^h(t_j)) \Big] \widetilde{N}(dz, ds)$$

$$+ \sum_{j=0}^{m-1} \int_{t_j}^{t_{j+1}} \int_{\mathcal{X}} (t_m - t_j)^{\alpha-1} S_{2h}(t_m - t_j)$$

$$\times P_h \Big[G(z, X^h(t_k)) - G(z, X_j^h) \Big] \widetilde{N}(dz, ds)$$

$$=: K_{31} + K_{32} + K_{33}. \tag{5.59}$$

Applying again the Itô isometry property (5.5), considering the fact that the variation of the compensated Poisson measure is independent, and inserting an appropriate power of $(-A)$, the discrete version of (5.17) with $t_2 = t_m - t_j$ and $t_1 = t_m - s$, $\nu = \frac{\beta}{2}$, and $\sigma = 0$, Assumption 5.4 with $\tau = 0$, the boundedness of P_h, the discrete version of (5.12), and the Cauchy–Schwartz inequality lead to

$$\|K_{31}\|_{L^2(\Omega;H)}^2 = \left\| \sum_{j=0}^{m-1} \int_{t_j}^{t_{j+1}} \int_{\mathcal{X}} (-A)^{\frac{\beta}{2}} \Big[(t_m - s)^{\alpha-1} S_{2h}(t_m - s) \right.$$

$$- (t_m - t_j)^{\alpha-1} S_{2h}(t_m - t_j) \Big]$$

$$\left. \times (-A)^{-\frac{\beta}{2}} P_h G(z, X^h(s)) \widetilde{N}(dz, ds) \right\|_{L^2(\Omega;H)}^2$$

$$\leq C \sum_{j=0}^{m-1} \int_{t_j}^{t_{j+1}} (s-t_j)^{2-(2-\beta)\alpha} \mathbb{E}$$
$$\times \left[\int_{\mathcal{X}} \left\| (-A)^{-\frac{\beta}{2}} P_h G(z, X^h(s)) \right\|^2 v(dz) \right] ds$$
$$\leq C \Delta t^{2-(2-\beta)\alpha} \left(\sum_{j=0}^{m-1} \int_{t_j}^{t_{j+1}} \left(1 + \mathbb{E}\left[\left\| X^h(s) \right\|^2 \right] \right) ds \right)$$
$$\leq C \Delta t^{2-(2-\beta)\alpha} t_m \left(1 + \mathbb{E}\left[\sup_{0 \leq s \leq T} \left\| X^h(s) \right\|^2 \right] \right)$$
$$\leq C \Delta t^{2-(2-\beta)\alpha}. \tag{5.60}$$

To estimate the second term $\|K_{32}\|_{L^2(\Omega;H)}^2$, applying the Itô isometry (5.5), using the boundedness of $S_{2h}(t)$ and P_h, Assumption 5.3, Theorem 5.2, the variable change $k = m - j$, and (5.55) with $\epsilon = 2\alpha - 1$ yields

$$\|K_{32}\|_{L^2(\Omega;H)}^2 = \left\| \sum_{j=0}^{m-1} \int_{t_k}^{t_{k+1}} \int_{\mathcal{X}} (t_m - t_j)^{\alpha-1} S_{2h}(t_m - t_j) \right.$$
$$\left. \times P_h \Big[G(z, X^h(s)) - G(z, X^h(t_k)) \Big] \widetilde{N}(dz, ds) \right\|_{L^2(\Omega;H)}^2$$
$$= \sum_{j=0}^{m-1} \mathbb{E}\left[\int_{t_j}^{t_{j+1}} \int_{\mathcal{X}} \left\| (t_m - t_j)^{\alpha-1} S_{2h}(t_m - t_j) \right. \right.$$
$$\left. \left. \times P_h \Big[G(z, X^h(s)) - G(z, X^h(t_j)) \Big] \right\|^2 v(dz) ds \right]$$
$$\leq \sum_{j=0}^{m-1} \mathbb{E}\left[\int_{t_j}^{t_{j+1}} (t_m - t_j)^{2\alpha-2} \| S_{2h}(t_m - t_j) P_h \|_{L(H)}^2 \right.$$
$$\left. \times \int_{\mathcal{X}} \left\| G(z, X^h(s)) - G(z, X^h(t_j)) \right\|^2 v(dz) ds \right]$$

$$\leq C \sum_{j=0}^{m-1} \int_{t_j}^{t_{j+1}} (t_m - t_j)^{2\alpha-2} \left\| X^h(s)) - X^h(t_j) \right\|^2_{L^2(\Omega;H)} ds$$

$$\leq C \sum_{j=0}^{m-1} \int_{t_j}^{t_{j+1}} (t_m - t_j)^{2\alpha-1} (s - t_j)^{\min(\alpha\beta,2\alpha-1)} ds$$

$$\leq C\Delta t^{1+\min(\alpha\beta,2\alpha-1)} \left(\sum_{j=0}^{m-1} (t_m - t_j)^{2\alpha-2} \right)$$

$$\leq C\Delta t^{\min(\alpha\beta,2-2\alpha)} \left(\sum_{k=1}^{m} t_k^{-1+2\alpha-1} \Delta t \right)$$

$$\leq C\Delta t^{\min(\alpha\beta,2\alpha-1)}. \tag{5.61}$$

To estimate the third term $\|K_{33}\|^2_{L^2(\Omega;H)}$, applying the Itô isometry (5.5), using the boundedness of $S_{2h}(t)$, P_h and Assumption 5.3 yields

$$\|K_{33}\|^2_{L^2(\Omega;H)} = \left\| \sum_{j=0}^{m-1} \int_{t_j}^{t_{j+1}} \int_{\mathcal{X}} (t_m - t_j)^{\alpha-1} S_{2h}(t_m - t_j) \right.$$

$$\left. \times P_h \left[G(z, X^h(t_j)) - G(z, X_j^h) \right] \widetilde{N}(dz, ds) \right\|^2_{L^2(\Omega;H)}$$

$$= \sum_{j=0}^{m-1} \mathbb{E} \left[\int_{t_j}^{t_{j+1}} \int_{\mathcal{X}} \left\| (t_m - t_j)^{\alpha-1} S_{2h}(t_m - t_j) \right.\right.$$

$$\left.\left. \times P_h \left[G(z, X^h(t_j)) - G(z, X_j^h) \right] \right\|^2 v(dz) ds \right]$$

$$\leq \sum_{j=0}^{m-1} \mathbb{E} \left[\int_{t_j}^{t_{j+1}} (t_m - t_j)^{2\alpha-2} \|S_{2h}(t_m - t_j) P_h\|^2_{L(H)} \right.$$

$$\left. \times \int_{\mathcal{X}} \left\| G(z, X^h(t_j)) - G(z, X_j^h) \right\|^2 v(dz) ds \right]$$

$$\leq C \sum_{j=0}^{m-1} (t_m - t_j)^{2\alpha-2} \int_{t_j}^{t_{j+1}} \left\| X^h(t_j) - X_j^h \right\|_{L^2(\Omega;H)}^2 ds$$

$$\leq C\Delta t \sum_{j=0}^{m-1} (t_m - t_j)^{2\alpha-2} \left\| X^h(t_j) - X_j^h \right\|_{L^2(\Omega;H)}^2$$

$$\leq C\Delta t^{2\alpha-1} \sum_{j=0}^{m-1} \left\| X^h(t_j) - X_j^h \right\|_{L^2(\Omega;H)}^2. \tag{5.62}$$

Substituting (5.60), (5.61), and (5.62) in (5.59) results in

$$\| K_3 \|_{L^2(\Omega;H)}^2$$

$$\leq C\Delta t^{\min(\alpha\beta, 2\alpha-1)} + C\Delta t^{2\alpha-1} \sum_{k=0}^{m-1} \left\| X^h(t_k) - X_k^h \right\|_{L^2(\Omega;H)}^2. \tag{5.63}$$

Using the similar procedure as for K_3, we obtain the following estimate of the norm of K_2:

$$\| K_2 \|_{L^2(\Omega;H)}^2$$

$$\leq C\Delta t^{\min(\alpha\beta, 2\alpha-1)} + C\Delta t^{2\alpha-1} \sum_{k=0}^{m-1} \left\| X^h(t_k) - X_k^h \right\|_{L^2(\Omega;H)}^2. \tag{5.64}$$

Substituting (5.58), (5.63), and (5.64) in (5.53) yields

$$\left\| X(t_m) - X_m^h \right\|_{L^2(\Omega;H)}^2$$

$$\leq C\Delta t^{\min(\alpha\beta, 2\alpha-1)} + C\Delta t^{2\alpha-1} \sum_{k=0}^{m-1} \left\| X^h(t_k) - X_k^h \right\|_{L^2(\Omega;H)}^2. \tag{5.65}$$

Applying discrete Gronwall's lemma to (5.65) and taking the square root lead to

$$\left\| X(t_m) - X_m^h \right\|_{L^2(\Omega;H)} \leq C\Delta t^{\frac{\min(\alpha\beta, 2\alpha-1)}{2}}. \tag{5.66}$$

Combining (5.66) and Theorem 5.3 completes the proof of Theorem 5.4. □

5.6. Numerical Simulations

We consider the following stochastic advection-diffusion-reaction SPDE:

$$\partial_t^\alpha X(t,x) = \left[\nabla \cdot (D\nabla X(t,x)) - q \cdot \nabla X(t,x) + \frac{X}{|X|+1} \right]$$
$$+ X(t,x) \frac{dW(t,x)}{dt} + X(t,x) \frac{d\widetilde{N}(t)}{dt},$$

where $D = \begin{pmatrix} 10^{-1} & 0 \\ 0 & 10^{-2} \end{pmatrix}$, $\widetilde{N}(t)$ is the compensated Poisson random measure with mean 0 and constant intensity parameter λ and the velocity field $q(x) = (q_i)_{1\leq i\leq d}$ is obtained exactly as in Refs. [17, 23, 31]. The covariance operator Q for $W(t)$ is given as in Refs. [17, 22, 23]. As mentioned in the introduction, the price to pay to simulate our scheme is the computation of ML matrix functions, which is more challenging than the standard exponential matrix functions [17, 31]. We have used the built-in Matlab function [27] to compute the ML matrix functions. As in Refs. [17, 31], we can easily check that the Assumptions 5.2–5.4 on noise and coefficients are satisfied. The grid has been chosen such that the size of the matrix A_h is 100. The final time is $T = 0.1$, and the reference samples solution, or "exact samples solution," is a numerical samples solution with a smaller time step of 0.0015. The errors are evaluated at the final time $T = 0.1$ with $\lambda = 0.1$. Figure 5.1(a) shows the streamline of the velocity field $q(x) = (q_i)_{1\leq i \leq d}$. A sample of the "exact solution" is shown in Figure 5.1(b). Figure 5.1(c) shows the convergence of the scheme for $\alpha = 0.51$, $\alpha = 0.75$, and $\alpha = 1$. We have used 50 samples in our error computation. The convergence orders are 0.501 for $\alpha = 1$, 0.341 for $\alpha = 0.75$, and 0.30 for $\alpha = 0.51$. We can observe that the order of convergence increases when α increase. The optimal order is 0.5 for $\alpha = 1$. We can also observe that our experimental convergence orders are in agreement with our theoretical results obtained using Theorem 5.4 for $\alpha = 0.75$ and $\alpha = 1$. When α is close to 0.5, our experimental convergence order is higher than that of our theoretical result obtained using Theorem 5.4.

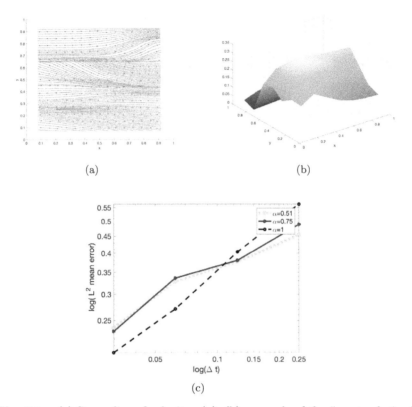

Fig. 5.1. (a) Streamline of velocity $q(x)$, (b) a sample of the "exact solution," and (c) convergence graphs for $\alpha = 0.51$, $\alpha = 0.75$, and $\alpha = 1$. The convergence orders are 0.501 for $\alpha = 1$, 0.341 for $\alpha = 0.75$, and 0.30 for $\alpha = 0.51$. We have used $\lambda = 0.1$ as the intensity parameter for the compensated Poisson random measure.

References

[1] M. Al-Maskari and S. Karaa, Numerical approximation of semilinear subdiffusion equations with nonsmooth initial data. *SIAM Journal on Numerical Analysis.* **57**(3), 1524–1544 (2019).
[2] R. Cont and P. Tankov, *Financial Modelling with Jump Process.* Financial Mathematics Series. CRC Press, Boca (2000).
[3] T. M. Elzaki, Y. Daoud, and J. Biazar, Decomposition method for fractional partial differential equations using modified integral transform. *World Applied Sciences Journal.* **37**(1), 18–24 (2019).
[4] N. J. Ford, J. Xiao, and Y. Yan, A finite element method for time fractional partial differential equation. *Fractional Calculus*

and *Applied Analysis.* **14**, 454–474 (2011). https://doi.org/10.2478/s13540-011-0028-2.

[5] F. Fujita and T. Suzuki, Evolution problems (part 1). In P. G. Ciarlet and J. L. Lions (eds.), *Handbook of Numerical Analysis*, Vol. 2. North-Holland, Amsterdam (1991), pp. 789–928.

[6] G. H. Gao, Z. Z. Sun, and H. W. Zhang, A new fractional numerical differentiation formula to approximate the Caputo fractional derivative and its applications. *Journal of Computational Physics.* **259**, 33–50 (2014).

[7] R. Garrappa, A family of Adams exponential integrators for fractional linear systems. *Computers & Mathematics with Applications.* **66**(5), 717–727 (2013).

[8] R. Garrappa and M. Popolizio, Computing the matrix Mittag-Leffler function with applications to fractional calculus. *Journal of Scientific Computing.* **77**, 129–153 (2018).

[9] M. Gunzburger, B. Li, and J. Wang, Sharp convergence rates of time discretization for stochastic time-fractional PDEs subject to additive space-time white noise (2018). arXiv:1704.02912v2 [math.NA], 8 August 2018.

[10] H. J. Haubold, A. M. Mathai, and R. K. Saxena, Mittag-Leffler functions and their applications. *Journal of Applied Mathematics.* **2011**, 298628, 51 pages (2011).

[11] D. Henry, *Geometric Theory of Semilinear Parabolic Equations*. Lecture Notes in Mathematics, Vol. 840. Springer, Berlin (1981).

[12] Y. Jianga and J. Ma, High-order finite element methods for time-fractional partial differential equations. *Journal of Computational and Applied Mathematics.* **235**, 3285–3290 (2011).

[13] R. Kruse, *Strong and Weak Approximation of Semilinear Stochastic Evolution Equations*. Springer, New-York (2014).

[14] X. Li, X. Yang, and Y. Zhang, Error estimates of mixed finite element methods for time-fractional Navier-Stokes equations. *Journal of Scientific Computing.* **70**, 500–515 (2017).

[15] Y. Lin and C. Xu, Finite difference/spectral approximations for the time-fractional diffusion equation. *Journal of Computational Physics.* **225**, 1533–1552 (2007).

[16] Y. Liu, H. Li, W. Gao, S. He, and Z. Fang, A new mixed element method for a class of time-fractional partial differential equations. *The Scientific World Journal* (2014). http://dx.doi.org/10.1155/2014/141467.

[17] G. J. Lord and A. Tambue, Stochastic exponential integrators for the finite element discretization of SPDEs for multiplicative & additive noise. *IMA Journal of Numerical Analysis.* **2**, 515–543 (2013).

[18] F. Mainardi, The fundamental solutions for the fractional diffusion-wave equation. *Applied Mathematics Letters.* **9**, 23–28 (1996).
[19] V. Mandreka and B. Rüdiger, *Stochastic Integration in Banach Space. Theory and Applications.* Probability Theory and Stochastic Modelling. Springer, Cham (2015).
[20] I. Moret and P. Novati, On the convergence of Krylov subspace methods for matrix Mittag-Leffler functions. *SIAM Journal on Numerical Analysis.* **49**(5), 2144–216 (2011).
[21] J. D. Mukam and A. Tambue, Optimal strong convergence rates of numerical methods for semilinear parabolic SPDE driven by Gaussian noise and Poisson random measure. *Computers & Mathematics with Applications.* **77**(10), 2786–2803 (2019).
[22] A. J. Noupelah and A. Tambue, Optimal strong convergence rates of some Euler-type timestepping schemes for the finite element discretization SPDEs driven by additive fractional Brownian motion and Poisson random measure. *Numerical Algorithms.* **88**, 315–363 (2021). https://doi.org/10.1007/s11075-020-01041-1.
[23] A. J. Noupelah, A. Tambue, and J. L. Woukeng, Strong convergence of a fractional exponential integrator scheme for finite element discretization of time-fractional SPDE driven by fractional and standard Brownian motions. *Communications in Nonlinear Science and Numerical Simulation.* **125**, 107371 (2023). https://doi.org/10.1016/j.cnsns.2023.107371.
[24] S. A. Osman and T. A. M. Langlands, An implicit Keller Box numerical scheme for the solution of fractional subdiffusion equations. *Applied Mathematics and Computation.* **348**, 609–626 (2019). https://doi.org/10.1016/j.amc.2018.12.015.
[25] A. Pazy, *Semigroups of Linear Operators and Applications to Partial Differential Equations.* Applied Mathematical Sciences, Vol. 44. Springer-Verlag, New York (1983).
[26] E. Platen and N. Bruti-Liberati, *Numerical Solution of Stochastic Differential Equations with Jumps in Finance.* Stochastic Modelling and Applied Probability, Vol. 64. Springer-Verlag, Berlin (2010).
[27] M. Popolizio, On the matrix Mittag-Leffler function: Theoretical properties and numerical computation. *Mathematics.* **7**(2), 1140 (2019). https://doi.org/10.3390/math7121140.
[28] D. Prato and G. J. Zabczyk, *Stochastic Equations in Infinite Dimensions*, Vol. 152. Cambridge University Press, Cambridge (2014).
[29] C. Prévôt and M. Röckner, *A Concise Course on Stochastic Partial Differential Equations.* Lecture Notes in Mathematics, Vol. 1905. Springer, Berlin (2007).

[30] G. S. Priya, P. Prakash, J. J. Nieto, and Z. Kayar, Higher order numerical scheme for the fractional heat equation with Dirichlet and Neumann boundary conditions. *Numerical Heat Transfer, Part B: Fundamentals.* **63**(6), 540–559 (2013).

[31] A. Tambue, Efficient Numerical Schemes for Porous Media Flow. PhD Thesis, Department of Mathematics, Heriot–Watt University (2010).

[32] Y. Yang and F. Zeng, Numerical analysis of linear and nonlinear time-fractional subdiffusion equations (2019). arXiv:1901.06814v1 [math.NA], 21 January 2019.

[33] H. Ye, J. Gao, and Y. Ding, A generalized Gronwall inequality and its application to a fractional differential equation. *Journal of Mathematical Analysis and Applications.* **328**, 1075–1081 (2007).

[34] L. Zhang, Y. Ding, K. Hao, and L. Hu, Moment stability of fractional stochastic evolution equations with Poisson jumps. *International Journal of Systems Science.* **45**(7), 1539–1547 (2014).

[35] G. A. Zou, B. Wang, and Y. Zhou, Existence and regularity of mild solution to fractional stochastic evolution equations. *Mathematical Modelling of Natural Phenomena.* **13** (2018). https://doi.org/10.1051/mmnp/2018004.

[36] G. A. Zou, Galerkin finite element method for time-fractional stochastic diffusion equations. *Computational and Applied Mathematics.* **37**, 4877–4898 (2018). https://doi.org/10.1007/s40314-018-0609-3.

Index

A

adjoint equation, 163, 165, 201, 203, 213
adjoint gradient representation, 162, 169, 189, 208
adjoint method, 204, 214
admissible control, 106–107
  finite-dimensional, 113
analytic semigroup, 222
analytical solutions, 223
asset price, 205

B

backward stochastic differential equation
  anticipating, 162, 164
  nonanticipating, 163
Banach space, 226, 230
Brownian motion, 155, 163, 204, 224–225

C

calibration, 207–208
Caputo, 225, 227
  derivative, 221
  fractional derivative, 222–223
Cauchy–Schwartz, 239, 247–248, 257
Chen's relation, 5, 29, 93
compact embedding, 110, 127, 145–146
complex Sobolev spaces, 109
control function, 172
control problem, 136
  ε-optimal solution, 140, 142–143
  existence of solution, 139
controlled rough path, 31, 33–34, 42–44, 46–47, 52, 56, 62, 70
convergence order, 225, 261
convergence analysis, 221
cost function, 107, 136
cost functional, 156, 160, 202, 208
covariation, generalized, 177

D

Dirichlet condition, 223, 231
discrete Gronwall's lemma, 260
discrete operator, 241

E

estimates for successive linearization method, 130
Euler scheme, 164, 166–167, 191, 201
exact solution, 261
exponential integrator scheme, 222, 225

F

filtered probability space, 222
filtration, 222
finite element method, 222–224
finite-dimensional optimal control problem, 136

fixed point, 72, 80, 86
flow of solutions, 98, 101
forward integral, 155–156, 162, 177
Fréchet derivative, 159, 183
fractional Brownian motion (fBm), 99, 115, 154, 204, 207
  infinite-dimensional, 26-27, 99
fractional calculus, 223
fractional derivative, 222, 225, 253
fractional Gronwall's lemma, 237, 250
Frobenius norm, 157

G

Galerkin, 224
Galerkin method, 106, 108
  convergence result, 121, 137
  existence of variational solution, 121
Gamma function, 227
Gaussian noise, 221, 224
Gaussian process, 157
geomteric rough path, 5
gradient descent, 156
greedy sequence, 175, 193
Gronwall lemma, 174, 192
Gubinelli
  derivative, 31–32
  integral, 56, 62

H

Hölder
  continuous, 158, 167
  rough path, 29
Heston model, fractional, 206
Hilbert space, 221–222, 226
  finite-dimensional, 112
  separable, 111
Hilbert–Schmidt, 226
Hurst index, 114–115
Hurst parameter, 154, 204, 209

I

infinitesimal generator, 36
integration-by-parts formula, 177, 180, 188

Itô
  formula for functionals of fBm, 115
  integral, 155, 162, 177
  isometry, 227, 235, 240, 249–250, 257–259

L

Lévy
  area, 29, 100
  process, 221, 225
Laplace transform, 225, 227
Lebesgue measure, 223
leverage effect, 155
linear operator, 221–222, 224–225, 229, 231–232
Lipschitz, 229–230
log-ODE method, 15–16
log-signature, 14
Love–Young estimate, 171

M

measurable space, 222
mild solution, 223, 225, 227–230, 237, 251, 253
Mittag–Leffler, 224
Monte Carlo, 202–203, 208

N

Neumann, 231
non-Gaussian noise, 221, 225
nonlinear functions, 223
nonlinear pathwise PDE, 107, 117
  convergence result, 125, 127
  finite-dimensional, 117–118
  variational solution, 117–118
numerical approximation, 221, 223, 251–252

O

orthogonal projection, 113
orthonormal basis, 226

P

p-variation
  estimate, 172, 183, 186
  finite, 157, 159

function, 157, 171
norm, 4, 157
process, 155–156, 204
seminorm, 157, 166
parameter set, 155, 207
partition, 157, 164
  mesh of, 157
  tagged, 171
Poisson random measure, 221, 223, 225

Q

quadratic variation, generalized, 177

R

random Poisson measure, 225, 261
regularity, 222, 225, 230, 234, 237, 245, 253
Riemann–Liouville, 223
Riemann–Stieltjes integral, 171
Riemann–Stieltjes sum, 171
Robin, 231
rough convolution, 26, 56, 62
rough differential equation, 7–9
rough integral, 6–7
Runge–Kutta method, 13, 16

S

second-order process, 29–30, 92, 100
self-adjoint, 221–225
semidiscrete problem, 225
semigroup, 232–233, 236
sensitivity equation, 159, 204, 213–214
signature, 3
space error, 253
spatial discretization, 225
spatial error, 225
standing assumptions, 158
stochastic differential equation
  matrix-valued, 159–160, 187
  parameter-dependent, 155
stochastic integral Wick-Itô-Skorohod type, 114

stochastic linear equation, 116
stochastic nonlinear Schrödinger equation, 114
  convergence result, 125, 134, 137
  estimate of variational solution, 125
  finite-dimensional, 124
  linearized and finite-dimensional, 129
  variational solution, 107, 124
stochastic partial differential equation, 222
stochastic process, 228
strong convergence, 225, 252
strong solution, 67–68, 70, 72, 85, 98, 101
strongly continuous semigroup, 35–36
successive linearization method, 106, 129
  convergence result, 134, 137
supremum norm, 159

T

Taylor method, 10–11
temporal discretization, 225
time
  error, 253
  regularity, 230

V

variation-of-constants formula, 161, 188
volatility, 154, 204–205

W

Wiener process, 88, 93, 97, 221–222
  infinite-dimensional, 88

Y

Young
  differential equation, 182, 187, 191, 201
  integral, 155, 170–171, 177, 182, 204

www.ingramcontent.com/pod-product-compliance
Lightning Source LLC
Chambersburg PA
CBHW050609010725
28797CB00004B/33

6